Genetics: Recent Developments

Genetics: Recent Developments

Edited by Ryan Jones

SYRAWOOD
PUBLISHING HOUSE

New York

Published by Syrawood Publishing House,
750 Third Avenue, 9th Floor,
New York, NY 10017, USA
www.syrawoodpublishinghouse.com

Genetics: Recent Developments
Edited by Ryan Jones

International Standard Book Number: 978-1-68286-844-7 (Hardback)

Cataloging-in-Publication Data

Genetics : recent developments / edited by Ryan Jones.
   p. cm.
Includes bibliographical references and index.
ISBN 978-1-68286-844-7
1. Genetics. 2. Genetics, Experimental. 3. Genomes. 4. Biology. I. Jones, Ryan.
QH430 .G46 2020
576.5--dc23

TABLE OF CONTENTS

PREFACE

Genetics is a field of science that is concerned with the study of genes, variation and heredity. It also involves the medical practice of diagnosing, treating and counseling patients with genetic disorders. Genetics overlaps with many other areas such as biotechnology, agriculture and medicine. Modern genetics focuses on the function and behavior of genes, gene structure and distribution from the context of cell and organism with respect to a population. Genetic processes interact with two factors - the environment and experiences of the organism. This influences the organism's behavior and development. Genetics is an upcoming field of science that has undergone rapid development over the past few decades. This book is a compilation of chapters that discuss the most vital concepts and emerging trends in the fields of plant and animal genetics. It will help the readers in keeping pace with the rapid changes in this field.

This book has been the outcome of endless efforts put in by authors and researchers on various issues and topics within the field. The book is a comprehensive collection of significant researches that are addressed in a variety of chapters. It will surely enhance the knowledge of the field among readers across the globe.

It gives us an immense pleasure to thank our researchers and authors for their efforts to submit their piece of writing before the deadlines. Finally in the end, I would like to thank my family and colleagues who have been a great source of inspiration and support.

Editor

Parameters for Successful Parental RNAi as An Insect Pest Management Tool in Western Corn Rootworm, *Diabrotica virgifera virgifera*

Ana M. Vélez [1,*], **Elane Fishilevich** [2], **Natalie Matz** [1], **Nicholas P. Storer** [2], **Kenneth E. Narva** [2] and **Blair D. Siegfried** [3]

[1]   Department of Entomology, University of Nebraska, 103 Entomology Hall, Lincoln, NE 68583, USA; nataliematz88@gmail.com

[2]   Dow AgroSciences, 9330 Zionsville Road, Indianapolis, IN 46268, USA; EFishilevich@dow.com (E.F.); nstorer@dow.com (N.P.S.); KNarva@dow.com (K.E.N.)

[3]   Entomology and Nematology Department, University of Florida, Charles Steinmetz Hall, PO Box 110620, Gainesville, FL 32611, USA; bsiegfried1@ufl.edu

*   Correspondence: avelezarango2@unl.edu

Academic Editor: Wenyi Gu

Abstract: Parental RNAi (pRNAi) is an RNA interference response where the gene knockdown phenotype is observed in the progeny of the treated organism. pRNAi has been demonstrated in female western corn rootworms (WCR) via diet applications and has been described as a potential approach for rootworm pest management. However, it is not clear if plant-expressed pRNAi can provide effective control of next generation WCR larvae in the field. In this study, we evaluated parameters required to generate a successful pRNAi response in WCR for the genes *brahma* and *hunchback*. The parameters tested included a concentration response, duration of the dsRNA exposure, timing of the dsRNA exposure with respect to the mating status in WCR females, and the effects of pRNAi on males. Results indicate that all of the above parameters affect the strength of pRNAi phenotype in females. Results are interpreted in terms of how this technology will perform in the field and the potential role for pRNAi in pest and resistance management strategies. More broadly, the described approaches enable examination of the dynamics of RNAi response in insects beyond pRNAi and crop pests.

Keywords: rootworm; *Diabrotica*; RNAi; parental RNAi; insect resistance; *brahma*; *hunchback*; chromatin-remodeling ATPase; transgenic crops

---

1. Introduction

Corn rootworms (CRW), *Diabrotica* species, are the most important pests of maize in the United States Corn Belt, with western corn rootworm (WCR), *Diabrotica virgifera virgifera*, being economically the most impactful [1]. The CRW larval stages cause economic damage by feeding on maize roots; root feeding by *D. v. virgifera* and the northern corn rootworm, *Diabrotica barberi* results in yield losses and costs of control that have been estimated to exceed $1 billion annually [2,3]. However, it is thought that this figure is underestimated today given the ongoing insecticide resistance problems, increased chemical costs, and technology fees associated with transgenic maize varieties [4]. Within the current management strategies, plant-expressed insecticidal proteins from *Bacillus thuringiensis* (Bt) have vastly changed the landscape of CRW control. Bt proteins expressed in maize provide significant root protection against CRW species, protecting yields, and are also believed to ameliorate the impacts of environmental stress conditions, such as drought, that are exacerbated by rootworm pressure [5,6].

RNA interference (RNAi) is recognized as a new potential management tool for this insect through feeding on plants producing long RNA hairpins (hpRNA) that suppress specific target genes in CRW [7]. WCR larvae and adults are generally susceptible to RNAi via ingestion of artificially produced double-stranded RNA (dsRNA) and hpRNA expressed in maize plants [7–9]. RNAi is likely to represent a next generation of biotechnological innovations for rootworm management [7,8,10,11]. Transgenic maize targeting WCR via RNAi will complement existing management practices including chemical insecticides and Bt traits [12,13].

To date, most genes proposed as RNAi targets for WCR cause lethality in the larval stage [7,8]. However, the sensitivity of WCR adults to RNAi was recently leveraged to produce phenotypes in larval progeny, referred to as parental RNAi (pRNAi). pRNAi has been described in coleopteran insects [13–16] as well as in other insect orders [17–25]. pRNAi functions by adult feeding on dsRNA that targets genes that regulate embryonic development resulting in reduced egg hatch rates or complete absence of viable larvae, while adults remain unaffected [16,26]. We recently evaluated the pRNAi effects of chromatin remodeling ATPase genes *brahma*, *mi-2*, *iswi-1*, and *iswi-2*, and the gap gene *hunchback* in WCR [16,26]. pRNAi has the potential to be used as part of integrated pest management (IPM) and insect resistance management (IRM) programs in combination with Bt toxins and related technologies to aid in slowing the emergence of alleles conferring resistance to the Bt toxins [26]. Moreover, by reducing the larval infestation in a maize field, pRNAi unique mode of action could potentially preserve the durability of other products used to manage WCR.

To fully evaluate the utility of pRNAi for pest and resistance management, key biological parameters need to be evaluated, including concentration-response, effects of the duration of the exposure, and effects of the timing of exposure within the adult lifecycle. With respect to the dose and exposure time, Bolognesi et al., 2012 [8] evaluated the effect of RNAi in WCR larvae: exposures of two to 24 h showed a response that was dependent on time of exposure and concentration in a 12 day assay. This concentration and time of exposure relationship illustrates that it is necessary to identify these parameters to achieve a consistent and effective RNAi response. Phenotypic responses to varying concentrations have also been documented in WCR larvae using genes critical for cuticle pigmentation [27]. While the effective concentration of dsRNA may be different depending on the target transcript, the sequence of dsRNA, and the stability of the encoded protein, the requirements for a pRNAi response in WCR progeny may further complicate the relationship between exposure and response. Further, since genes targeted by pRNAi affect embryonic development, it is necessary to determine the stage of female reproductive development at which the pRNAi effect is most successful. Ultimately, these parameters may be correlated to the conditions in the field when females are actively consuming maize tissues. With adults being much more mobile than larvae, establishing exposure requirements is critical for ensuring effective pRNAi concentrations in plants expressing hpRNA.

WCR have a univoltine life cycle; eggs are typically laid from late to July to early September and diapause in the soil [3,28]. Egg hatch varies depending on soil temperature; in the Midwest larvae typically hatch between May and early June [29], and feed underground on maize roots. Adults emerge during the summer and are present in and around maize fields from late June to autumn frost [28,30]. The larvae feed continuously for three to four weeks only on the roots of grasses (Graminae), especially maize [31,32]. Adults are strongly attracted to pollen and reproductive plant parts. They feed mainly on maize [4] as well as on other crops such as cucurbits, alfalfa, and soybeans [33–35], and pollen of non-crop flowers including *Ambrosia*, *Helianthus*, and *Amaranthus* [36,37].

WCR males emerge approximately five days before females [38]. However, the male emergence period overlaps with the females', since approximately five to seven days are required for the males to reach sexual maturity [39], while the females are sexually mature upon emergence [40]. In the field, males often intercept teneral virgin females (within 12–24 h of emergence, with pale and soft bodies) shortly after emergence from the soil [41–43]; mating couples are commonly observed at the base of maize plants. WCR females usually mate only once and they do not mate again as long as they are actively laying eggs [43]. Females feed on maize tissues available in the field where they emerged before mating or immediately after mating [4]. Female post-emergence dispersal prior to mating is believed to be minimal (1–5 m) and is dependent of whether there are sufficient numbers of males present at emergence [4,44], whereas dispersal after mating can be significant (<1 m/flight to as long as 24 km/flight) [45,46]. Several days after mating, 15% to 24% of the females engage in "sustained" or migratory flights [45,46]. Later studies showed that approximately 70% of females take flights ("trivial or "sustained") after mating, most of the flights (85%–90%) are of less than 1 min in duration, with only 0.5% of the female flights lasting longer than 20 min [47]. Campbell and Meinke [36] reported that WCR adults frequently move between a maize-prairie interface primarily after corn pollination, when it becomes less attractive than the adjoining prairie. WCR movement is also affected by changes in crop phenology within and among fields [48]. WCR female movement increases in later maize vegetative stages [49,50] and adults tend to move from early-planted maize to late planted maize [46]; adult movement is also density-dependent [51]. The above studies suggest that movement and feeding behavior could influence adult exposure to pRNAi in asynchronous fields with different traits or pRNAi fields adjacent to prairies. Based on the WCR behaviors described above, interplay between exposure duration and parental effect could also affect the success of refuge-based resistance management strategies that are intended to delay the onset of resistance to the pRNAi and Bt proteins in WCR populations [13]. This highlights the importance of identifying the duration of exposure to dsRNA or hpRNA, necessary to generate a pRNAi response and how adult movement will affect this exposure. WCR female feeding and mating behaviors also suggest that females could be exposed to hpRNA at different times of the reproductive cycle indicating the importance of evaluating the timing of exposure required for a successful pRNAi response.

This study aimed to identify the parameters required for a successful pRNAi for two genes in WCR, the chromatin remodeling gene *brahma* (*brm*), and the gap gene *hunchback* (*hb*). The parameters explored in the current work included: (1) a dsRNA concentration response; (2) duration of the dsRNA exposure; and (3) timing of the dsRNA exposure with respect to the mating status in WCR females. The concentration required to generate a pRNAi response with six exposures over twelve days was 0.2 µg/pellet or higher for both *brm* and *hb*. An exposure of four days for *brm* and eight days of *hb* of 2 µg of dsRNA/food pellet (highest amount used; equivalent to ~1.1 µg/insect/day) were necessary to achieve pRNAi responses in WCR. Further, recent work demonstrates that exposure of WCR females to *brm* homologs or *hb* dsRNA significantly affects larval emergence [16,26], however, the effect of *brm* and *hb* on the fecundity and fertility of adult WCR males has not been determined. In this study, we evaluated the effects of *brm* and *hb* dsRNA on male sperm viability and fecundity. Exposure of WCR males to *brm* and *hb* dsRNA had a subtle effect on sperm counts but no detectable effect on the number of offspring produced. The results obtained in this study further characterize the potential effectiveness of in planta pRNAi expression as a pest management tool for rootworm.

2. Materials and Methods

2.1. Gene Identification

WCR transcriptome sequencing and gene identification was described previously [16,52]. The amino acid sequences of *brahma* (*brm*) and *hunchback* (*hb*) from *Tribolium* were used as query sequences to search the WCR transcriptome. The GenBank accession numbers for WCR sequences for *brm* and *hb* are KR152260 and KR152261, respectively [16].

2.2. cDNA Preparation and dsRNA Synthesis

cDNA preparation and dsRNA synthesis was performed as previously described [16,26]. Briefly, total RNA was isolated from non-diapausing WCR adults (Crop Characteristics Inc., Farmington, MN, USA) using RNeasy Mini Kit (Qiagen, Valencia, CA, USA). Total RNA (1 µg) was used to synthesize first strand cDNA using the Quantitech Reverse Transcription Kit (Qiagen) and DNA was amplified using Takara Taq DNA Polymerase (Clontech Laboratories, Inc. Mountain View,

CA, USA). All primers contained a T7 promoter sequence at their 5' ends to enable T7 transcription (Supplementary Materials Table S1) [16]. *Green Fluorescent Protein* (*GFP*) dsRNA was used as a negative control. *Brm, hb,* and *GFP* PCR products were used as templates for in vitro synthesis of dsRNAs using the MEGAscript™ T7 RNAi Kit (Ambion, Life Technologies, Carlsbad, CA, USA) and purified using the RNeasy Mini Kit (Qiagen). The dsRNA products were quantified using a NanoDrop™ 100 spectrophotometer (Thermo Scientific, Franklin, MA, USA) at 260 nm and analyzed by gel electrophoresis to determine purity.

2.3. pRNAi Phenotypes in Embryos and Ovaries

WCR embryos from females fed with *hb* dsRNA and ovaries of females fed with diet treated with water, GFP, *brm* and *hb* dsRNA for 12 days before or after mating, were dissected under a Leica Zoom 200 stereomicroscope (Leica, Wetzlar, Germany) and stored in 70% ethanol. Images were captured with an Olympus SZX16 microscope, Olympus SDF PLAPO 2X PFC lens and the Olympus CellSens Dimensions software (Olympus, Tokyo, Japan).

2.4. brahma and hunchback Concentration Response

Test insects were purchased from Crop Characteristics (Farmington, MN, USA). In each treatment, ten females and ten males (24–48 h old) were maintained on untreated artificial diet and allowed to mate for four days in 16-well trays (5.1 cm long × 3.8 cm wide × 2.9 high) with vented lids. The artificial diet was adapted from Branson and Jackson [53] to provide the consistency necessary to cut diet plugs that could be treated with dsRNA. Diet was poured into Petri dishes to a depth of approximately 0.5 cm and after solidification the diet plugs (~4 mm in diameter × 2 mm height) were cut from the diet with a #1 (4 mm) cork borer. Trays were held in a growth chamber at 23 ± 1 °C, relative humidity >80%, and 16:8 L:D photoperiod [16]. Four replications of ten females and ten males were completed per treatment.

Four days after mating, males were removed and the remaining females were provided with eleven diet plugs surface-treated with gene specific dsRNA. WCR females were exposed to four concentrations of *brm* or *hb* dsRNA, 2 μg, 0.2 μg, 0.02 μg, and 0.002 μg per diet plug. Water and 2 μg of *GFP* dsRNA served as the controls. Freshly-treated diet was provided every other day, for a total of six exposures over twelve days. On Day 10 of exposure, females were transferred to polystyrene oviposition egg boxes (7.5 cm × 5.5 cm × 5.5 cm) (ShowMan box, Althor Products, Wilton, CT, USA) using the design of Campbell and Meinke [54]. The boxes contained moistened silty clay loam soil, pre-sifted through a 60-mesh sieve and autoclaved [55]. Females were allowed to lay eggs for four days, then were removed and flash frozen for qRT-PCR. Eggs were incubated in soil within the oviposition boxes for ten days at 27 °C, relative humidity >80% and 24 h dark. Eggs were removed from the soil by washing through a 60-mesh sieve. Harvested eggs were held in Petri dishes on moistened filter paper at 28 °C, relative humidity >80%, 24 h dark. The Petri dishes were photographed and total eggs counted using the cell counter function of ImageJ software [56]. The number of larvae hatching from each plate was recorded daily for fifteen days to determine egg viability [16].

2.5. Duration of brahma and hunchback dsRNA Exposure

To identify the duration of exposure necessary to generate a pRNAi response, females were exposed to 2 μg of dsRNA/plug one, two, four, or six times. The methodology used for this experiment was similar to that described for the concentration response experiment. Briefly, ten females and ten males (24–48 h old) were maintained on untreated artificial diet and allowed to mate for four days. After mating, males were removed and females were transferred to new trays with eleven dsRNA- or control-treated diet plugs. Freshly treated diet was provided every other day for eleven days but unlike the concentration response experiment, females were exposed one, two, four, or six times to 2 μg of *brm* or *hb* dsRNA per diet plug. Untreated artificial diet was provided for the remaining days. The controls, water and 2 μg of *GFP* dsRNA, were provided six times. After four days in the oviposition boxes, females were flash frozen for qRT-PCR. Eggs were washed, placed in Petri dishes, imaged, and analyzed with ImageJ, as described above. Larval hatching was monitored daily for fifteen days to determine egg viability. Four replicates of ten females and ten males were completed per treatment.

2.6. Timing of brahma and hunchback dsRNA Exposure with Respect to Mating Status

Previous experiments evaluated the pRNAi response for *brm* and *hb* in females exposed to dsRNA immediately after mating [16]. To determine if WCR female sensitivity to pRNAi varies with age and mating status, females were exposed to dsRNA six times prior to mating, immediately after mating, and six days after mating. The methodology used for this experiment was similar to that described for the concentration and duration response experiments.

Four replications of ten females and ten males per replication were completed for each type of exposure. The evaluation of the pRNAi effects immediately after mating was used as a reference and was performed using the methods described for the concentration response experiment. Briefly, artificial diet was surface-treated with water or 2 μg of *brm*, *hb* or *GFP* dsRNA six times over eleven days. After oviposition in oviposition boxes, females were flash frozen for qRT-PCR. Eggs were washed, placed in Petri dishes, imaged, and analyzed with ImageJ. Larval hatching was monitored daily for fifteen days to determine egg viability.

To determine the pRNAi effects in females before mating, ten virgin females (24–48 h old) were fed artificial diet treated with water or 2 μg of *brm*, *hb* or *GFP* dsRNA six times over eleven days. On Day 12 females were paired with ten virgin males and provided with untreated diet. Four days after mating, males were removed and females were transferred to trays with untreated diet. Females were transferred to oviposition boxes after six days, allowed to lay eggs for four days then removed and flash frozen for qRT-PCR. Eggs were washed, placed in Petri dishes, imaged, and analyzed with ImageJ. Larval hatching was monitored daily for fifteen days to determine egg viability.

To evaluate the pRNAi effect after mating, ten females and ten males (24–48 h old) were allowed to mate for four days. After mating, males were removed and females were transferred to trays with untreated diet. Females were provided untreated diet every other day for five days. Six days after mating, females were transferred to trays with artificial diet surface-treated with water or 2 μg of *brm*, *hb* or *GFP* dsRNA six times over eleven days. Females were transferred to oviposition boxes the day of the second exposure to dsRNA. One day after the last exposure, females were removed and flash frozen for qRT-PCR. Eggs were washed, placed in Petri dishes, imaged, and analyzed with ImageJ. Larval hatching was monitored daily for fifteen days to determine egg viability.

2.7. Effects of brahma and hunchback on Males

The effect of pRNAi in males was evaluated by exposing virgin males to artificial diet treated with dsRNA before mating. Ten virgin males (24–48 h old) were fed eleven pellets of artificial diet treated with water or 2 μg of specific *brm*, *hb* or *GFP* dsRNA. Freshly treated diet was provided every other day for seven days for a total of four exposures. On Day 8, three males per replication per treatment were flash frozen for qRT-PCR and the remaining males were paired with ten virgin females. Males were removed after four days and females were transferred to trays with untreated diet. Six days after mating females were transferred to oviposition boxes and allowed to lay eggs for four days. Eggs were washed, placed in Petri dishes, imaged, and analyzed with ImageJ. Larval hatching was monitored daily for fifteen days to determine egg viability. Three replications per treatment were performed. A second experiment with six exposures to dsRNA and three replications was performed to evaluate the effect on oviposition, egg hatching, and relative gene expression.

Sperm viability was evaluated in live males after four exposures to dsRNA over eight days. One day after the last exposure four males per replication were evaluated. Sperm viability was assessed using the Live/Dead Sperm Viability Kit (Invitrogen, Carlsbad, CA, USA) to discriminate between living and dead sperm [57]. WCR males were anesthetized on ice, testes and seminal vesicles were dissected under a stereomicroscope, placed in 10 µl of buffer (HEPES 10 mM, NaCl 150 mM, BSA 10%, pH 7.4) and crushed with a toothpick. Immediately after dissection, 1 µL of SYBR 14 (0.1 mM in dimethyl sulfoxide (DMSO)) was added and incubated at room temperature for ten minutes, followed by 1 µL of propidium iodine (2.4 mM) and incubated again at room temperature for ten minutes. Ten microliters of the sperm stained solution was transferred to a glass microslide and evaluated using a Nikon Eclipse 90i microscope with a Nikon A1 confocal and NIS-Elements Software (Melville, NY, USA). Samples were visualized at $10\times$ with 488 excitation, a 500–550 nm band pass for live sperm (SYBR 14) and 663–738 nm band pass for dead sperm (propidium iodine) simultaneously. Digital images were recorded for five fields of view per sample. The numbers of live (green) and dead (red) sperm were evaluated using the cell counter function of ImageJ [56].

2.8. Quantitative Real-Time PCR (qRT-PCR)

WCR qRT-PCR was performed using SYBR green and the 7500 Fast System Real-Time PCR System (Applied Biosystems, Foster City, CA, USA). Total RNA isolation and cDNA preparation was performed as described in the previous section. cDNA was diluted 50-fold for use as template. β-actin was selected as the reference gene based on its stability of expression across different life stages of WCR [58]. Primers used for qRT-PCR were designed using Beacon Designer software (Premier Biosoft International, Palo Alto, CA, USA) and are provided in Supplementary Materials Table S1. The 7500 Fast System SDS v.2.0.6 Software was used to determine the slope, correlation coefficients, and efficiencies (Supplementary Materials Table S1). Primer efficiencies were evaluated using 5-fold serial dilutions (1: 1/5: 1/25: 1/125: 1:625) in triplicate. Amplification efficiencies were higher than 96.1% for all the qRT-PCR primer pairs used in this study (Supplemental Materials Table S1). qRT-PCR analysis was performed with three to six biological replicates; each biological replicate had two technical replications. qRT-PCR cycling parameters were set as described in the supplier's protocol. At the end of each PCR reaction, a melting curve was generated to confirm single peaks and rule out the possibility of primer–dimer and nonspecific product formation. Relative quantifications of the transcripts were calculated using the comparative $2^{-\Delta\Delta CT}$ method [59] and were normalized to β-actin [9].

2.9. Statistical Analysis

Statistical analyses were performed with JMP® Pro 11 [60]. Data were analyzed with a one-way analysis of variance (ANOVA) and the means of the treatments were compared using a Student's t-test with Dunnett's adjustment (α, 0.05).

3. Results

3.1. brahma and hunchback Concentration Response

As previously described, feeding of adult WCR with *brm* or *hb* dsRNA leads to significant reductions in egg hatch rates [16,26]. While the eggs of *brm* dsRNA-fed females showed no signs of embryonic development and appeared as undeveloped or unfertilized eggs; embryos produced by WCR *hb* dsRNA-fed females had missing segments and deformed mouthparts (Supplementary Materials Figure S1) [16]. To determine the lowest concentration necessary to generate a pRNAi response, mated WCR females were exposed a range of concentrations of *brm* and *hb* dsRNA from 0.002 µg to 2 µg of dsRNA per artificial diet pellet six times over twelve days. No significant reduction in the number of eggs per female was observed after females were fed with *brm* dsRNA, although 2 µg of *brm* dsRNA produced a downward trend in oviposition (Figure 1a). Egg production in females exposed to any of the *hb* dsRNA concentrations was unaffected (Figure 1a). A significant reduction in egg hatching was observed with six feedings over a period of 12 days with 0.2 and 2 µg dsRNA for both *brm* and *hb* (Figure 1b). Significant reductions of *brm* transcript levels were detected when females were fed 0.2 and 2 µg of *brm* dsRNA (Figure 1c), while significant reductions of *hb* transcript levels were observed at all three *hb* dsRNA exposure concentrations (Figure 1d). Although a reduction in *hb* expression was observed at the 0.02 µg exposure, there was no reduction in the hatch rate.

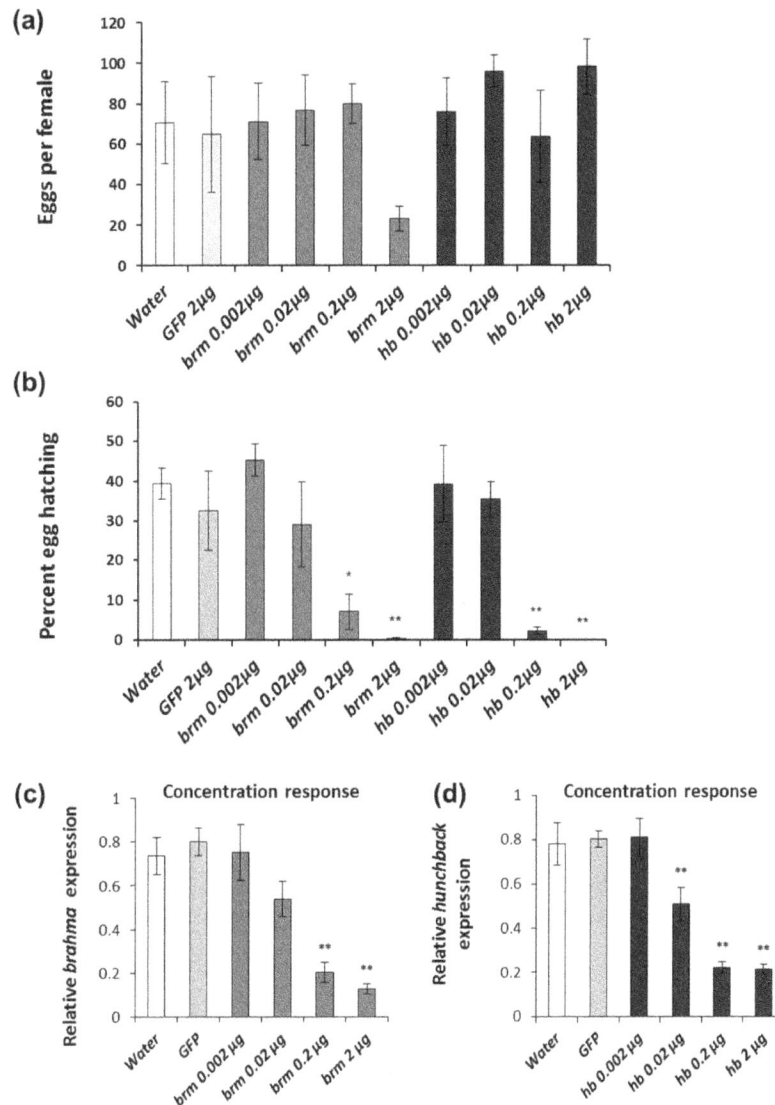

Figure 1. pRNAi concentration response to *brahma* (*brm*) and *hunchback* (*hb*) dsRNA in *Diabrotica virgifera virgifera*. Females were fed with diet treated with 0.002 µg, 0.02 µg, 0.2 µg and 2 µg *brm* or *hb* dsRNA. Diet treated with water or 2 µg *GFP* dsRNA were used as controls. Treatments were applied six times every other day for twelve days. (**a**) Eggs collected from dsRNA-fed females after last feeding exposure; (**b**) Eggs hatched based on numbers oviposited; (**c,d**) Relative transcript level for *brahma* (*brm*) and *hunchback* (*hb*) in *D. v. virgifera* females. Comparisons performed with Dunnett's test, * significance at $p < 0.05$. ** significance at $p < 0.001$.

3.2. Duration of brahma and hunchback dsRNA Exposure

Since females feed on a variety of plant material [36,46], it is important to determine the minimal duration of the exposure to dsRNA that generates a pRNAi response. For this purpose, we exposed females to artificial diet treated with 2 µg of dsRNA once, twice, four, or six times, providing freshly treated or untreated diet every other day. Although *brm* dsRNA-fed females showed a downward-trend in oviposition, the number of eggs per female was not significantly different from the water control. Similarly, for the results observed in the concentration response experiment, the egg production of females exposed to *hb* dsRNA was unaffected (Figure 2a). The percentage of larvae hatching was significantly reduced compared to water for all exposures with both genes (Figure 2b). For *brm*, the pRNAi effect was stronger with two to six feedings of 2 μg dsRNA; and for *hb*, at least four feedings were necessary to generate over 50% reduction in egg hatching (Figure 2b). Relative *brm* transcript levels were significantly reduced with all the exposures (Figure 2c), while for *hb* four dsRNA feedings were necessary to observe a significant reduction in *hb* transcript levels (Figure 2d).

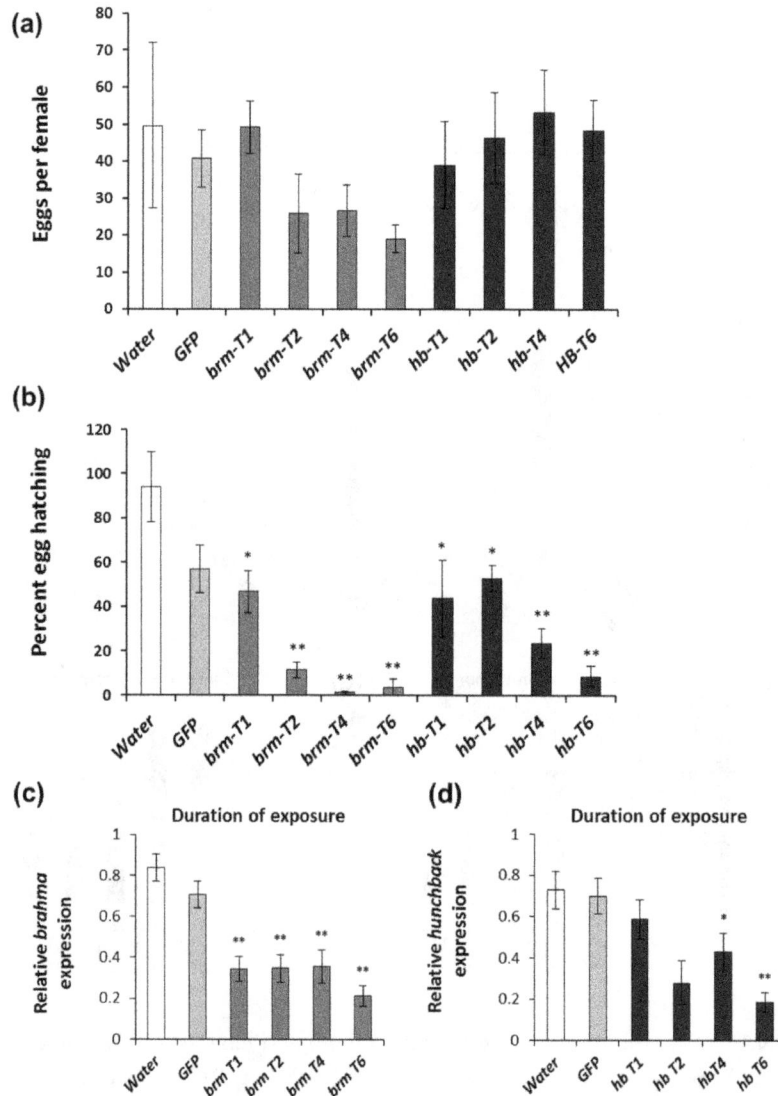

Figure 2. Duration of exposure effects on pRNAi response using *brahma* (*brm*) and *hunchback* (*hb*) dsRNA in *D. v. virgifera*. Females were fed with diet treated with dsRNA; the number following T indicates the number of times that females received dsRNA (2 μg per diet pellet), diet provided every other day for twelve days. Diet treated with water and *GFP* dsRNA provided six times were used as controls. (**a**) Eggs collected from dsRNA-fed females after last feeding exposure; (**b**) Eggs hatched based on numbers oviposited; (**c,d**) Relative transcript level for *brahma* (*brm*) and *hunchback* (*hb*) in *D. v. virgifera* females. T indicates the number of times that females received dsRNA (2 μg per diet pellet), diet provided every other day for twelve days. Comparisons performed with Dunnett's test (control group = water), * significance at $p < 0.05$. ** significance at $p < 0.001$.

3.3. Timing of brahma and hunchback dsRNA Exposure with Respect to Mating Status

In a field setting, females will be exposed to dsRNA at different times of their reproductive cycle. We evaluated females exposed to 2 μg of *brm* and *hb* dsRNA six times before mating, immediately after mating, and six days after mating to determine the impact of reproductive status on gene expression and phenotypic response. As in the concentration response experiment (Figure 1a), the number of eggs per female was reduced in females exposed to *brm* dsRNA but it was not significantly different from the water control (Figure 3a). In females exposed to *hb* dsRNA egg production was not significantly affected (Figure 3a). The percent eggs hatching was significantly reduced when females were fed *brm* dsRNA immediately after mating and six days after mating; this effect was stronger in females that fed immediately after mating (Figure 3b). Even though the percent of eggs hatching from females fed with *brm* dsRNA before mating was not significantly different from females fed with water treated diet, egg hatching was based on five emerging larvae from a total of 43 eggs, while the total number of eggs for the controls were 2128 and 1406 for the water and *GFP,* respectively (Figure 3b). Low egg hatch was observed in females fed with *hb* dsRNA at any time of their reproductive cycle. The total egg hatch rate was lower when females were fed *hb* dsRNA before mating and immediately after mating (Figure 3b). Relative transcript levels were significantly reduced in females fed *brm* (Figure 4a) and *hb* (Figure 4b) dsRNA before and immediately after mating. Overall, gene knockdown was stronger for *brm* compared to *hb*. To determine if *brm* and *hb* dsRNA treatments caused phenotypic changes in WCR ovary, ovaries of females that were treated with dsRNA before and after mating were dissected; no morphological differences were observed between dsRNA and control treatments (Supplementary Materials Figure S2).

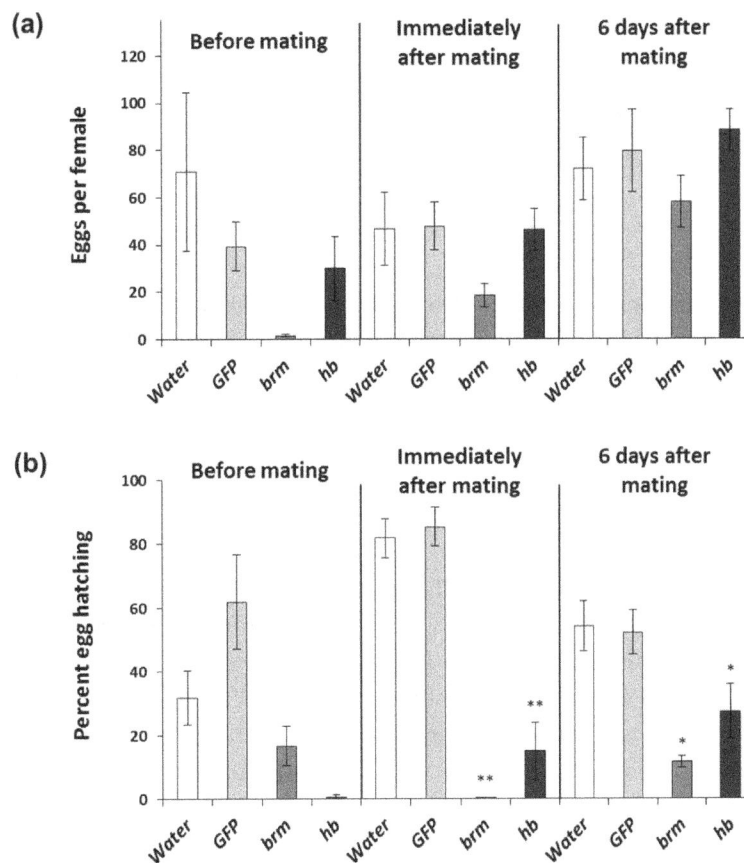

Figure 3. Timing of exposure effects on pRNAi response using *brahma* (*brm*) and *hunchback* (*hb*) dsRNA in *D. v. virgifera*. Females were fed diet with 2 µg dsRNA six times before mating, six times immediately after mating, and six times six days after mating. Diet treated with water and *GFP* dsRNA were used as controls. (**a**) Eggs collected from dsRNA-fed females after last feeding exposure; (**b**) Eggs hatched based on the numbers oviposited. Comparisons performed with Dunnett's test (control group = water), * significance at $p < 0.05$. ** significance at $p < 0.001$.

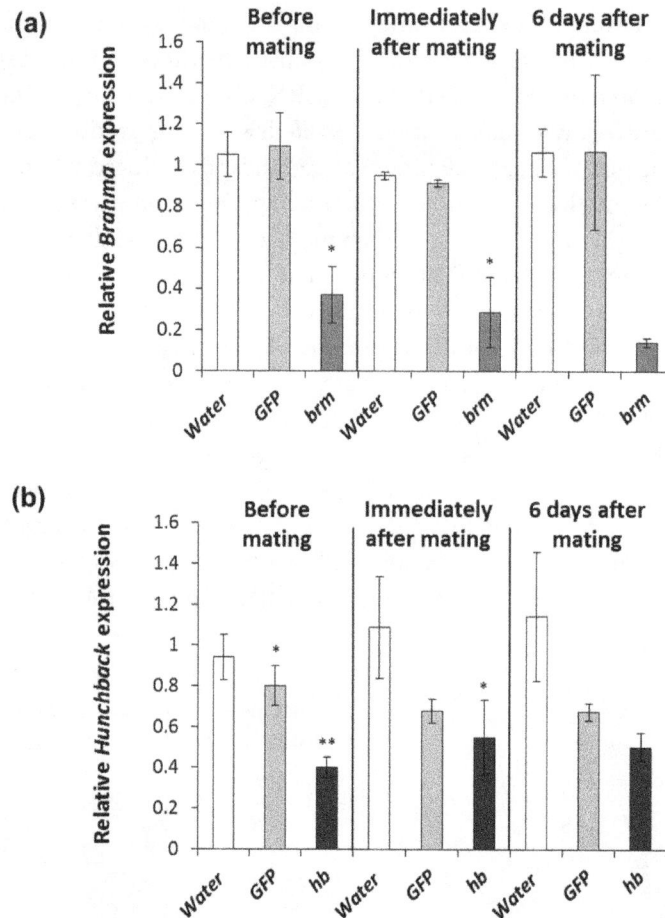

Figure 4. Relative transcript level of *brahma* (*brm*) and *hunchback* (*hb*) in *D. v. virgifera* females: (**a**) relative *brahma* transcript expression for timing of exposure; and (**b**) relative *hunchback* transcript expression from timing of exposure. Comparisons performed with Dunnett's test (control group = water), * significance at $p < 0.05$. ** significance at $p < 0.001$.

3.4. Effects of brahma and hunchback on Males

Fertility of males exposed to *brm* and *hb* dsRNA was assessed by testing sperm viability using fluorescent staining techniques [57] and viability of the offspring of males exposed to dsRNA. Sperm cells stained using Live/Dead Viability Kit yielded green fluorescence (500–550 nm) if live and red fluorescence (663–738 nm) if dead (Figure 5a). Results indicated that the overall sperm count was significantly lower in males fed four times with *brm* dsRNA. Whereas males treated with *hb* dsRNA showed a lower live and total number of sperm compared to the controls, although these numbers were not significantly different from water-treated controls (Figure 5b). Even though we observed an overall reduction in the number of sperm, there was no impact on the number of eggs and egg viability from females mated with males exposed to *brm* and *hb* dsRNA after four exposures over eight days (Figure 6a,b) and six exposures over 12 days (Figure 6c,d). A significant reduction of transcript levels was achieved when males were fed *hb* dsRNA but not *brm* when fed dsRNA four times (Supplementary Materials Figure S3a,b); yet significant reduction in transcript abundance was observed when fed dsRNA six times over 12 days for both genes (Supplementary Materials Figure S3c,d). To determine if the differences in gene knockdown between females and males were sex-specific, females and males were fed 2 μg *brm* and *hb* dsRNA four and six times. Gene knockdown for both genes in males was higher or similar to females (Supplementary Materials Figure S3).

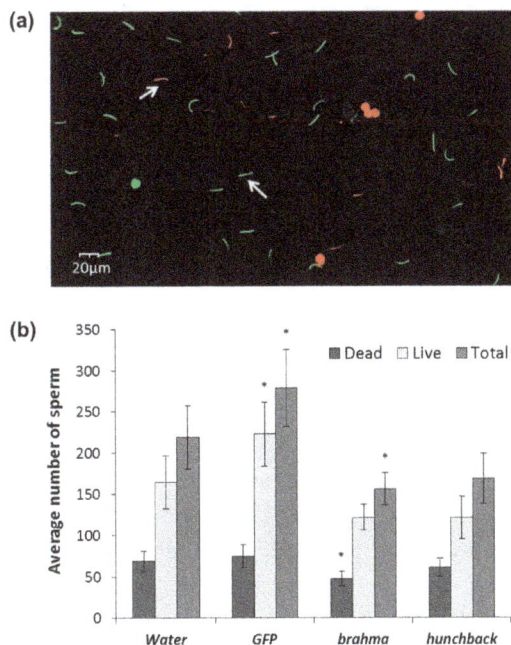

Figure 5. Sperm viability of *D. v. virgifera* males exposed to *brahma* (*brm*) and *hunchback* (*hb*) dsRNA: (**a**) sample composite digital image of live (green) and dead (red) sperm from half of a single field of view (10X); and (**b**) total number of sperm (live + dead) of males exposed four times to *brm* and *hb* dsRNA. Comparisons performed with Dunnett's test (control group = water), * significance at $p < 0.05$.

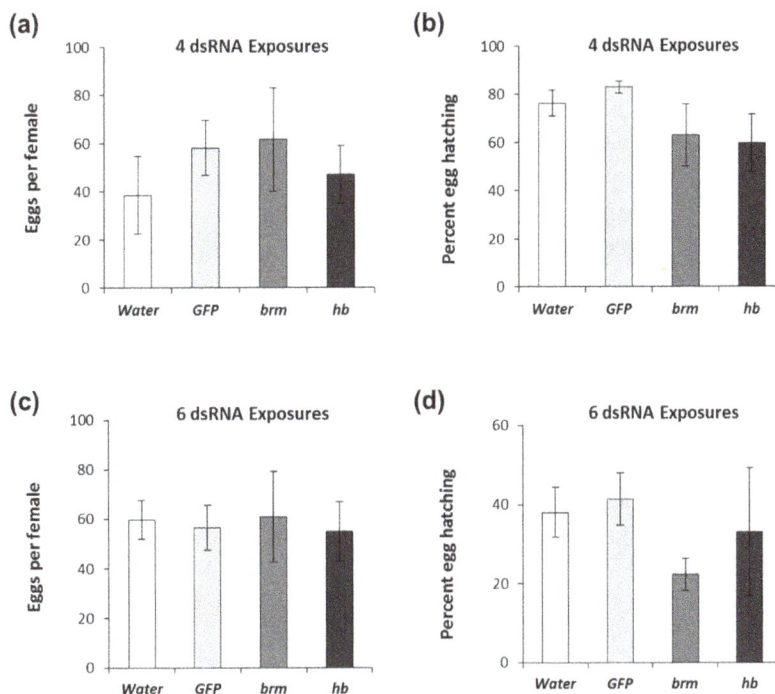

Figure 6. pRNAi response to *brahma* (*brm*) and *hunchback* (*hb*) dsRNA of *D. v. virgifera* males. Males were fed with diet treated with 2 μg dsRNA. Treatment was provided four and six times every other day and mated with females immediately after receiving all dsRNA treatments. (**a**) Eggs collected from females mated with males with four dsRNA exposures; (**b**) Eggs hatched based on numbers oviposited; (**c**) Eggs collected from females mated with males with six dsRNA exposures; (**d**) Eggs hatched based on numbers oviposited. Comparisons performed with Dunnett's test (control group = water), * significance at $p < 0.05$.

4. Discussion

The information obtained in this study informs the discussion on dsRNA exposure as it relates to achieving pRNAi responses that may be applied to management of corn rootworm populations. Moreover, this work began to examine the dynamics of the RNAi response in insects that may go beyond pRNAi and crop pests. Adult insects appear to be unaffected by pRNAi targets, allowing estimation of parameters such as the duration of gene knockdown. These parameters are more difficult to measure with genes that affect pigmentation or generate lethality, given that the effects may be confounded by the phenotype itself (i.e., one cannot monitor a recovery of the response once the treated insects are dead; parameters such as recovery in pigmentation may be slower and more difficult to quantify over time). While qRT-PCR may be used to accurately measure gene knockdown [61], it does not take into account protein turnover, which will have profound effects on the outcome or phenotype of the RNAi treatment. Therefore, the use of pRNAi as a model RNAi system may enable a better understanding of the concentration-over-time exposures, onset of the RNAi effect, and interactions of different dsRNA treatments given that changes in both the transcript levels and the phenotype can be quantified. Based on our observations, we postulate that the pRNAi could be used as a model to better understand the RNAi response in insects in general.

The experiments performed with WCR allowed us to quantify the level of exposure to dsRNA that consistently produces a pRNAi response in females. Our results suggest that there is a correlation between the response and the concentration of dsRNA and the exposure time. We observed that the concentration required to generate a reduction in egg hatching for both *brm* and *hb* was at least 0.2 µg per diet pellet with six exposures over twelve days of feeding (Figure 1b). When the dsRNA amount is fixed at 2 µg per diet pellet, the duration of feeding should be of at least four days (two exposures) for *brm* and eight days (four exposures) for *hb* (Figure 2b). In the above experiments, ten WCR females were provided with eleven diet pellets with various amounts of dsRNA every other day. Thus, approximately, 1.1 pellets were provided for each female and the diet pellets were consumed in their entirety in most of the experiments. This setup provides rough estimates of dsRNA consumption per insect and over time (e.g., 1.1 µg of dsRNA per day per female when the dsRNA amount was 2 µg/pellet over two days). Extrapolating from these artificial diet-based observations, females would need to consume approximately 1.1 µg of dsRNA per day over a four-day period (4.4 µg total/female), 0.11 µg per day over a twelve-day period (1.32 µg total/female) or a combination of dose and duration that equals these parameters. Interestingly, we observed that six exposures at 0.2 µg dsRNA (1.32 µg/female) (Figure 2b) were more efficacious than a single exposure of 2 µg of dsRNA (2.2 µg/female) (Figure 2b). Since the single-exposure experiments lasted for the same period of time as the three-exposure experiments, the protein half-life is not the likely explanation for the difference. To determine the benefits of prolonged low-concentration dsRNA exposure vs. acute high-dose dsRNA application more detailed studies need to be performed.

Earlier studies have demonstrated robust and highly sensitive lethal RNAi response in WCR adults [9,62]. In the aforementioned studies, a similar WCR adult feeding approach was used. The LD_{50} for *v-ATPase A* was found to be ~500 ng/diet pellet of dsRNA, applied six times over twelve days [9,62]. For pRNAi, the 2 µg/diet pellet application is four times higher than LC_{50} of a lethal gene. However, is important to consider that *v-ATPase* genes are highly expressed in the WCR midgut [63], while *brahma* and *hunchback* are expressed in the ovaries. The RNAi response in WCR has been found to be systemic [63], hence the movement of dsRNA from the midgut to the ovaries could explain the higher amount of dsRNA needed for pRNAi genes. Additionally, the dose needed to trigger a pRNAi response in this study may reflect the lower sensitivity of the ovary to RNAi or the dose-response of the specific genes used to probe pRNAi in WCR.

It was recently postulated that there could be competition of siRNA and miRNA pathways [64]. Interestingly, in WCR, even at high doses, application of pRNAi or non-lethal dsRNA targets does not produce observable fitness effects [62]. It is also possible that even if the miRNA pathway is affected in response to dsRNA, the miRNA pathway may not be essential during the adult stages of WCR. The observations that *brm* and *hb* dsRNA treatments cause primarily egg hatch defects and no or low-level reduction in oviposition is consistent with no observable changes seen in the morphology of ovaries (Supplementary Materials Figure S2). In an earlier study, we observed *brahma* dsRNA-induced oviposition and ovary development phenotypes in the stink bug, *Euschistus heros* [26], hence the low-level oviposition phenotype in WCR was not surprising. Brahma and other chromatin remodeling ATPases are known to play various roles in oogenesis, early and late embryogenesis [65–69]. The difference in the effects of *brahma* dsRNA on oviposition in WCR and *E. heros* may stem from the differences in the function of these genes between different insect orders. Further, the parental RNAi approach for pest insect control does not necessarily exclude lethality. In the present study, the absence of strong morphological or lethal phenotypes in the adult insect enables a more accurate characterization of the pRNAi response. However, the best plant protection may be achieved by an RNAi trait that confers both lethal and parental effects.

The amount of plant material consumed by WCR is likely to vary depending on the nutritional value of the plant tissue and other biotic and abiotic factors that may affect rates of consumption. Therefore, the best studies to estimate the minimum in-plant dsRNA concentrations for a robust pRNAi response should be performed directly with hpRNA-expressing plant materials. In addition to dose and duration, important factors that need to be considered for successful field exposure include feeding, mating, and dispersal behaviors. Considering that adult rootworms can utilize a variety of plant materials as their food sources [36,37] and can readily move between transgenic, non-transgenic fields [70] as well as native weed species [36], it is likely that adults will not feed exclusively on a single plant. However, the strong fidelity of WCR to maize fields, and the fact that WCR females usually feed on maize tissue after emergence, during mating, or immediately after mating [30] suggest that pRNAi could potentially reduce fecundity in a field setting.

Our results suggest that females were more sensitive to pRNAi before mating and the sensitivity of the response seemed to decrease as the adult females aged. This suggests that for a stronger pRNAi response females should preferably feed on dsRNA before mating and immediately after mating, although a decline in egg hatching was still observed in females that fed after mating. Based on these results, pRNAi would be most successful if females feed on dsRNA before or immediately after mating. This will align well to the behaviors observed in the field, given that females feed on maize tissues immediately after mating to stimulate egg development [4,71]. In addition, after mating, WCR females tend to remain in their natal maize field for several days before dispersal [72]; this would be in the range of the four-day exposure that we tested in the lab. Females emerging from adjacent maize fields that do not express pRNAi and migrate to a field expressing a pRNAi trait will be exposed later in their reproductive development. Since we observed pRNAi phenotypes in females that were exposed to pRNAi six days after mating, even shorter exposures or exposure several days after mating may produce pRNAi effects. Given that only 5%–10% of eggs successfully establish in the field [73], greenhouse or field-based testing will be suited best to answer these more complex scenarios. In practice, the concept of pRNAi would best be implemented in maize plants in combination with Bt toxins and/or RNAi lethal genes so any emerging larvae will be potentially killed by maize expressing a Bt toxin and/or lethal RNAi. In the above pyramid, pRNAi would serve the function of an added control measure to extend durability [13,26].

Unlike the robust fecundity phenotypes observed in females, no egg viability defects were detected after dsRNA treatments of males. A decrease in the total number of sperm was observed after exposure of males to *brm* and *hb* dsRNA, however this decrease may not be enough to result in measurable changes in male fertility.

The experiments performed in this study provided a means to quantify the level of exposure to dsRNA that consistently produces a pRNAi response in exposed WCR females; this will assist in establishing a baseline for the potential efficacy of transgenic maize plants expressing hpRNA for pRNAi target sequences. The next step for the validation of this technology will be testing the efficacy of maize plants expressing long hairpin RNA for *brm* and *hb* and to correlate the effects of the successful pRNAi exposure parameters to behaviors of WCR adults. Furthermore, because females have been reported to oviposit for up to 60 days during their lifespan [42,74], it will be important to validate the onset and the longevity of the pRNAi response in WCR females. Further research evaluating the effects of *brm* and *hb* on larval survival, development, and the ability of larvae exposed to parental RNAi to produce offspring will provide a better understanding of the pRNAi response and its potential use for corn rootworm management.

The benefits of pRNAi for crop protection reach beyond WCR. In a recent publication, we demonstrated that dsRNA can generate a strong pRNAi response in the Neotropical brown stink bug *E. heros*, by injection [26]. For insects like stink bugs that have multiple generations per year, the use of a pRNAi strategy will have most benefit since it will control insects within the same season. However, there is no oral response in stink bugs. Once the barriers to oral delivery to lepidopteran and hemipteran insects are overcome, multiple pest management areas, particularly for multivoltine pests, may benefit from pRNAi.

5. Conclusions

In conclusion, this study has probed the concentration, duration, and timing of the exposure needed for a successful pRNAi response in WCR. Described herein, diet-based RNAi studies have an advantage in that the amount of dsRNA is tightly controlled, enabling basic research pertaining to the concentration and the exposure parameters of pRNAi and RNAi in general. These experiments provide a framework for plant-based testing of pRNAi, allowing for a more accurate assessment of the potential of pRNAi, when applied at the field level. A path forward for pRNAi as a pest management tool will build on this work to ascertain efficacy in transgenic plants and the longevity of the pRNAi effect. Furthermore, *brm*, *hb*, and potentially other pRNAi also provide a platform to better understand RNAi in insects.

Supplementary Materials: The following are available online at www.mdpi.com/2073/4425/8/1/7/s1, Figure S1: Parental *hunchback* dsRNA phenotypes in *D. v. virgifera*, Figure S2: Ovaries of *hunchback* and *brahma* dsRNA-fed *D. v. virgifera*, Figure S3: Comparison of relative transcript level for *brahma* (*brm*) and *hunchback* (*hb*) between *D. v. virgifera* females and males, Table S1: Primer pairs used to amplify DNA templates for *D. v. virgifera* dsRNA synthesis and qRT-PCR.

Acknowledgments: The authors thank Lance Meinke for providing insights on the ecology of rootworms. The authors also thank Ronda Hamm and Miles Lepping for a critical reading of the manuscript. John Wang is thanked for bioinformatics advice and contributions to study design. This research was supported through the University of Nebraska-Lincoln, Life Sciences Industry Partnership Grant Program with Dow AgroSciences.

Author Contributions: A.M.V., B.D.S., E.F. and K.E.N. conceived and designed the experiments; A.M.V. and N.M. performed the experiments; A.M.V. and E.F. analyzed the data; A.M.V., E.F., N.P.S. and B.D.S. interpreted the results; and A.M.V., E.F. and N.P.S. composed the manuscript.

References

1. Chandler, L.D.; Coppedge, J.R.; Edwards, C.R.; Tollefson, J.J.; Wilde, G.R.; Faust, R.M. Corn rootworm areawide pest management in the midwestern USA. In *Areawide Pest Management: Theory and Implementation*; CABI Publishing: Wallingford, UK, 2008; pp. 191–207.

2. Metcalf, R.L. Foreword. In *Methods for the Study of Pest Diabrotica*; Krysan, J.L., Miller, T.A, Eds.; Springer: New York, NY, USA, 1986.

3. Gray, M.E.; Sappington, T.W.; Miller, N.J.; Moeser, J.; Bohn, M.O. Adaptation and invasiveness of western corn rootworm: Intensifying research on a worsening pest. *Annu. Rev. Entomol.* **2009**, *54*, 303–321. [CrossRef] [PubMed]

4. Spencer, J.L.; Levine, E.; Isard, S.A.; Mabry, T.R. Movement, dispersal and behavior of western corn rootworm adults in rotated maize and soybean fields. In *Western Corn Rootworm: Ecology and Management*; Vidal, S., Kuhlmann, U., Edwards, C.R., Eds.; CABI Publishing: Wallingford, UK, 2005; pp. 121–144.

5. Ma, B.L.; Meloche, F.; Wei, L. Agronomic assessment of Bt trait and seed or soil-applied insecticides on the control of corn rootworm and yield. *Field Crops Res.* **2009**, *111*, 189–196. [CrossRef]

6. Marra, M.C.; Piggott, N.E.; Goodwin, B.K. The impact of corn rootworm protected biotechnology traits in the United States. *AgBioForum* **2012**, *15*, 217–230.

7. Baum, J.A.; Bogaert, T.; Clinton, W.; Heck, G.R.; Feldmann, P.; Ilagan, O.; Johnson, S.; Plaetinck, G.; Munyikwa, T.; Pleau, M.; et al. Control of coleopteran insect pests through RNA interference. *Nat. Biotechnol.* **2007**, *25*, 1322–1326. [CrossRef] [PubMed]

8. Bolognesi, R.; Ramaseshadri, P.; Anderson, J.; Bachman, P.; Clinton, W.; Flannagan, R.; Ilagan, O.; Lawrence, C.; Levine, S.; Moar, W.; et al. Characterizing the mechanism of action of double-stranded RNA activity against western corn rootworm (*Diabrotica virgifera virgifera* LeConte). *PLoS ONE* **2012**, *7*, e47534. [CrossRef] [PubMed]

9. Rangasamy, M.; Siegfried, B.D. Validation of RNA interference in western corn rootworm *Diabrotica virgifera virgifera* LeConte (Coleoptera: Chrysomelidae) adults. *Pest Manag. Sci.* **2012**, *68*, 587–591. [CrossRef] [PubMed]

10. Ahmad, A.; Negri, I.; Oliveira, W.; Brown, C.; Asiimwe, P.; Sammons, B.; Horak, M.; Jiang, C.; Carson, D. Transportable data from non-target arthropod field studies for the environmental risk assessment of genetically modified maize expressing an insecticidal double-stranded RNA. *Transgenic Res.* **2016**, *25*, 1–17. [CrossRef] [PubMed]

11. Petrick, J.S.; Frierdich, G.E.; Carleton, S.M.; Kessenich, C.R.; Silvanovich, A.; Zhang, Y.; Koch, M.S. Corn rootworm-active RNA DvSnf7: Repeat dose oral toxicology assessment in support of human and mammalian safety. *Regula. Toxicol. Pharm.* **2016**. [CrossRef] [PubMed]

12. Levine, S.L.; Tan, J.; Mueller, G.M.; Bachman, P.M.; Jensen, P.D.; Uffman, J.P. Independent action between DvSnf7 RNA and Cry3Bb1 protein in southern corn rootworm, *Diabrotica undecimpunctata howardi* and Colorado potato beetle, *Leptinotarsa decemlineata*. *PLoS ONE* **2015**, *10*, e0118622. [CrossRef] [PubMed]

13. Fishilevich, E.; Vélez, A.M.; Storer, N.P.; Li, H.; Bowling, A.J.; Rangasamy, M.; Worden, S.E.; Narva, K.E.; Siegfried, B.D. RNAi as a pest management tool for the western corn rootworm, *Diabrotica virgifera virgifera*. *Pest Manag. Sci.* **2016**. [CrossRef] [PubMed]

14. Bucher, G.; Scholten, J.; Klingler, M. Parental RNAi in *Tribolium* (Coleoptera). *Curr. Biol.* **2002**, *12*, R85–R86. [CrossRef]

15. Shukla, J.N.; Palli, S.R. Production of all female progeny: Evidence for the presence of the male sex determination factor on the Y chromosome. *J. Exp. Biol.* **2014**, *217*, 1653–1655. [CrossRef] [PubMed]

16. Khajuria, C.; Vélez, A.M.; Rangasamy, M.; Wang, H.; Fishilevich, E.; Frey, M.L.F.; Carneiro, N.; Premchand, G.; Narva, K.E.; Siegfried, B.D. Parental RNA interference of genes involved in embryonic development of the western corn rootworm, *Diabrotica virgifera virgifera* LeConte. *Insect Biochem. Mol. Biol.* **2015**, *63*, 54–62. [CrossRef] [PubMed]

17. Konopova, B.; Akam, M. The *hox* genes *ultrabithorax* and *abdominal-a* specify three different types of abdominal appendage in the springtail *Orchesella cincta* (Collembola). *Evodevo* **2014**. [CrossRef] [PubMed]

18. Mito, T.; Okamoto, H.; Shinahara, W.; Shinmyo, Y.; Miyawaki, K.; Ohuchi, H.; Noji, S. *Kruppel* acts as a gap gene regulating expression of *hunchback* and even-skipped in the intermediate germ cricket *Gryllus bimaculatus*. *Dev. Biol.* **2006**, *294*, 471–481. [CrossRef] [PubMed]

19. Liu, P.Z.; Kaufman, T.C. *Hunchback* is required for suppression of abdominal identity, and for proper germband growth and segmentation in the intermediate germband insect *Oncopeltus fasciatus*. *Development* **2004**, *131*, 1515–1527. [CrossRef] [PubMed]

20. Paim, R.M.; Araujo, R.N.; Lehane, M.J.; Gontijo, N.F.; Pereira, M.H. Long-term effects and parental RNAi in the blood feeder *Rhodnius prolixus* (Hemiptera: Reduviidae). *Insect Biochem. Mol. Biol.* **2013**, *43*, 1015–1020. [CrossRef] [PubMed]

21. Coleman, A.D.; Wouters, R.H.; Mugford, S.T.; Hogenhout, S.A. Persistence and transgenerational effect of plant-mediated RNAi in aphids. *J. Exp. Bot.* **2015**, *66*, 541–548. [CrossRef] [PubMed]

22. Mao, J.J.; Liu, C.Y.; Zeng, F.R. *Hunchback* is required for abdominal identity supression and germband growth in the parthenogenetic embryogenesis of the pea aphis, *Acyrthosiphon pisum*. *Arch. Insect Biochem.* **2013**, *84*, 209–221. [CrossRef] [PubMed]

23. Yoshiyama, N.; Tojo, K.; Hatakeyama, M. A survey of the effectiveness of non-cell autonomous RNAi throughout development in the sawfly, *Athalia rosae* (Hymenoptera). *J. Insect Physiol.* **2013**, *59*, 400–407. [CrossRef] [PubMed]

24. Piulachs, M.D.; Pagone, V.; Belles, X. Key roles of the broad-complex gene in insect embryogenesis. *Insect Biochem. Mol.* **2010**, *40*, 468–475. [CrossRef] [PubMed]

25. Nakao, H. Anterior and posterior centers jointly regulate *Bombyx embryo* body segmentation. *Dev. Biol.* **2012**, *371*, 293–301. [CrossRef] [PubMed]

26. Fishilevich, E.; Velez, A.M.; Khajuria, C.; Frey, M.L.; Hamm, R.L.; Wang, H.; Schulenberg, G.A.; Bowling, A.J.; Pence, H.E.; Gandra, P.; et al. Use of chromatin remodeling atpases as RNAi targets for parental control of western corn rootworm (*Diabrotica virgifera virgifera*) and neotropical brown stink bug (*Euschistus heros*). *Insect Biochem. Mol. Biol.* **2016**, *71*, 58–71. [CrossRef] [PubMed]

27. Miyata, K.; Ramaseshadri, P.; Zhang, Y.J.; Segers, G.; Bolognesi, R.; Tomoyasu, Y. Establishing an in vivo assay system to identify components involved in environmental RNA interference in the western corn rootworm. *PLoS ONE* **2014**, *9*, e101661. [CrossRef] [PubMed]

28. Branson, T.F.; Krysan, J.L. Feeding and oviposition behavior and life cycle strategies of *Diabrotica*: An evolutionary view with implications for pest management (Coleoptera, Chrysomelidae). *Environ. Entomol.* **1981**, *10*, 826–831. [CrossRef]

29. Meinke, L.J.; Sappington, T.W.; Onstad, D.W.; Guillemaud, T.; Miller, N.J.; Komáromi, J.; Levay, N.; Furlan, L.; Kiss, J.; Toth, F. Western corn rootworm (*Diabrotica virgifera virgifera* LeConte) population dynamics. *Agric. Forest Entomol.* **2009**, *11*, 29–46. [CrossRef]

30. Spencer, J.L.; Hibbard, B.E.; Moeser, J.; Onstad, D.W. Behaviour and ecology of the western corn rootworm (*Diabrotica virgifera virgifera* LeConte). *Agric. Forest Entomol.* **2009**, *11*, 9–27. [CrossRef]

31. Krysan, J.L. Introduction: Biology, distribution, and identification of pest *Diabrotica*. In *Methods for the Study of Pest Diabrotica*; Krysan, J.L., Miller, T.A., Eds.; Springer: New York, NY, USA, 1986; pp. 1–23.

32. Oyediran, I.O.; Hibbard, B.E.; Clark, T.L. Prairie grasses as hosts of the western corn rootworm (Coleoptera: Chrysomelidae). *Environ. Entomol.* **2004**, *33*, 740–747. [CrossRef]

33. Mabry, T.R.; Spencer, J.L.; Levine, E.; Isard, S.A. Western corn rootworm (Coleoptera: Chrysomelidae) behavior is affected by alternating diets of corn and soybean. *Environ. Entomol.* **2004**, *33*, 860–871. [CrossRef]

34. Siegfried, B.D.; Mullin, C.A. Effects of alternative host plants on longevity, oviposition, and emergence of western and northern corn rootworms (Coleoptera, Chrysomelidae). *Environ. Entomol.* **1990**, *19*, 474–480. [CrossRef]

35. Shaw, J.T.; Paullus, J.H.; Luckmann, W.H. Corn rootworm (Coleoptera Chrysomelidae) oviposition in soybeans. *J. Econ. Entomol.* **1978**, *71*, 189–191. [CrossRef]

36. Campbell, L.A.; Meinke, L.J. Seasonality and adult habitat use by four *Diabrotica* species at prairie-corn interfaces. *Environ. Entomol.* **2006**, *35*, 922–936. [CrossRef]

37. Moeser, J.; Vidal, S. Nutritional resources used by the invasive maize pest *Diabrotica virgifera virgifera* in its new south-east-european distribution range. *Entomol. Exp. Appl.* **2005**, *114*, 55–63. [CrossRef]

38. Branson, T.F. The contribution of prehatch and posthatch development to protandry in the Chrysomelid, *Diabrotica virgifera virgifera*. *Entomol. Exp. Appl.* **1987**, *43*, 205–208. [CrossRef]

39. Guss, P.L. The sex pheromone of the western corn rootworm (*Diabrotica virgifera*). *Environ. Entomol.* **1976**, *5*, 219–223. [CrossRef]

40. Hammack, L. Calling behavior in female western corn rootworm beetles (Coleoptera, Chrysomelidae). *Ann. Entomol. Soc. Am.* **1995**, *88*, 562–569. [CrossRef]

41. Ball, H.J. On the biology and egg-laying habits of the western corn rootworm. *J. Econ. Entomol.* **1957**, *50*, 126–128. [CrossRef]

42. Hill, R.E. Mating, oviposition patterns, fecundity and longevity of the western corn rootworm. *J. Econ. Entomol.* **1975**, *68*, 311–315. [CrossRef]

43. Branson, T.F.; Guss, P.L.; Jackson, J.J. Mating frequency of the western corn rootworm. *Ann. Entomol. Soc. Am.* **1977**, *70*, 506–508. [CrossRef]

44. Quiring, D.T.; Timmins, P.R. Influence of reproductive ecology on feasibility of mass trapping *Diabrotica virgifera virgifera* (Coleoptera, Chrysomelidae). *J. Appl. Ecol.* **1990**, *27*, 965–982. [CrossRef]

45. Coats, S.A.; Tollefson, J.J.; Mutchmor, J.A. Study of migratory flight in the western corn rootworm (Coleoptera, Chrysomelidae). *Environ. Entomol.* **1986**, *15*, 620–625. [CrossRef]

46. Naranjo, S.E. Movement of corn rootworm beetles, *Diabrotica* spp. (Coleoptera, Chrysomelidae), at cornfield boundaries in relation to sex, reproductive status, and crop phenology. *Environ. Entomol.* **1991**, *20*, 230–240. [CrossRef]

47. Stebbing, J.A.; Meinke, L.J.; Naranjo, S.E.; Siegfried, B.D.; Wright, R.J.; Chandler, L.D. Flight behavior of methyl-parathion-resistant and -susceptible western corn rootworm (Coleoptera: Chrysomelidae) populations from Nebraska. *J. Econ. Entomol.* **2005**, *98*, 1294–1304. [CrossRef] [PubMed]

48. Darnell, S.J.; Meinke, L.T.; Young, L.J.; Gotway, C.A. Geostatistical investigation of the small-scale spatial variation of western corn rootworm (Coleoptera: Chrysomelidae) adults. *Environ. Entomol.* **1999**, *28*, 266–274. [CrossRef]

49. Pierce, C.M.F.; Gray, M.E. Western corn rootworm, *Diabrotica virgifera virgifera* LeConte (Coleoptera: Chrysomelidae), oviposition: A variant's response to maize phenology. *Environ. Entomol.* **2006**, *35*, 423–434. [CrossRef]

50. Cagan, L.; Rosca, I. Seasonal dispersal of the western corn rootworm (*Diabrotica virgifera virgifera*) adults in Bt and non-Bt maize fields. *Plant Prot. Sci.* **2012**, *48*, S36–S42.

51. Steffey, K.L.; Tollefson, J.J. Spatial dispersion patterns of northern and western corn rootworm adults in Iowa cornfields. *Environ. Entomol.* **1982**, *11*, 283–286. [CrossRef]

52. Eyun, S.I.; Wang, H.C.; Pauchet, Y.; Ffrench-Constant, R.H.; Benson, A.K.; Valencia-Jimenez, A.; Moriyama, E.N.; Siegfried, B.D. Molecular evolution of glycoside hydrolase genes in the western corn rootworm (*Diabrotica virgifera virgifera*). *PLoS ONE* **2014**, *9*, e94052. [CrossRef] [PubMed]

53. Branson, T.F.; Jackson, J.J. An improved diet for adult *Diabrotica virgifera vrgifera* (Coleoptera, Chrysomelidae). *J. Kansas Entomol. Soc.* **1988**, *61*, 353–355.

54. Campbell, L.A.; Meinke, L.J. Fitness of *Diabrotica barberi*, *Diabrotica longicornis*, and their hybrids (Coleoptera: Chrysomelidae). *Ann. Entomol. Soc. Am.* **2010**, *103*, 925–935. [CrossRef]

55. Jackson, J.J. Rearing and handling of *Diabrotica virgifera* and *Diabrotica undecimpunctata howardi*. In *Methods for the Study of Pest Diabrotica*; Krysan, J.L., Miller, T.A., Eds.; Springer: New York, NY, USA, 1986; pp. 25–47.

56. Schneider, C.A.; Rasband, W.S.; Eliceiri, K.W. NIH image to ImageJ: 25 years of image analysis. *Nat. Methods* **2012**, *9*, 671–675. [CrossRef] [PubMed]

57. Johnson, R.M.; Dahlgren, L.; Siegfried, B.D.; Ellis, M.D. Effect of in-hive miticides on drone honey bee survival and sperm viability. *J. Apic. Res.* **2013**, *52*, 88–95. [CrossRef]

58. Rodrigues, T.B.; Khajuria, C.; Wang, H.; Matz, N.; Cunha Cardoso, D.; Valicente, F.H.; Zhou, X.; Siegfried, B. Validation of reference housekeeping genes for gene expression studies in western corn rootworm (*Diabrotica virgifera virgifera*). *PLoS ONE* **2014**, *9*, e109825. [CrossRef] [PubMed]

59. Livak, K.J.; Schmittgen, T.D. Analysis of relative gene expression data using real-time quantitative PCR and the $2^{-\Delta\Delta CT}$ method. *Methods* **2001**, *25*, 402–408. [CrossRef] [PubMed]

60. *JMP®*, version 11 pro; SAS Institute Inc.: Cary, NC, USA, 1989–2007.

61. Wynant, N.; Verlinden, H.; Breugelmans, B.; Simonet, G.; Vanden Broeck, J. Tissue-dependence and sensitivity of the systemic rna interference response in the desert locust, *Schistocerca gregaria*. *Insect Biochem. Mol. Biol.* **2012**, *42*, 911–917. [CrossRef] [PubMed]

62. Vélez, A.M.; Khajuria, C.; Wang, H.; Narva, K.E.; Siegfried, B.D. Knockdown of RNA interference pathway genes in western corn rootworms (*Diabrotica virgifera virgifera* LeConte) demonstrates a possible mechanism of resistance to lethal dsRNA. *PLoS ONE* **2016**, *11*, e0157520. [CrossRef] [PubMed]

63. Li, H.; Bowling, A.J.; Gandra, P.; Rangasamy, M.; Pence, H.E.; McEwan, R.; Khajuria, C.; Siegfried, B.; Narva, K.E. Systemic RNAi in western corn rootworm, *Diabrotica virgifera virgifera* leconte, does not involve transitive pathways. *Insect Sci.* **2016**. [CrossRef] [PubMed]

64. Khan, A.A.; Betel, D.; Miller, M.L.; Sander, C.; Leslie, C.S.; Marks, D.S. Transfection of small RNAs globally perturbs gene regulation by endogenous microRNAs. *Nat. Biotechnol.* **2009**, *27*, 549–555. [CrossRef] [PubMed]

65. Brizuela, B.J.; Elfring, L.; Ballard, J.; Tamkun, J.W.; Kennison, J.A. Genetic analysis of the *brahma* gene of *Drosophila melanogaster* and polytene chromosome subdivisions 72ab. *Genetics* **1994**, *137*, 803–813. [PubMed]

66. Deuring, R.; Fanti, L.; Armstrong, J.A.; Sarte, M.; Papoulas, O.; Prestel, M.; Daubresse, G.; Verardo, M.; Moseley, S.L.; Berloco, M.; et al. The ISWI chromatin-remodeling protein is required for gene expression and the maintenance of higher order chromatin structure in vivo. *Mol. Cell* **2000**, *5*, 355–365. [CrossRef]

67. McDaniel, I.E.; Lee, J.M.; Berger, M.S.; Hanagami, C.K.; Armstrong, J.A. Investigations of CHD1 function in transcription and development of *Drosophila melanogaster*. *Genetics* **2008**, *178*, 583–587. [CrossRef] [PubMed]

68. Kehle, J.; Beuchle, D.; Treuheit, S.; Christen, B.; Kennison, J.A.; Bienz, M.; Muller, J. dMi-2, a hunchback-interacting protein that functions in polycomb repression. *Science* **1998**, *282*, 1897–1900. [CrossRef] [PubMed]

69. He, J.; Xuan, T.; Xin, T.; An, H.; Wang, J.; Zhao, G.; Li, M. Evidence for chromatin-remodeling complex pbap-controlled maintenance of the *Drosophila* ovarian germline stem cells. *PLoS ONE* **2014**, *9*, e103473. [CrossRef] [PubMed]

70. Marquardt, P.T.; Krupke, C.H. Dispersal and mating behavior of *Diabrotica virgifera virgifera* (Coleoptera: Chrysomelidae) in Bt cornfields. *Environ. Entomol.* **2009**, *38*, 176–182. [CrossRef] [PubMed]

71. Elliott, N.C.; Gustin, R.D.; Hanson, S.L. Influence of adult diet on the reproductive biology and survival of the western corn rootworm, *Diabrotica virgifera virgifera*. *Entomol. Exp. Appl.* **1990**, *56*, 15–21. [CrossRef]

72. Isard, S.A.; Spencer, J.L.; Mabry, T.R.; Levine, E. Influence of atmospheric conditions on high-elevation flight of western corn rootworm (Coleoptera: Chrysomelidae). *Environ. Entomol.* **2004**, *33*, 650–656. [CrossRef]

73. Hibbard, B.E.; Higdon, M.L.; Duran, D.P.; Schweikert, Y.M.; Ellersieck, M.R. Role of egg density on establishment and plant-to-plant movement by western corn rootworm larvae (Coleoptera: Chrysomelidae). *J. Econ. Entomol.* **2004**, *97*, 871–882. [CrossRef] [PubMed]

74. Fisher, J.R.; Sutter, G.R.; Branson, T.F. Influence of corn planting date on the survival and on some reproductive parameters of *Diabrotica virgifera virgifera* (Coleoptera, Chrysomelidae). *Environ. Entomol.* **1991**, *20*, 185–189. [CrossRef]

Neurotoxic Doses of Chronic Methamphetamine Trigger Retrotransposition of the Identifier Element in Rat Dorsal Dentate Gyrus

Anna Moszczynska [1,*], Kyle J. Burghardt [2] and Dongyue Yu [1]

[1] Department of Pharmaceutical Sciences, Eugene Applebaum College of Pharmacy and Health Sciences, Wayne State University, Detroit, MI 48201, USA; dongyue.yu@wayne.edu

[2] Department of Pharmacy Practice, Eugene Applebaum College of Pharmacy and Health Sciences, Wayne State University, Detroit, MI 48201, USA; kburg@wayne.edu

* Correspondence: amosz@wayne.edu

Academic Editor: Dennis R. Grayson

Abstract: Short interspersed elements (SINEs) are typically silenced by DNA hypermethylation in somatic cells, but can retrotranspose in proliferating cells during adult neurogenesis. Hypomethylation caused by disease pathology or genotoxic stress leads to genomic instability of SINEs. The goal of the present investigation was to determine whether neurotoxic doses of binge or chronic methamphetamine (METH) trigger retrotransposition of the identifier (ID) element, a member of the rat SINE family, in the dentate gyrus genomic DNA. Adult male Sprague-Dawley rats were treated with saline or high doses of binge or chronic METH and sacrificed at three different time points thereafter. DNA methylation analysis, immunohistochemistry and next-generation sequencing (NGS) were performed on the dorsal dentate gyrus samples. Binge METH triggered hypomethylation, while chronic METH triggered hypermethylation of the CpG-2 site. Both METH regimens were associated with increased intensities in poly(A)-binding protein 1 (PABP1, a SINE regulatory protein)-like immunohistochemical staining in the dentate gyrus. The amplification of several ID element sequences was significantly higher in the chronic METH group than in the control group a week after METH, and they mapped to genes coding for proteins regulating cell growth and proliferation, transcription, protein function as well as for a variety of transporters. The results suggest that chronic METH induces ID element retrotransposition in the dorsal dentate gyrus and may affect hippocampal neurogenesis.

Keywords: methamphetamine; DNA methylation; identifier element; short interspersed elements; rat brain; dentate gyrus; retrotransposition

1. Introduction

Transposable elements are non-coding pieces of DNA with the ability to "jump" to different locations in the genome upon activation [1]. They can control genes epigenetically when inserted (transposed) into genes or gene regulatory regions. Activation of transposable elements most often depends on DNA hypomethylation at promoter regions [2–4]. Their transposition to another location in the genome takes place during both embryonic and adult neurogenesis [1,5–7]. Short interspersed elements (SINEs) are non-autonomous transposable elements known as short retrotransposons [8]. They are short DNA sequences with a length of 80–400 bp that are dispersed over the eukaryotic genome, are amplified by a reverse transcription and retrotranspose by a copy-and-paste mechanism [9]. In contrast to long retrotransposons (LINEs), SINEs do not code for a reverse transcriptase [4,10] and, therefore, they must interact with endogenous cellular factors to retrotranspose. Specifically, SINEs are reverse-transcribed by partner LINE-1s [10–12] and interact with cellular proteins, including poly(A)-binding proteins (PABPs) to form RNA-protein (RNP) complexes in the cytoplasm [13] that later translocate to the nucleus for the insertion of SINEs into DNA.

Identifier (ID) elements belong to the group of simple SINEs and are widespread in the rat genome [14]. The human counterpart to the ID element is the *Alu* element, while its mouse counterparts are the B1 and B2 elements. ID elements are about 100 bp long and consists of a core domain containing an internal RNA polymerase III promoter, a poly(A) region and 5'- and 3'-flanking regions [14,15]. Due to their widespread presence in the rat genome, ID elements have been used as determinants of global methylation in rat tissues [16]. Human and mouse counterparts of the rat ID element were demonstrated to be activated by hypomethylation [17]; therefore, ID element transcription is most likely activated by removal of methyl groups within its sequence. Exposure to heat shock, genotoxic agents, mechanical damage or ischemia increases SINEs transcription, which can lead to their retrotransposition [10,11,18–21]. Increases in SINE transcripts, as well as SINE retrotransposition can regulate gene expression, with the former having a short-term and the latter a long-term effect. To date, no studies have examined ID element methylation status or its retrotransposition after neurotoxic doses of methamphetamine (METH).

METH is a widely-abused central nervous system (CNS) psychostimulant, which reduces hippocampal volume and induces apoptosis in the hippocampus in experimental animals and humans, particularly when administered at high doses [22–26]. METH also affects adult neurogenesis in the hippocampus [22,27,28]. These molecular events are thought to underlie a variety of cognitive impairments observed in chronic human METH users [29]. Relatively little is known about epigenetic changes induced by neurotoxic doses of METH in adult hippocampus. To our knowledge, only our laboratory investigated METH effects on transposable elements in the hippocampus and found increased LINE-1 expression in the dentate gyrus [30]. The aim of the present investigation was to determine whether binge or chronic administration of neurotoxic METH doses leads to retrotransposition of the ID element in the dentate gyrus of adult male rats. Toward this goal, we measured ID element methylation over time, as well as the diversity in amplification of the element in DNA samples from the dentate gyrus of saline- and METH-treated rats. We also assessed the levels of PABP1, a putative ID element-binding protein with a high affinity for the poly(A) tail of mRNA [31] and regulator of SINE retrotransposition [13].

We have found that binge METH causes hypomethylation of CpG-2 site within the ID element sequence at 1 h and 24 h after the last injection of the drug, while chronic METH causes hypermethylation of this site at 1 h after the last METH injection as compared to saline-treated control rats. The methylation status of the ID element returned to basal levels by the seventh day after binge METH and by 24 h after chronic METH. Seven days after chronic METH, the CpG-2 site displayed small, but statistically significant hypermethylation. An increase in the levels of PABP1 was detected at 24 h after binge METH and after two days of chronic METH administration. Regarding METH-induced ID element retrotransposition, the loci with the greatest difference in relative ID element amplification between METH and control samples were mapped to genes encoding for proteins regulating cell growth and proliferation, transcription, protein function as well as for a variety of transporters.

2. Materials and Methods

2.1. Animals

Adult male Sprague-Dawley rats (Harlan, Indianapolis, IN, USA) (weighing 250–300 g on arrival) were pair-housed under a 12-h light/dark cycle in a temperature- (20–22 °C) and humidity-controlled room. Food and water were available *ad libitum*. The animals were allowed to acclimate for a week before the start of the study. All animal procedures were conducted between 7:00 a.m. and 7:00 p.m. in strict accordance with the National Institutes of Health (NIH) Guide for Care and Use of Laboratory Animals and approved by the Institutional Animal Care and Use Committee (IACUC) at Wayne

State University (Animal Protocol #A 05-07-13). The description of the animal procedures meets the Animal Research: Reporting of in Vivo Experiments (ARRIVE) recommended guidelines described by The National Centre for the Replacement, Refinement and Reduction of Animals in Research [32].

2.2. Administration of Methamphetamine

(+)-Methamphetamine hydrochloride (METH) (Sigma-Aldrich, St. Louis, MO, USA) or saline was administered to rats in a binge (10 mg/kg, every 2 h in four successive intraperitoneal (i.p.) injections) or chronically (20 mg/kg, daily, for 10 days, i.p.). Both paradigms are established as models of METH neurotoxicity in rats and other experimental animals. METH neurotoxicity (neurodegeneration or damage to neuronal components within neuronal terminals or cell bodies causing dysregulation of neuronal function) is associated with hyperthermia [33], which peaks at approximately at 1 h after i.p. injection of METH. Therefore, core body temperatures were measured via a rectal probe (Thermalert TH-8; Physitemp Instruments, Clifton, NJ, USA) before the treatments (baseline temperatures) and 1 h after each METH or saline injection. Rats were sacrificed by decapitation at 1 h, 24 h or 7 days after the last injection of the drug or saline. For some analyses, rats were sacrificed 24 h after the first two days of chronic METH administration, at which point the rats were injected with the same total dose of METH as the binge METH rats (2×20 mg/kg and 4×10 mg/kg, respectively). The experimental design is shown in Figure 1.

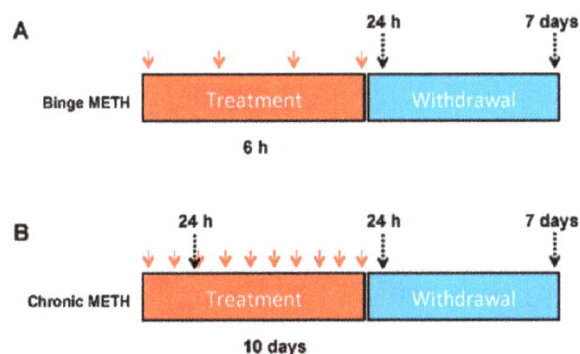

Figure 1. Experimental design. Adult male Sprague-Dawley rats were administered saline (1 mL/kg), binge METH (4×10 mg/kg, i.p. every 2 h) (**A**) or chronic METH (20 mg/kg/day for 10 days, i.p.) (**B**). Core body temperatures (°C) were measured before treatments and 1 h after each METH or saline injection. The rats were sacrificed 24 h or 7 days after the last dose of saline or METH or on the 3rd day of chronic METH regimen (24 h after the 2nd dose). The red arrows indicate the injection times, whereas the black arrows indicate sacrifice times.

2.3. Tissue Collection

The brains were removed and dissected out into discrete brain areas (striatum, dentate gyrus, Ammon's horn, prefrontal cortex and cerebellum) and stored at −80 °C until assayed. The subgranular zone (SGZ) was dissected out together with the dentate gyrus. The dentate gyrus was dissected out according to a previously described protocol [34] modified for the rat brain. Briefly, the brain was cut sagittally to divide the hemispheres, which were then placed medial side up after removal of the regions posterior to lambda. The dentate gyrus and Ammon's horn (CA1, CA2 and CA3) (which were visible upon removal of thalamus and hypothalamus) were dissected out using fine tip surgical instruments. Muscle tissues (negative control for METH effects) were also collected and stored at −80 °C until analyzed.

2.4. Polymerase Chain Reaction and Pyrosequencing

Each ID element core contains four CpG nucleotides (Figure 2). DNA methylation of the first two CpG sites within the ID element sequence was determined by pyrosequencing at EpigenDx Inc.

(Hopkinton, MA, USA), using glyceraldehyde-3-phosphate dehydrogenase (GADPH) as a reference gene [35]. Briefly, 500 ng of DNA from each rat brain tissue sample were first treated with bisulfite and subsequently purified using Zymo Research DNA columns (Zymo Research, Irvine, CA, USA). A 1/20th eluted solution was used for each PCR. Biotinylated PCR products were bound to Streptavidin Sepharose HP (GE Healthcare, Waukesha, WI, USA). Immobilized PCR products were purified using Pyrosequencing Vacuum Prep Tool (Qiagen, Valencia, CA, USA) according to the manufacturer's instructions. The 0.2 μM pyrosequencing primer was annealed to purified single-strained PCR product. The PCR products (10 uL) were sequenced using the Pyrosequencing PSQ96 HS System (Biotage AB, Charlotte, NC, USA). The methylation status of each locus was analyzed individually as a T/C SNP using QCpG software (Pyrosequencing, Qiagen). The analysis was performed on duplicate samples, and the samples were averaged. The data are expressed as percent (%) of methylation (mean ± SEM).

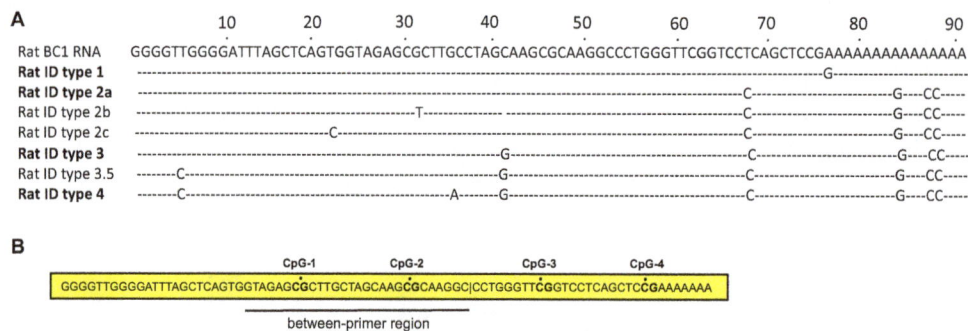

Figure 2. Schematic illustration of ID element sequences in the rat. (**A**) Alignment of seven consensus sequences of rat ID subfamilies with the rat BC1 RNA gene. The dot indicates the base identity at each position. The lower case letters indicate the ambiguity at that position. Each subfamily is named in sequential order, types 1–4, based on the age of each subfamily. Three major subfamilies are in bold type. The diagnostic changes of these major subfamilies and one minor subfamily, type 3.5, accumulate progressively from older to younger subfamilies (bold-type). The two minor subfamilies, types 2b and 2c, are derived from the previous major subfamily, type 2 [14]; (**B**) Rat BC1 RNA with CpG sites and the between-primers region.

2.5. Immunohistochemistry

Brain tissue from the rats sacrificed by decapitation at 24 h after the last injection of METH or saline was fixed in 4% paraformaldehyde for 24 h, then incubated in 20% and 30% buffered glycerol concentrations for 24 h each (4°C). Every other of the coronal sections (20 μm, 3 sections/rat) from the dentate gyrus (−3.12−−4.68 mm from Bregma) was examined for the levels of PABP1, using the immunofluorescence technique. Sections were first pretreated with 1× citrate buffer for 40 min at 70 °C, then allowed to cool to room temperature before being blocked in blocking buffer (phosphate-buffered saline (PBS) pH 7.4, 0.1% Triton X-100, and 5% bovine serum albumin (BSA)) for 1 h at room temperature. The sections were then incubated overnight at 4 °C with a rabbit anti-PABP1 primary antibody (1:100, #4992, Cell Signaling Technology, Danvers, MA, USA). The next day, the sections were incubated for 3 h at room temperature with anti-rabbit Alexa-594 secondary antibody (1:2000, Life Technologies, Carlsbad, CA, USA). The incubations with primary and secondary antibodies were separated by three 5 min-long washes with PBS that contained 0.1% Triton and 5% BSA. The nuclei were labeled with DRAQ5 dye (Life Technologies). The sections were mounted on slides using Flouromount mounting medium (Sigma-Aldrich). Images (taken along the dentate gyrus) were captured using a Leica TCS SPE-II confocal microscope under the 63 × oil objective (Leica, Buffalo Grove, IL, USA United States. PABP1 immunofluorescence, measured in 3 non-overlapping areas on each slice, was first averaged per slice and then averaged per rat before statistical comparison of the controls and METH groups. The data are expressed as PABP1 immunofluorescence normalized to saline controls (mean ± SEM).

2.6. Next-Generation Sequencing and Sequence Diversity Analysis

Analysis of dynamic transposable element activity via next-generation sequencing (NGS) technologies has been difficult in the past because it is challenging to align sequencing reads to repetitive regions of the genome [36]. Computational approaches to profiling the sequence diversity of ID elements have since been introduced to help overcome these barriers, and our group has applied these strategies for detection of ID elements. This method first characterizes the dynamic ID element activity by quantifying the relative proportion of clonal population subtypes, using the BC1 consensus sequence as a reference, within this ID element family, followed by mapping of the most significant subtypes back to their genomic locations [15]. This approach has an advantage over the traditional alignment approach because it considers all ID elements in a population and is agnostic to mapping, whereas the traditional approach must focus on the rare subset of sequences that mapped uniquely to the reference genome. We were able to estimate the change in total copy number associated with ID element activity in METH-treated rats relative to controls. Briefly, dorsal dentate gyrus samples were sequenced using Ion Torrent Sequencing technology after amplification of the target BC1 ID element master gene (BC1 reference sequence: GGTTGGGGATTTAGCTCAGTGGTAGAGCGCTTGCCTAGCAAGCGCAAGGCCCTGGGTTCGGTC CTCA). In order to describe the diversity of the aligned ID elements after sequencing, a set of mutations, or discrepancies from the reference, were determined for each read, and the total number of reads was determined for each unique ID element subtype. To avoid intrinsic error associated with sequencing, subtypes with 100 or more normalized per million reads were then grouped according to treatment condition, and the ID element subtypes with the greatest change relative to controls were identified. The top sequences (defined by a p-value <0.05) from the identified subtypes were aligned to a reference genome (Rat Genome Sequencing Consortium 6.0/ Rattus norvegicus 6 (RGSC 6.0/rn6) using the Bowtie2 read aligner tool [37]. The number of times each sequence aligned to the genome and its location in the genome was also determined, and only genomic alignments with zero mismatches were kept for interpretation. ID subtypes that did not uniquely match at least once to the reference genome were not included in the results table.

2.7. Statistical Analysis

All data were tested with the Levene test for the heterogeneity of variances and with the Shapiro-Wilk test for normality. Two-way repeated-measures ANOVA followed by the Student-Newman-Keuls post hoc test was performed on the temperature data. Two-way ANOVA followed by the Fisher LSD post hoc test was performed on the methylation data. As our sample sizes were small and group comparisons were preplanned, the Fisher LSD test was chosen as a post hoc test to avoid making the type II error (accepting a null hypothesis that is actually false), as the Student-Newman-Keuls or Tukey's HSD test may be too conservative for this study. This strategy will maximize our ability to detect differences in a small, pilot sample, and so, findings from using this strategy will likely need to be replicated in further samples in the future. Student's unpaired two-tailed t-test was performed on the immunochemistry data. The NGS data were analyzed by Student t-tests comparing the number of reads in ID element subtypes between METH- and saline-treated rats. The results are expressed as the mean ± SEM. Statistical significance was set at $p < 0.05$.

3. Results

3.1. The ID Element Is Similarly Methylated in Different Rat Brain Regions

To our knowledge, the methylation status of ID element CpG sites in different brain areas has not been determined. To assess whether there are regional difference in methylation status of the ID element, we assessed methylation of two CpG sites, CpG-1 and CpG-2, in five different brain areas, namely the striatum, dentate gyrus, Ammon's horn, frontal cortex and cerebellum, as well as in muscle tissue of the rat. The methylation percentages for each rat brain area were very similar, 57%–60%

at CpG-1 and 29%–31% at CpG-2. Muscle tissue displayed similar methylation status as the brain. The data are summarized in Figure 3.

ID element methylation in rat brain

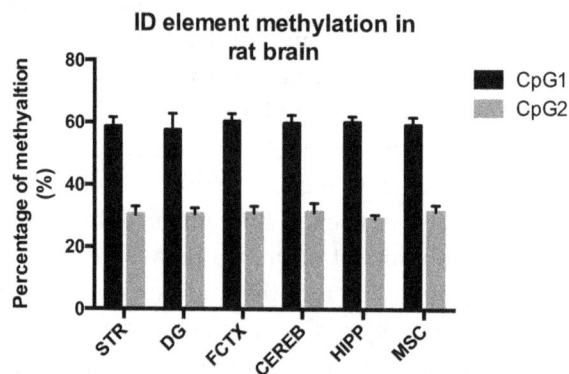

Figure 3. Methylation status of ID element CpG-1 and CpG-2 sites in several brain areas and muscle tissue in the rat. The methylation percentage of CpG-1 and CpG-2 in five different brain areas, namely the striatum, dentate gyrus, Ammon's horn, frontal cortex and cerebellum, as well as in muscle tissue of the rat, was similar: 57%–60% at CpG-1 and 29%–31% at CpG-2. The data are expressed as the mean $\pm$ SEM (n = 12). Abbreviations: CEREB, cerebellum; CpG, C-phosphate-G; DG, dentate gyrus; FCTX, frontal cortex; HIPP, CA1-CA3 of the hippocampus (Ammon's horn); MSC, muscle, STR, striatum.

3.2. Binge and Chronic High-Dose METH-Induced Hyperthermia in the Rat

Hyperthermia is an important contributing factor in METH neurotoxicity. A decrease in core body temperature decreases METH neurotoxicity, whereas an increase in core body temperature increases toxicity [38,39]. To assess body temperature profiles in both METH regimens, core body temperature was measured prior to the administration of METH or saline and 1 h after each METH or saline injection. As shown if Figure 4A, binge METH administration significantly increased core body temperatures over time. Chronic METH administration initially resulted in a hyperthermic response, which decreased by the seventh day of drug administration (Figure 4B). The chronic METH-induced hyperthermia positively correlated with CpG-2 hypermethylation (Pearson r^2 = 0.99, $p < 0.01$, two-tailed test, n = 4); however, group sizes have to be increased to make this result conclusive.

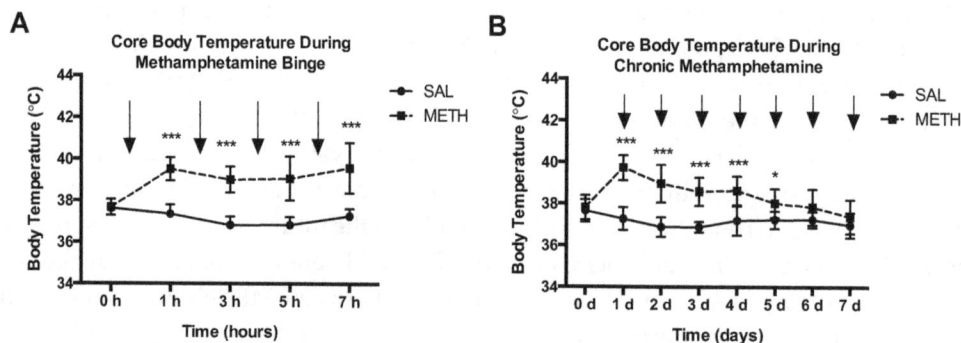

Figure 4. METH-induced hyperthermia. Adult male Sprague-Dawley rats were administered saline (1 mL/kg), binge METH (4 $\times$ 10 mg/kg, i.p. every 2 h) or chronic METH (20 mg/kg/day for 10 days, i.p.). Core body temperatures ($^\circ$C) were measured before treatments and 1 h after each METH or saline injection. The black arrows indicate the injection times. (**A**) Binge METH induced significant hyperthermia during the treatment. (**B**) During chronic METH regiment, rats developed tolerance to METH-induced hyperthermia. Saline vs. METH: * $p < 0.05$, *** $p < 0.0001$ (two-way ANOVA with repeated measures, followed by the Student-Neuman-Keuls post hoc test). Data are expressed as the mean $\pm$ SEM. Abbreviations: METH, methamphetamine; SAL, saline.

3.3. Binge and Chronic METH Differentially Alter the Methylation Status of the ID Element

Retrotransposition of transposable elements does not take place in non-neurogenic areas of adult brain; it can occur in neurogenic niches [6,7,40]. Of note, environmental stressors, including substance abuse, can alter methylation and expression (RNA transcripts) of these elements in non-neurogenic brain areas [7]. To determine whether high-dose binge and chronic METH regimens induce short-term or long-term hypomethylation or hypermethylation of CpG-1 and CpG-2 sites in the dentate gyrus, rats were treated with METH or saline and killed 1 h, 24 h or 7 days after the last injection. As compared to binge saline, binge METH significantly decreased methylation of CpG-2 in the dental gyrus at 1 h after the last METH injection (-2.7%, $p < 0.05$, two-way ANOVA followed by the Fisher LSD post hoc test, $n = 4–7$) (Figure 5A). In contrast to binge METH, chronic METH increased methylation of the CpG-2 site at the 1-h time point ($+5.1\%$, $p < 0.0001$, two-way ANOVA followed by the Fisher LSD post hoc test, $n = 4–7$) (Figure 5B). By Day 7, the methylation of CpG-2 returned to the control values after both METH administrations. There was no change in methylation status of the CpG-1 site at any time point either after binge or chronic METH. Of note, binge METH-treated rats with higher core body temperatures tended to have lower CpG-2 methylation.

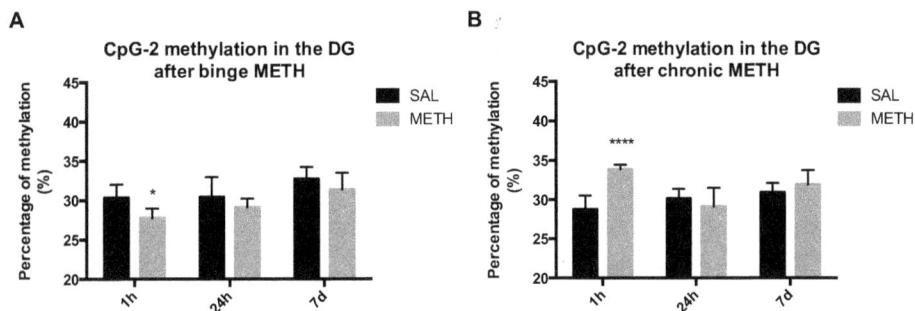

Figure 5. The effects of binge and chronic METH on methylation status of the ID element in the dentate gyrus at 1 h, 24 h and 7 days after drug administration. Adult male Sprague-Dawley rats were administered saline (binge or chronic, 1 mL/kg), binge METH (4×10 mg/kg, i.p. every 2 h) or chronic METH (20 mg/kg/day for 10 days, i.p.) and killed 1 h, 24 h or 7 days later. As compared to controls, binge METH triggered significant hypomethylation of CpG-2 site at 1 h (-2.7%, * $p < 0.05$) in the dentate gyrus (**A**), whereas chronic METH triggered significant hypermethylation of CpG-2 at 1 h in the dentate gyrus (5.1%, **** $p < 0.0001$); (**B**). The CpG-2 methylation status returned to the control values within seven days after METH administration. In the binge METH paradigm, there was a significant main effect of time ($F(2,27) = 7.3$, $p < 0.01$) and treatment condition (saline or METH) ($F(1,27) = 7.6$, $p < 0.05$). In the chronic METH paradigm, there was a significant main effect of time ($F(2,27) = 4.09$, $p < 0.05$) and treatment condition (saline or METH) ($F(1,27) = 8.9$, $p < 0.01$), as well as a significant time $\times$ treatment interaction ($F(2,27) = 8.9$, $p < 0.01$). Analysis was performed by two-way ANOVA followed by the Fisher LSD post hoc test ($n = 4–7$). Data are expressed as the mean $\pm$ SEM. Abbreviations: CpG, C-phosphate-G; DG, dentate gyrus; METH, methamphetamine; SAL, saline.

3.4. PABP1 Protein Levels Are Increased by METH in the Dentate Gyrus

PABP1 is important for the regulation of mRNA translation and stability, and its levels increase during recovery from a heat shock [41]. As other PABP proteins, PABP1 may bind to SINEs and contribute to the SINE retrotransposition process [13]. To determine whether binge METH-induced ID element hypomethylation at the CpG-2 is accompanied by increased expression of PABP1 in the dentate gyrus, hippocampal slices from rats killed at 24 h after the last dose of binge METH were probed with the anti-PABP1 antibody. Compared to controls, PABP1 immunofluorescence in the dentate gyrus was significantly increased in binge METH-treated rats ($+26\%$, $p < 0.05$, Student's t-test, $n = 4$) (Figure 6A). Based on the methylation data, we hypothesized that tolerance to METH effects developed over the course of its chronic administration. To determine whether PABP1 protein levels are altered in the beginning of the chronic METH regimen, rats were treated with two injections of saline or 20 mg/kg/day METH and killed 24 h after the second injection. Compared to controls, two doses of 20 mg/kg METH administered over a period of two days markedly increased PABP1 immunofluorescence in the dentate gyrus (+88%, $p < 0.05$, Student's t-test, $n = 4$) (Figure 6B).

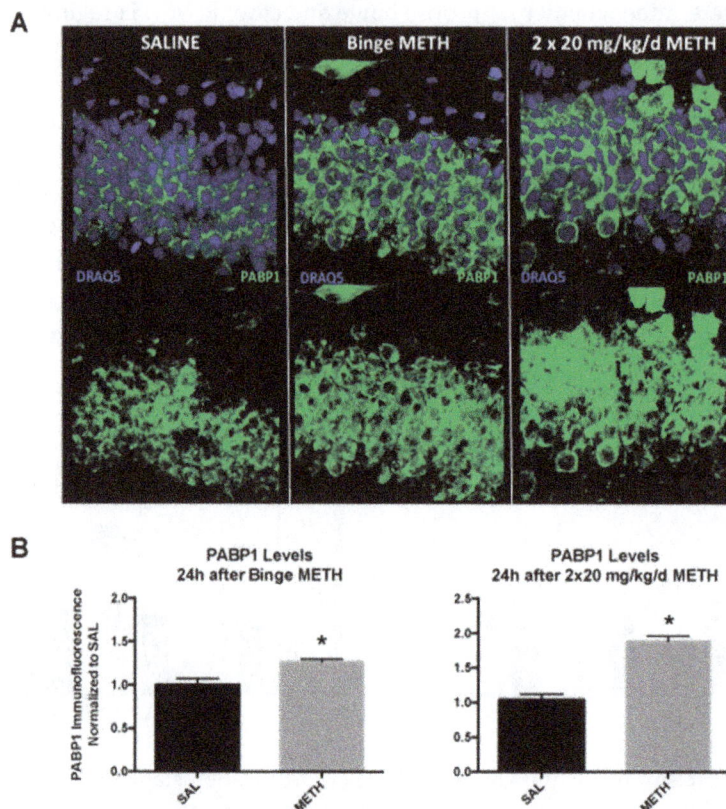

Figure 6. The effects of METH binge and two high doses of daily METH on PABP1 protein levels in the dentate gyrus at 24 h after drug administration. Adult male Sprague-Dawley rats were administered saline (1 mL/kg per injection), binge METH (4 × 10 mg/kg, i.p. every 2 h) or treated with two injections of 20 mg/kg/day METH and killed 24 h after the last injection of saline or METH. (**A**) Compared to saline controls, four doses of 10 mg/kg METH administered every 2 h increased PABP1 immunofluorescence in the dentate gyrus by 26% ($p < 0.05$, Student's t-test, $n = 4$). (**B**) Compared to saline controls, two doses of 20 mg/kg METH administered over a period of two days markedly increased PABP1 immunofluorescence in the dentate gyrus (+88%, $p < 0.05$, Student's t-test, $n = 4$). Data were normalized and expressed as the mean ± SEM. Abbreviations: METH, methamphetamine; SAL, saline; PABP1, poly(A)-binding protein 1.

3.5. Chronic METH Induces a Persistent Difference in ID Element Amplification in the Dentate Gyrus

It is known that SINEs can retrotranspose in response to injury [11] and that the ID element can be activated by injury to the CNS [18,19]. However, the hypomethylation of ID element and increase in PABP1 protein levels in the dentate gyrus do not indicate that ID elements retrotransposed upon exposure to binge or chronic METH; these changes only suggest ID element activation. Therefore, we next compared amplification, in other words, the number of copies across the entire genome, of ID element species within gDNA isolated from dentate gyri of the control and chronic METH-treated rats. We detected several loci with significant sequence diversity of clonal ID element populations between the controls and METH-treated samples. The BC1 consensus sequence and ID element type 3 displayed the highest number of reads in the METH and the control samples; however, there was no significant difference between the groups (BC1, SAL: 9303 ± 2026, METH: 8855 ± 764; ID element type 3, SAL: 4763 ± 587, METH: 5120 ± 1060; normalized reads per million; $p > 0.1$, Student's unpaired two-tailed t-test, $n = 4$–5). For ID element type 2b, the detected numbers of reads were: SAL: 598 ± 197; METH: 650 ± 235 (normalized reads per million; $p > 0.1$, Student's unpaired two-tailed t-test, $n = 4$–5). ID element type 1, type 2b and type 2c could not be reliably identified. The ID sequence 29G > A | 38G > C | 43C > T | 52C > T was the sequence that showed the highest, but not significantly different, amplification in the METH group as compared to the control group (six-fold increase, normalized reads per million; SAL: 7 ± 7, METH: 51 ± 22; $p = 0.14$, Student's unpaired two-tailed t-test, $n = 4$–5). Six sequences that could be matched at least once to the reference rat genome were significantly amplified ($p < 0.05$, Student's unpaired two-tailed t-test, $n = 4$–5): 28C > T | 38G > C (+61%, SAL: 34 ± 6, METH: 55 ± 2), 29G > A | 44G > A | 52_53insT (+122%, SAL: 19 ± 6, METH: 42 ± 5), 33G > A | 38G > C | 39C > G | 40A > G (+23%, SAL: 67 ± 4, METH: 82 ± 3), 41A > G (+95%, SAL: 25 ± 9, METH: 49 ± 5), 23delT | 29G > A (+162, SAL: 17 ± 4, METH: 45 ± 9) and 28C > G | 29G > C | 31_32insC | 34C > G (+76%, SAL: 30 ± 10, METH: 52 ± 2). Some of these sequences were mapped to the intergenic regions in the rat genome, while some were mapped to genes. Genes associated with preferential amplification and their roles are shown in Table 1. Two ID element sequences that could be matched at least once to the reference rat genome were significantly decreased in the METH group in terms of proportions as compared to the control group: 1G > A | 33G > A | 39C > G | 3T > G | 3_4insT and 1G > A | 3T > G | 3_4insT | 43 C> T | 52C > T (−55% and −56%, respectively, $p < 0.05$, Student's unpaired to-tailed t-test, $n = 4$–5), suggesting that proliferation of dentate gyrus cells containing these sequences was suppressed by METH.

It is known that SINEs can alter the expression of genes near their insertion site [42]. The precise location of chronic METH-triggered ID element insertions was not possible to determine with the current analysis; however, we detected thirteen ID sequences that did not match even once to the reference genome, but were significantly amplified in the METH group compared to the controls. These ID elements could have been new ID element types or ID element sequences that retrotransposed during chronic METH treatment.

Table 1. Top six preferentially amplified ID element subtypes and their associated genomic locations. Dorsal dentate gyrus samples were analyzed using next-generation sequencing technology to identify ID element subtypes that were preferentially amplified in rats administered methamphetamine (METH) relative to rats administered saline.

ID Element Subtype [a]	% Change [b]	p-Value	Genes Associated with Preferential Amplification [c]
28C > T	38G > C	61	0.00828	Cipc (Chr6:110986686), Slc9a9 (Chr8:102453600), LOC680227 (ChrX:74311735), Ift80 (Chr2:165544369), Mfsd14b (Chr17:2643624), Slc12a2 (Chr18:52921730), Uba6 (Chr14:23511767), Lasp1 (Chr10:85756102), Dcakd (Chr10:91116391), Smpdl3b (Chr5:150930976), Atp2b2 (Chr4:145726282), Smad9 (Chr2:144054261), Asic2 (Chr10:68890335), Abca17 (Chr10:13680892), Rab3c (Chr2:41772443), Ralgps2 (Chr13:74477445), Gtf2e2 (Chr16:62133089), Map3k12 (Chr7:144103811)
29G > A	44G > A	52_53insT	122	0.01785	Veph1 (Chr2:157964489), Mapk8ip3 (Chr10:14288892)

Table 1. *Cont.*

ID Element Subtype [a]	% Change [b]	*p*-Value	Genes Associated with Preferential Amplification [c]
33G > A | 38G > C | 39C > G | 40A > G	23	0.01951	Proximal to Cep162 (Chr8:94937230)
41A > G	95	0.03976	Hdac4 (Chr9:99135435), Atg3 (Chr11:60587858), Nutf2 (Chr19:37841169), Urm1 (Chr3:8399609), Rsl1d1 (Chr10:4498654), Kpna3 (Chr15:41704423), Cnga3 (Chr9:43837574), Tfap2d (Chr9:25355932), Crkl (Chr11:87792847), Atp10d (Chr14:38491114), Vegfd (ChrX:31818688), Slc24a3 (Chr3:139780502), Pbx1 (Chr13:86476988), Prpsap1 (Chr10:105447151), Mylk (Chr11:69140146), Synpo Chr18:55910398), Zbtb46 (Chr3:176916873), Rtcd1 (Chr2:219556431), Slit3 (Chr10:20429924), Lcmt1 (Chr1:193375030), Rab30 (Chr1:157633572), Sepw1 (Chr1:77806751), Tjap1 (Chr9:17089159), Laptm4b (Chr7:72935301), Optn (Chr17:77215077), Nup210l (Chr2:189478449), Aaas (Chr7:143939751), Astn2 (Chr5:82045057), Rb1cc1 (Chr5:13068263), Smarcal1 (Chr9:79976755), Gsto2 (Chr1:267621506), Sppl3 (Chr12:47334193), Tctn2 (Chr12:37366571), Impdh1 (Chr4:56485763), Tbc1d14 (Chr14:79314924), Tbc1d8 (Chr9:46159589), Fcgr2b (Chr13:89356250), Scarb1 (Chr12:36712849), Rab3ip (Chr7:59939610), Prpf40a (Chr3:38695276), Utp20 (Chr7:29314493), Arsb (Chr2:23413525), Entpd7 (Chr1:263467744), Capn2 (Chr13:100899435), Txndc11 (Chr10:4613068), Gspt1 (Chr10:4386117), Six4 (Chr6:95989936), Susd5 (Chr8:122359614), Il23r (Chr4:98243720), Casc4 (Chr3:113853687), Ranbp17 (Chr10:18194793), Gabrb1 (Chr14:38814344), Cdc37 (Chr8:22163945), Fam172a (Chr2:5339706), Gtf2f2 (Chr15:57997967), Adra1a Chr15:43354439), Pum2 (Chr6:33799082), Gpr137b (Chr17:90744301), Atp1b3 (Chr8:104217176), Ubap1 (Chr5:57741520), Sgpl1 (Chr20:30724101), Ptprk (Chr1:17762449), Wwp2 (Chr19:39600430), Sbno1 (Chr12:37628774), Zfand3 (Chr20:8810745), Wrn (Chr16:62554494), Kyat3 (Chr2:248661110), Aldh2 (Chr12:40485627), Dpp6 (Chr4:4219884), Plekho2 (Chr8:71098823)
23delT | 29G > A	162	0.04303	Taar8c (Chr1:22334448)
28C > G | 29G > C | 31_32insC | 34C > G	76	0.04654	Ythdf2 (Chr5:150375359)

[a] The reference sequence on which ID element subtypes were aligned and identified was the BC1 consensus sequence: GGTTGGGGATTTAGCTCAGTGGTAGAGCGCTTGCCTAGCAAGCGCAAGGCCCTGGGTTCGGTCCTCA. Each subtype name is represented by the changes it contains compared to the reference sequence. For example, for the ID element subtype, 28C > T \| 38G > C; this indicates that the 28th and 38th positions with respect to the consensus sequence have been changed to T and C, respectively. #_#ins indicates an insertion of a nucleotide between the given positions. #del indicates the nucleotide position that was deleted. [b] METH relative to controls. [c] Gene regions to which each ID element subtype mapped. Intergenic regions are not shown. Genomic coordinates using Rat Genome Sequencing Consortium (RGSC) 6.0/rn6.

4. Discussion

The present study demonstrates that neurotoxic doses of binge and chronic METH regimen differentially affected the CpG methylation within the ID element sequence at 1 h after METH; binge METH triggered hypomethylation, while chronic METH triggered hypermethylation of the CpG-site 2 in the ID element sequence. We also demonstrate that high doses of METH increase the levels of SINE regulatory protein PABP1 in this brain region. Using a new method in NGS analysis, we are the first to demonstrate significant differences in the sequence diversity of clonal ID element populations between controls and chronic METH-exposed samples of dentate gyri. We also identify several ID element sequences that amplified preferentially in certain genomic loci within the reference rat genome.

DNA methylation at the carbon 5 position of the cytosine ring of the CpG nucleotides is the most common epigenetic modification within vertebrate genomes [43,44]. SINEs and LINEs are frequently methylated in mouse, rat and human tissues [44–46]. The methylation percentage of CpG sites within the ID element in gDNA from six tissues (liver, lung, kidney, spleen, ovary, testis) from two-month-old male and female Sprague-Dawley rats was found to be about 54% at the CpG-1 and about 60% at the CpG-2 [16]. We found a similar methylation percentage at the CpG-1 and about 30% at the CpG-2 in several brain areas in two-month-old male Sprague-Dawley control rats. The reason for the discrepancy is currently unknown. It might relate to age, tissue specificity and/or methodology. Our data suggest that ID methylation in young adult rats is partial and does not differ between brain regions.

Activation of transposable elements, including SINEs, usually depends on their DNA methylation status and takes place during both embryonic and adult neurogenesis. These elements are silenced by methylation in somatic cells in adults [2,3]. Activation of transposable elements is observed in several diseases and can be triggered by environmental stimuli. Thus, hypomethylation and increased expression of SINEs was reported in cancer, autoimmune and neurodegenerative diseases, substance abuse [1,3,7] and upon exposure to several cellular stressors, including thermal stress and hypoxia [47–50]. An acute low dose of METH (4 mg/kg) was shown to decrease the expression of DNA methyltransferase 2 in the dentate gyrus of Wistar rats when measured 24 h later [51]. This finding and our finding of ID element hypomethylation at 24 h after the high-dose binge METH administration suggest that both low and high METH doses might be able to trigger hypomethylation of the element. The extent of ID element hypomethylation at the CpG-2 site was small (2.7%). Similarly, less than 10% hypomethylation of *Alu* was detected in cancer cells [52,53] and Klinefelter syndrome [54]. Less than 10% hypomethylation of the B1/B2 element was found after exposure to soil dust and traffic exhaust [55]. As in our experiment, the weak increase in mouse SINEs, detected at the 24-h time-point, was followed by their hypermethylation later on [55]. Activation of transposable elements is more complex than simultaneous hypomethylation of all CpG sites [2,50,56]. This fact could explain, in part, the lack of changes in the methylation state at the CpG-1 site.

At 24 h after chronic METH, the CpG-2 site of the ID element was hypermethylated and returned to the control levels by the seventh day of withdrawal from METH, suggesting that the ID element was silenced by de novo methylation sometime during chronic METH administration. This conclusion is supported by other published studies that observed increases in global DNA methylation after chronic administration of low doses of amphetamines. Thus, in adult rats, chronic exposure to amphetamine (1 mg/kg) led to increases in global DNA methylation in the nucleus accumbens, prefrontal cortex and olfactory tubercle [57], while chronic exposure to METH (4 mg/kg or 0.5–3 mg/kg) resulted in increased expression of DMT1 (DNA-methylating enzyme 1) in the striatum and nucleus accumbens [58,59] at two weeks after drug administration.

It has been proposed that methylation of histones rather than methylation of gDNA plays a dominant role in transcription of SINEs [56]. Therefore, it is possible that the hypermethylation of the ID element did not suppress its activity, but histone methylation did. Along these lines, neurotoxic self-administered METH increased the levels of histone-3 methylation in the striatum at 1 h and 24 h after cessation of METH admministration [60]. Similarly, hypomethylation of CpG-2 might have played a small or no role in ID element activation. Of note, binge METH-treated rats with higher core body temperatures tended to have lower CpG-2 methylation, which together with the data on increased expression of SINEs from several different species following thermal shock [47–49] suggest that binge METH-induced hyperthermia leads to hypomethylation of the ID element. In addition to thermal shock, ID element activation and increased expression can be induced by inflammatory response [61], inhibition of protein synthesis [48] or DNA breakage, with the latter being known to also induce SINE retrotransposition [11]. Consequently, the increased expression of the ID element in the dentate gyrus may have occurred not only due to the thermal shock, but also due to METH-induced inflammatory response [62], DNA breakage [63–65] and/or inhibition of protein synthesis [66], with at least some of these elements retrotransposing within gDNA.

PABPs are multifunctional RNA-binding proteins that regulate multiple aspects of mRNA translation and stability [67], including mediating retrotransposition of SINEs. SINEs are non-autonomous elements [8]. They use PABPs and proteins encoded by endogenous LINE-1s [4,9–11] to form RNP complexes that are needed for the transport of SINEs to the nucleus for reverse transcription [42]. PABP1 is a protein suggested to positively regulate reverse transcription of the ID element-containing BC1 gene [13] and LINE-1 [68], as well as to have a role in cellular recovery from thermal stress [41]. Consequently, the increase in PAB1 levels observed by us at 24 h after binge METH and on the third day of chronic METH treatment could have mediated either stabilization of ID element RNA and ID element retrotransposition, or re-initiation of protein synthesis after METH-induced thermal shock, or both. The more pronounced increase in PABP1 levels in the dentate gyrus after two daily doses of 20 mg/kg METH than after binge METH (4 × 10 mg/kg, every 2 h) could have been due to the difference in the timeline, i.e., more PABP1 protein was synthesized over three days than one day since the first METH injection.

ID element amplification within gDNA was detected after METH treatment, which is the most important finding of our study. The most abundant subtypes of ID element (e.g., types 3, 4, etc.) were detected in our analyses; however, they were not significantly different between the groups, suggesting that the significantly amplified sequences originated (copied) from different locations than these sequences' loci within the rat genome [69,70] and that BC1, ID element type 2b, type 3 and type 4 activities were suppressed during METH administration.

We identified several subtypes of ID elements, closely related to the master BC1 sequence, that were significantly affected by METH treatment. Certain subtypes, like 28C > T | 38G > C, were associated with a broader set of genes compared to other subtypes, such as 29G > A | 44G > A | 52_53insT. This may indicate that the distribution of ID element subtypes could be substantially different and that this further may be influenced by METH treatment, although further work is needed to test these hypotheses. Although not described in Table 1, ID element insertion in the same gene at multiple locations was also present with several subtypes. The reason for this apparent targeting of certain genes by ID element subtypes is unknown, but could suggest the possible importance of a gene or even non-specific effects by METH. In addition to ID element subtypes being found within genes, many mapped locations were in intergenic locations not associated with a particular gene. Intergenic transposable elements could be important when considering genomic stability and nearby gene regulation. One explanation for the lack of increases in the number of BC1 sequences and increases in other ID elements is the fact that the older subfamilies of this element appear to be incapable of retrotransposition, with younger families being capable of amplifying [71] from numerous source genes [70]. In agreement with the latter finding, we identified multiple potential gene sources of the ID element in the rat dentate gyrus. Secondly, different cell populations within the rat dentate gyrus might have reacted differently to METH [22]. Thirdly, duplications and deletions of the element may have taken place [72].

Of the most significant ID element subtypes that were preferentially amplified in the METH group, the associated genomic locations had a diverse range of functions. For example, ID element subtype 28C > T | 38G > C was found in genes coding for transporters (*SLC12A2*, *SLC9A9*), protein modification (*UBA6*, *Map3k12*) and transcriptional regulation (*CIPC*). Others, such as *GPC5* [73,74], have been implicated in addiction, while *UBA6* hypomethylation has been found with METH treatment [75]. Overall, the genes associated with the identified ID element subtypes require further, targeted investigation to understand what effects METH has on their expression levels. The current investigation was limited by a low power to detect statistical significance and the ability to see absolute differences, making future work in this area important. We believe the novel approach to identify ID element changes with METH treatment overcomes these limitations and has identified important effects that can be further investigated.

5. Strengths and Limitations

Strengths of the current work include a novel and unbiased approach to characterizing ID element changes in the brain. Additionally, our work assessed both pre- and post-translational changes (e.g., epigenetic to protein expression). The limitations of our study are relatively small sample sizes, the inability to detect all ID element subtypes and the pilot character of the study due to the primer design (the primers were designed to be contained within the consensus sequence in order to improve NGS performance). Finally, the lack of a targeted qPCR and sequencing analysis of ID elements makes comparisons to other studies difficult. Nevertheless, this work provides candidate genes and locations from which our identified ID element subtypes may have originated in the reference genome. We plan to improve and validate the method in the near future through replication and a better understanding of the destination of ID element retrotransposition.

6. Conclusions

In summary, our results suggest that high-dose chronic METH triggers a chain of events leading to ID element retrotransposition in the dentate gyrus and that the ID element may play a role in METH-mediated changes in hippocampal neurogenesis. The hippocampus is involved in the relapse effect of drug abuse and memory tasks commonly impaired in human chronic METH users [23,76,77]. Consequently, determination of METH effects on the ID element and other SINEs is clinically relevant to abuse of amphetamines and the related decline of cognitive skills. Identification of ID-containing genes affected by high METH doses adds to knowledge on SINE-related epigenetic events induced by neurotoxic doses of binge and chronic METH. Further work will aid in understanding the origin of the ID elements that change their position in the genome after chronic METH treatment.

Acknowledgments: Supported by National Institutes of Health (NIH)/National Institute of Drug Abuse (NIDA) RO1 Grant DA034783. Anna Moszczynska had received funds for covering the costs to publish in open access.

Author Contributions: Research design: Anna Moszczynska. Contributed to data generation: Dongyue Yu. Performed data analysis: Anna Moszczynska and Kyle J. Burghardt. Wrote the manuscript: Anna Moszczynska and Kyle J. Burghardt.

Abbreviations

The following abbreviations are used in this manuscript:

DMT	DNA methyltransferase
ID element	Identifier element
LINEs	Long interspersed elements
METH	Methamphetamine
PABP	Poly(A)-binding protein
SAL	Saline
SINEs	Short interspersed elements

References

1. Hancks, D.C.; Kazazian, H.H., Jr. Active human retrotransposons: Variation and disease. *Curr. Opin. Genet. Dev.* **2012**, *22*, 191–203. [CrossRef] [PubMed]

2. Hata, K.; Sakaki, Y. Identification of critical CpG sites for repression of L1 transcription by DNA methylation. *Gene* **1997**, *189*, 227–234. [CrossRef]

3. Schulz, W.A.; Steinhoff, C.; Florl, A.R. Methylation of endogenous human retroelements in health and disease. *Curr. Top. Microbiol. Immunol.* **2006**, *310*, 211–250. [PubMed]

4. Liu, W.M.; Maraia, R.J.; Rubin, C.M.; Schmid, C.W. Alu transcripts: Cytoplasmic localisation and regulation by DNA methylation. *Nucleic Acids Res.* **1994**, *22*, 1087–1095. [CrossRef] [PubMed]

5. Beck, C.R.; Garcia-Perez, J.L.; Badge, R.M.; Moran, J.V. LINE-1 elements in structural variation and disease. *Annu. Rev. Genom. Hum. Genet.* **2011**, *12*, 187–215. [CrossRef] [PubMed]

6. Muotri, A.R.; Zhao, C.; Marchetto, M.C.; Gage, F.H. Environmental influence on L1 retrotransposons in the adult hippocampus. *Hippocampus* **2009**, *19*, 1002–1007. [CrossRef] [PubMed]

7. Erwin, J.A.; Marchetto, M.C.; Gage, F.H. Mobile DNA elements in the generation of diversity and complexity in the brain. *Nat. Rev. Neurosci.* **2014**, *15*, 497–506. [CrossRef] [PubMed]

8. Deininger, P.L.; Batzer, M.A. Mammalian retroelements. *Genome Res.* **2002**, *12*, 1455–1465. [CrossRef] [PubMed]

9. Kramerov, D.A.; Vassetzky, N.S. Short retroposons in eukaryotic genomes. *Int. Rev. Cytol.* **2005**, *247*, 165–221. [CrossRef]

10. Dewannieux, M.; Esnault, C.; Heidmann, T. LINE-mediated retrotransposition of marked Alu sequences. *Nat. Genet.* **2003**, *35*, 41–48. [CrossRef] [PubMed]

11. Hagan, C.R.; Sheffield, R.F.; Rudin, C.M. Human Alu element retrotransposition induced by genotoxic stress. *Nat. Genet.* **2003**, *35*, 219–220. [CrossRef] [PubMed]

12. Okada, N.; Hamada, M.; Ogiwara, I.; Ohshima, K. SINEs and LINEs share common 3′ sequences: A review. *Gene* **1997**, *205*, 229–243. [CrossRef]

13. West, N.; Roy-Engel, A.M.; Imataka, H.; Sonenberg, N.; Deininger, P. Shared protein components of SINE RNPs. *J. Mol. Biol.* **2002**, *321*, 423–432. [CrossRef]

14. Kass, D.H.; Kim, J.; Deininger, P.L. Sporadic amplification of ID elements in rodents. *J. Mol. Evol.* **1996**, *42*, 7–14. [CrossRef] [PubMed]

15. Kim, J.; Martignetti, J.A.; Shen, M.R.; Brosius, J.; Deininger, P. Rodent BC1 RNA gene as a master gene for ID element amplification. *Proc. Natl. Acad. Sci. USA* **1994**, *91*, 3607–3611. [CrossRef] [PubMed]

16. Kim, H.H.; Park, J.H.; Jeong, K.S.; Lee, S. Determining the global DNA methylation status of rat according to the identifier repetitive elements. *Electrophoresis* **2007**, *28*, 3854–3861. [CrossRef] [PubMed]

17. Ichiyanagi, K. Epigenetic regulation of transcription and possible functions of mammalian short interspersed elements, SINEs. *Genes Genet. Syst.* **2013**, *88*, 19–29. [CrossRef] [PubMed]

18. Chen, D.; Jin, K.; Kawaguchi, K.; Nakayama, M.; Zhou, X.; Xiong, Z.; Zhou, A.; Mao, X.O.; Greenberg, D.A.; Graham, S.H.; et al. Ero1-L, an ischemia-inducible gene from rat brain with homology to global ischemia-induced gene 11 (Giig11), is localized to neuronal dendrites by a dispersed identifier (ID) element-dependent mechanism. *J. Neurochem.* **2003**, *85*, 670–679. [CrossRef] [PubMed]

19. Gillen, C.; Gleichmann, M.; Spreyer, P.; Muller, H.W. Differentially expressed genes after peripheral nerve injury. *J. Neurosci. Res.* **1995**, *42*, 159–171. [CrossRef] [PubMed]

20. Rudin, C.M.; Thompson, C.B. Transcriptional activation of short interspersed elements by DNA-damaging agents. *Genes Chromosomes Cancer* **2001**, *30*, 64–71. [CrossRef]

21. Li, T.; Spearow, J.; Rubin, C.M.; Schmid, C.W. Physiological stresses increase mouse short interspersed element (SINE) RNA expression in vivo. *Gene* **1999**, *239*, 367–372. [CrossRef]

22. Mandyam, C.D.; Wee, S.; Crawford, E.F.; Eisch, A.J.; Richardson, H.N.; Koob, G.F. Varied access to intravenous methamphetamine self-administration differentially alters adult hippocampal neurogenesis. *Biol. Psychiatry* **2008**, *64*, 958–965. [CrossRef] [PubMed]

23. Thompson, P.M.; Hayashi, K.M.; Simon, S.L.; Geaga, J.A.; Hong, M.S.; Sui, Y.; Lee, J.Y.; Toga, A.W.; Ling, W.; London, E.D. Structural abnormalities in the brains of human subjects who use methamphetamine. *J. Neurosci.* **2004**, *24*, 6028–6036. [CrossRef] [PubMed]

24. Deng, X.; Wang, Y.; Chou, J.; Cadet, J.L. Methamphetamine causes widespread apoptosis in the mouse brain: Evidence from using an improved TUNEL histochemical method. *Brain Res. Mol. Brain Res.* **2001**, *93*, 64–69. [CrossRef]

25. Hori, N.; Kadota, M.T.; Watanabe, M.; Ito, Y.; Akaike, N.; Carpenter, D.O. Neurotoxic effects of methamphetamine on rat hippocampus pyramidal neurons. *Cell. Mol. Neurobiol.* **2010**, *30*, 849–856. [CrossRef] [PubMed]

26. Thanos, P.K.; Kim, R.; Delis, F.; Ananth, M.; Chachati, G.; Rocco, M.J.; Masad, I.; Muniz, J.A.; Grant, S.C.; Gold, M.S.; et al. Chronic Methamphetamine Effects on Brain Structure and Function in Rats. *PLoS ONE* **2016**, *11*, e0155457. [CrossRef] [PubMed]

27. Teuchert-Noodt, G.; Dawirs, R.R.; Hildebrandt, K. Adult treatment with methamphetamine transiently decreases dentate granule cell proliferation in the gerbil hippocampus. *J. Neural Transm.* **2000**, *107*, 133–143. [CrossRef] [PubMed]

28. Yuan, C.J.; Quiocho, J.M.; Kim, A.; Wee, S.; Mandyam, C.D. Extended access methamphetamine decreases immature neurons in the hippocampus which results from loss and altered development of neural progenitors without altered dynamics of the S-phase of the cell cycle. *Pharmacol. Biochem. Behav.* **2011**, *100*, 98–108. [CrossRef] [PubMed]

29. Hart, C.L.; Marvin, C.B.; Silver, R.; Smith, E.E. Is cognitive functioning impaired in methamphetamine users? A critical review. *Neuropsychopharmacology* **2012**, *37*, 586–608. [CrossRef] [PubMed]

30. Moszczynska, A.; Flack, A.; Qiu, P.; Muotri, A.R.; Killinger, B.A. Neurotoxic Methamphetamine Doses Increase LINE-1 Expression in the Neurogenic Zones of the Adult Rat Brain. *Sci. Rep.* **2015**, *5*, 14356. [CrossRef] [PubMed]

31. Gorlach, M.; Burd, C.G.; Dreyfuss, G. The mRNA poly(A)-binding protein: Localization, abundance, and RNA-binding specificity. *Exp. Cell Res.* **1994**, *211*, 400–407. [CrossRef] [PubMed]

32. Kilkenny, C.; Browne, W.J.; Cuthill, I.C.; Emerson, M.; Altman, D.G. Improving bioscience research reporting: The ARRIVE guidelines for reporting animal research. *PLoS Biol.* **2010**, *8*, e1000412. [CrossRef] [PubMed]

33. Yamamoto, B.K.; Moszczynska, A.; Gudelsky, G.A. Amphetamine toxicities: Classical and emerging mechanisms. *Ann. N. Y. Acad. Sci.* **2010**, *1187*, 101–121. [CrossRef] [PubMed]

34. Hagihara, H.; Toyama, K.; Yamasaki, N.; Miyakawa, T. Dissection of hippocampal dentate gyrus from adult mouse. *J. Vis. Exp.* **2009**. [CrossRef] [PubMed]

35. Tost, J.; Gut, I.G. DNA methylation analysis by pyrosequencing. *Nat. Protoc.* **2007**, *2*, 2265–2275. [CrossRef] [PubMed]

36. Treangen, T.J.; Salzberg, S.L. Repetitive DNA and next-generation sequencing: Computational challenges and solutions. *Nat. Rev. Genet.* **2011**, *13*, 36–46. [CrossRef] [PubMed]

37. Langmead, B.; Salzberg, S.L. Fast gapped-read alignment with Bowtie 2. *Nat. Methods* **2012**, *9*, 357–359. [CrossRef] [PubMed]

38. Ali, S.F.; Newport, G.D.; Holson, R.R.; Slikker, W., Jr.; Bowyer, J.F. Low environmental temperatures or pharmacologic agents that produce hypothermia decrease methamphetamine neurotoxicity in mice. *Brain Res.* **1994**, *658*, 33–38. [CrossRef]

39. Bowyer, J.F.; Davies, D.L.; Schmued, L.; Broening, H.W.; Newport, G.D.; Slikker, W., Jr.; Holson, R.R. Further studies of the role of hyperthermia in methamphetamine neurotoxicity. *J. Pharmacol. Exp. Ther.* **1994**, *268*, 1571–1580. [PubMed]

40. Kurnosov, A.A.; Ustyugova, S.V.; Nazarov, V.I.; Minervina, A.A.; Komkov, A.Y.; Shugay, M.; Pogorelyy, M.V.; Khodosevich, K.V.; Mamedov, I.Z.; Lebedev, Y.B. The evidence for increased L1 activity in the site of human adult brain neurogenesis. *PLoS ONE* **2015**, *10*, e0117854. [CrossRef] [PubMed]

41. Ma, S.; Bhattacharjee, R.B.; Bag, J. Expression of poly(A)-binding protein is upregulated during recovery from heat shock in HeLa cells. *FEBS J.* **2009**, *276*, 552–570. [CrossRef] [PubMed]

42. Deininger, P. Alu elements: Know the SINEs. *Genome Biol.* **2011**, *12*, 236. [CrossRef] [PubMed]

43. Xing, J.; Hedges, D.J.; Han, K.; Wang, H.; Cordaux, R.; Batzer, M.A. Alu element mutation spectra: Molecular clocks and the effect of DNA methylation. *J. Mol. Biol.* **2004**, *344*, 675–682. [CrossRef] [PubMed]

44. Meissner, A.; Mikkelsen, T.S.; Gu, H.; Wernig, M.; Hanna, J.; Sivachenko, A.; Zhang, X.; Bernstein, B.E.; Nusbaum, C.; Jaffe, D.B.; et al. Genome-scale DNA methylation maps of pluripotent and differentiated cells. *Nature* **2008**, *454*, 766–770. [CrossRef] [PubMed]

45. Sati, S.; Tanwar, V.S.; Kumar, K.A.; Patowary, A.; Jain, V.; Ghosh, S.; Ahmad, S.; Singh, M.; Reddy, S.U.; Chandak, G.R.; et al. High resolution methylome map of rat indicates role of intragenic DNA methylation in identification of coding region. *PLoS ONE* **2012**, *7*, e31621. [CrossRef] [PubMed]

46. Lister, R.; Pelizzola, M.; Dowen, R.H.; Hawkins, R.D.; Hon, G.; Tonti-Filippini, J.; Nery, J.R.; Lee, L.; Ye, Z.; Ngo, Q.M.; et al. Human DNA methylomes at base resolution show widespread epigenomic differences. *Nature* **2009**, *462*, 315–322. [CrossRef] [PubMed]

47. Liu, W.M.; Chu, W.M.; Choudary, P.V.; Schmid, C.W. Cell stress and translational inhibitors transiently increase the abundance of mammalian SINE transcripts. *Nucleic Acids Res.* **1995**, *23*, 1758–1765. [CrossRef] [PubMed]

48. Kimura, R.H.; Choudary, P.V.; Schmid, C.W. Silk worm Bm1 SINE RNA increases following cellular insults. *Nucleic Acids Res.* **1999**, *27*, 3380–3387. [CrossRef] [PubMed]

49. Allen, T.A.; von Kaenel, S.; Goodrich, J.A.; Kugel, J.F. The SINE-encoded mouse B2 RNA represses mRNA transcription in response to heat shock. *Nat. Struct. Mol. Biol.* **2004**, *11*, 816–821. [CrossRef] [PubMed]

50. Pal, A.; Srivastava, T.; Sharma, M.K.; Mehndiratta, M.; Das, P.; Sinha, S.; Chattopadhyay, P. Aberrant methylation and associated transcriptional mobilization of Alu elements contributes to genomic instability in hypoxia. *J. Cell. Mol. Med.* **2010**, *14*, 2646–2654. [CrossRef] [PubMed]

51. Numachi, Y.; Yoshida, S.; Yamashita, M.; Fujiyama, K.; Naka, M.; Matsuoka, H.; Sato, M.; Sora, I. Psychostimulant alters expression of DNA methyltransferase mRNA in the rat brain. *Ann. N. Y. Acad. Sci.* **2004**, *1025*, 102–109. [CrossRef] [PubMed]

52. Sirivanichsuntorn, P.; Keelawat, S.; Danuthai, K.; Mutirangura, A.; Subbalekha, K.; Kitkumthorn, N. LINE-1 and Alu hypomethylation in mucoepidermoid carcinoma. *BMC Clin. Pathol.* **2013**, *13*, 10. [CrossRef] [PubMed]

53. Park, S.Y.; Seo, A.N.; Jung, H.Y.; Gwak, J.M.; Jung, N.; Cho, N.Y.; Kang, G.H. Alu and LINE-1 hypomethylation is associated with HER2 enriched subtype of breast cancer. *PLoS ONE* **2014**, *9*, e100429. [CrossRef] [PubMed]

54. Viana, J.; Pidsley, R.; Troakes, C.; Spiers, H.; Wong, C.C.; Al-Sarraj, S.; Craig, I.; Schalkwyk, L.; Mill, J. Epigenomic and transcriptomic signatures of a Klinefelter syndrome (47, XXY) karyotype in the brain. *Epigenetics* **2014**, *9*, 587–599. [CrossRef] [PubMed]

55. Miousse, I.R.; Chalbot, M.C.; Aykin-Burns, N.; Wang, X.; Basnakian, A.; Kavouras, I.G.; Koturbash, I. Epigenetic alterations induced by ambient particulate matter in mouse macrophages. *Environ. Mol. Mutagen.* **2014**, *55*, 428–435. [CrossRef] [PubMed]

56. Varshney, D.; Vavrova-Anderson, J.; Oler, A.J.; Cowling, V.H.; Cairns, B.R.; White, R.J. SINE transcription by RNA polymerase III is suppressed by histone methylation but not by DNA methylation. *Nat. Commun.* **2015**, *6*, 6569. [CrossRef] [PubMed]

57. Mychasiuk, R.; Muhammad, A.; Ilnytskyy, S.; Kolb, B. Persistent gene expression changes in NAc, mPFC, and OFC associated with previous nicotine or amphetamine exposure. *Behav. Brain Res.* **2013**, *256*, 655–661. [CrossRef] [PubMed]

58. Numachi, Y.; Shen, H.; Yoshida, S.; Fujiyama, K.; Toda, S.; Matsuoka, H.; Sora, I.; Sato, M. Methamphetamine alters expression of DNA methyltransferase 1 mRNA in rat brain. *Neurosci. Lett.* **2007**, *414*, 213–217. [CrossRef] [PubMed]

59. Jayanthi, S.; McCoy, M.T.; Chen, B.; Britt, J.P.; Kourrich, S.; Yau, H.J.; Ladenheim, B.; Krasnova, I.N.; Bonci, A.; Cadet, J.L. Methamphetamine downregulates striatal glutamate receptors via diverse epigenetic mechanisms. *Biol. Psychiatry* **2014**, *76*, 47–56. [CrossRef] [PubMed]

60. Krasnova, I.N.; Chiflikyan, M.; Justinova, Z.; McCoy, M.T.; Ladenheim, B.; Jayanthi, S.; Quintero, C.; Brannock, C.; Barnes, C.; Adair, J.E.; et al. CREB phosphorylation regulates striatal transcriptional responses in the self-administration model of methamphetamine addiction in the rat. *Neurobiol. Dis.* **2013**, *58*, 132–143. [CrossRef] [PubMed]

61. Gelfand, B.D.; Wright, C.B.; Kim, Y.; Yasuma, T.; Yasuma, R.; Li, S.; Fowler, B.J.; Bastos-Carvalho, A.; Kerur, N.; Uittenbogaard, A.; et al. Iron Toxicity in the Retina Requires Alu RNA and the NLRP3 Inflammasome. *Cell Rep.* **2015**, *11*, 1686–1693. [CrossRef] [PubMed]

62. Goncalves, J.; Baptista, S.; Martins, T.; Milhazes, N.; Borges, F.; Ribeiro, C.F.; Malva, J.O.; Silva, A.P. Methamphetamine-induced neuroinflammation and neuronal dysfunction in the mice hippocampus: Preventive effect of indomethacin. *Eur. J. Neurosci.* **2010**, *31*, 315–326. [CrossRef] [PubMed]

63. Venkatesan, A.; Uzasci, L.; Chen, Z.; Rajbhandari, L.; Anderson, C.; Lee, M.H.; Bianchet, M.A.; Cotter, R.; Song, H.; Nath, A. Impairment of adult hippocampal neural progenitor proliferation by methamphetamine: Role for nitrotyrosination. *Mol. Brain* **2011**, *4*, 28. [CrossRef] [PubMed]

64. Baptista, S.; Lasgi, C.; Benstaali, C.; Milhazes, N.; Borges, F.; Fontes-Ribeiro, C.; Agasse, F.; Silva, A.P. Methamphetamine decreases dentate gyrus stem cell self-renewal and shifts the differentiation towards neuronal fate. *Stem Cell Res.* **2014**, *13*, 329–341. [CrossRef] [PubMed]

65. Smith, K.J.; Butler, T.R.; Self, R.L.; Braden, B.B.; Prendergast, M.A. Potentiation of *N*-methyl-D-aspartate receptor-mediated neuronal injury during methamphetamine withdrawal in vitro requires co-activation of IP3 receptors. *Brain Res.* **2008**, *1187*, 67–73. [CrossRef] [PubMed]

66. Baliga, B.S.; Zahringer, J.; Trachtenberg, M.; Moskowitz, M.A.; Munro, H.N. Mechanism of D-amphetamine inhibition of protein synthesis. *Biochim. Biophys. Acta* **1976**, *442*, 239–250. [CrossRef]

67. Gray, N.K.; Hrabalkova, L.; Scanlon, J.P.; Smith, R.W. Poly(A)-binding proteins and mRNA localization: Who rules the roost? *Biochem. Soc. Trans.* **2015**, *43*, 1277–1284. [CrossRef] [PubMed]

68. Dai, L.; Taylor, M.S.; O'Donnell, K.A.; Boeke, J.D. Poly(A) binding protein C1 is essential for efficient L1 retrotransposition and affects L1 RNP formation. *Mol. Cell. Biol.* **2012**, *32*, 4323–4336. [CrossRef] [PubMed]

69. Kim, J.; Kass, D.H.; Deininger, P.L. Transcription and processing of the rodent ID repeat family in germline and somatic cells. *Nucleic Acids Res.* **1995**, *23*, 2245–2251. [CrossRef] [PubMed]

70. Johnson, L.J.; Brookfield, J.F. A test of the master gene hypothesis for interspersed repetitive DNA sequences. *Mol. Biol. Evol.* **2006**, *23*, 235–239. [CrossRef] [PubMed]

71. Deininger, P.L.; Batzer, M.A.; Hutchison, C.A., 3rd; Edgell, M.H. Master genes in mammalian repetitive DNA amplification. *Trends Genet.* **1992**, *8*, 307–311. [CrossRef]

72. Kazazian, H.H., Jr.; Goodier, J.L. LINE drive. retrotransposition and genome instability. *Cell* **2002**, *110*, 277–280. [CrossRef]

73. Joslyn, G.; Wolf, F.W.; Brush, G.; Wu, L.; Schuckit, M.; White, R.L. Glypican Gene GPC5 Participates in the Behavioral Response to Ethanol: Evidence from Humans, Mice, and Fruit Flies. *G3 (Bethesda)* **2011**, *1*, 627–635. [CrossRef] [PubMed]

74. Robison, A.J.; Nestler, E.J. Transcriptional and epigenetic mechanisms of addiction. *Nat. Rev. Neurosci.* **2011**, *12*, 623–637. [CrossRef] [PubMed]

75. Itzhak, Y.; Ergui, I.; Young, J.I. Long-term parental methamphetamine exposure of mice influences behavior and hippocampal DNA methylation of the offspring. *Mol. Psychiatry* **2015**, *20*, 232–239. [CrossRef] [PubMed]

76. Recinto, P.; Samant, A.R.; Chavez, G.; Kim, A.; Yuan, C.J.; Soleiman, M.; Grant, Y.; Edwards, S.; Wee, S.; Koob, G.F.; et al. Levels of neural progenitors in the hippocampus predict memory impairment and relapse to drug seeking as a function of excessive methamphetamine self-administration. *Neuropsychopharmacology* **2012**, *37*, 1275–1287. [CrossRef] [PubMed]

77. Volkow, N.D.; Chang, L.; Wang, G.J.; Fowler, J.S.; Leonido-Yee, M.; Franceschi, D.; Sedler, M.J.; Gatley, S.J.; Hitzemann, R.; Ding, Y.S.; et al. Association of dopamine transporter reduction with psychomotor impairment in methamphetamine abusers. *Am. J. Psychiatry* **2001**, *158*, 377–382. [CrossRef] [PubMed]

3

Alternative Splicing Profile and Sex-Preferential Gene Expression in the Female and Male Pacific Abalone *Haliotis discus hannai*

Mi Ae Kim [1,†]**, Jae-Sung Rhee** [2,†]**, Tae Ha Kim** [1]**, Jung Sick Lee** [3]**, Ah-Young Choi** [4]**,
Beom-Soon Choi** [4]**, Ik-Young Choi** [5,*] **and Young Chang Sohn** [1,*]

[1] Department of Marine Molecular Bioscience, Gangneung-Wonju National University, Gangneung 25457, Korea; kimmiaecho@gmail.com (M.A.K.); kio1231@naver.com (T.H.K.)
[2] Department of Marine Science, College of Natural Sciences, Incheon National University, Incheon 22012, Korea; jsrhee@inu.ac.kr
[3] Department of Aqualife Medicine, Chonnam National University, Yeosu 59626, Korea; ljs@chonnam.ac.kr
[4] Phyzen Genomics Institute, Seongnam 13558, Korea; ahyoung@phyzen.com (A.-Y.C.); bschoi@phyzen.com (B.-S.C.)
[5] Department of Agriculture and Life Industry, Kangwon National University, Chuncheon 24341, Korea
* Correspondence: choii@kangwon.ac.kr (I.-Y.C.); ycsohn@gwnu.ac.kr (Y.C.S.)

† These authors contributed equally to this work.

Academic Editor: J. Peter W. Young

Abstract: In order to characterize the female or male transcriptome of the Pacific abalone and further increase genomic resources, we sequenced the mRNA of full-length complementary DNA (cDNA) libraries derived from pooled tissues of female and male *Haliotis discus hannai* by employing the Iso-Seq protocol of the PacBio RSII platform. We successfully assembled whole full-length cDNA sequences and constructed a transcriptome database that included isoform information. After clustering, a total of 15,110 and 12,145 genes that coded for proteins were identified in female and male abalones, respectively. A total of 13,057 putative orthologs were retained from each transcriptome in abalones. Overall Gene Ontology terms and Kyoto Encyclopedia of Genes and Genomes (KEGG) pathways analyzed in each database showed a similar composition between sexes. In addition, a total of 519 and 391 isoforms were genome-widely identified with at least two isoforms from female and male transcriptome databases. We found that the number of isoforms and their alternatively spliced patterns are variable and sex-dependent. This information represents the first significant contribution to sex-preferential genomic resources of the Pacific abalone. The availability of whole female and male transcriptome database and their isoform information will be useful to improve our understanding of molecular responses and also for the analysis of population dynamics in the Pacific abalone.

Keywords: abalone; *Haliotis discus hannai*; transcriptome; isoform; PIS system

1. Introduction

The abalone (Gastropoda; Haliotidae) species is a marine gastropod that is herbivorous, single-shelled, and reef-dwelling. It is widely distributed throughout temperate and tropical coastal regions (~100 species) [1]. Of them, approximately 20 species are important for commercial aquaculture and wild fisheries worldwide as highly prized seafood items containing bioactive molecules (e.g., polysaccharides, proteins, fatty acids) that support health benefits beyond basic nutrition [2]. Global abalone production from farms reached 63,245 metric tons (MT) in 2010 after rapid development of aquaculture techniques [3]; total production increased to 85,344 MT in 2011 and 103,464 MT in 2013 [4]. Asian countries such as China, Korea, and Japan are currently major suppliers of abalone in the global market, while abalone culture is also growing in Europe (e.g., the UK, the Channel Islands, Ireland, France, and Spain) [4]. Particularly, Southeast Asia, along with China (80%) and Korea (12%), are the largest producers of farmed abalone in the world, whereas Japan and China are the major consumer regions [2,4].

Of the *Haliotis* species, the majority of Korean abalone production is composed of *Haliotis discus hannai*; its production is estimated to have increased over 60 times during the past 10 years and is predicted to reach over 10,000 MT by 2015 [4]. Because the abalone *H. discus hannai* is an organism of major economic interest, genomic resources and applications have recently focused on the understanding of the physiology, molecular adaptation, genetic selection, disease, defense mechanisms, and ecological genetic diversity of this species. Genes involved in numerous metabolic pathways such as innate immunity, cell stress and repair system, and antioxidant defense systems have been cloned and tested to determine their mRNA expression, translational inducibility, and molecular functions. Recently, the use of a genomics platform was successfully applied in *H. discus hannai*, although many research groups have continuously employed '-omics' platforms (e.g., genomics, proteomics, linkage map development, Bacterial Artificial Chromosome (BAC) library construction) to understand the growth, development, reproduction, molecular adaptation, or innate immune response mechanisms upon environmental stressors or pathogen challenges in the *Haliotis* genus [5–14]. For *H. discus hannai*, a genetic linkage map was constructed as a potential application of marker-assisted selection in breeding programs using amplified fragment length polymorphism (AFLP) markers [15]. The genome size of *H. discus hannai* was measured as 1.84 pg of C value by using flow cytometry [16]. Recently, an RNA-sequencing application revealed various innate immune challenge-responsive transcriptomes in the *Vibrio parahaemolyticus*-infected *H. discus hannai* [17]. However, the public availability of the gene/genome information of *H. discus hannai* and the global response profiles of transcriptomes or proteomes are still scarce.

In this study, we analyzed the sex-preferential transcriptome of *H. discus hannai* and identified genes that are differentially expressed in female or male abalone. Specifically, we analyzed entire isoforms of female- or male-specific transcripts that are alternatively spliced at transcriptional control. To validate transcriptome profiles, quantitative real-time (RT)-PCR (qPCR) was employed using several sex-related genes in both female and male abalones. This transcriptome information will be useful as a first significant contribution to genomic resources for each sex of the Pacific abalone. Furthermore, it will provide valuable data for numerous forms of future research studies abalones.

2. Materials and Methods

2.1. Sample Preparation

Sexually mature male (131.3 g, 10.2 cm) and female (119.1 g, 9.5 cm) *H. discus hannai* were obtained from a public fish market on July 4th, 2015 at Wando-gun (Jeollanam-do, Korea), and identified to the subspecies using morphological characteristics [18,19]. The abalones were anesthetized with $MgCl_2$ before dissection. To maximize the discovery of *H. discus hannai* gene pools, multiple tissues including ganglia, gills, intestine, hepatopancreas, muscle, and gonads were pooled, quickly frozen in liquid nitrogen, and stored at $-80\,°C$ until total RNA extraction. The sampling was performed in accordance with relevant institutional and national guidelines.

2.2. RNA Extraction and Library Construction

Pooled tissues of each sex were frozen in liquid nitrogen and homogenized with a glass pestle. Total RNA was extracted using the Hybrid-RTM kit (GeneAll Biotechnology Co., Seoul, Korea) according to the manufacturer's protocol. Total RNA was quantified by absorption of light at A260 and quality checked by analyzing the RNA Integrity Number (RIN) using a 2100 Bi system (Agilent Technologies Inc., Santa Clara, CA, USA).

Construction of full-length complementary DNA (cDNA) libraries and sequencing with the PacBio RSII platform (Pacific Bio-science Inc., Menlo Park, CA, USA) were performed at the National Instrumentation Center for Environmental Management (NICEM, Seoul National University, Seoul, Korea). Briefly, the first-strand cDNA was synthesized using the Clontech SMARTer PCR cDNA Synthesis kit (Takara Bio USA, Inc., Mountain View, CA, USA). The large-scale double-strand cDNA was generated with the optimal number of PCR cycles (for amplification) by the PCR cycle optimization test. The large-scale PCR products were purified and eluted to the optimal size (approximately 1–6 Kb) using the BluepippinTM System (Sage Science Inc., Beverly, MA, USA) and were re-purified using AMPure PB Beads (Pacific Biosciences Inc.) two additional times. A SMRTbell template library was prepared by repairing DNA damage, blunt end ligation, and annealing the SMRTbell adapter (according to the manufacturer's protocol) using the SMRTbell library kit. The template was eluted to construct libraries of approximately 1–2 Kb, 2–3 Kb, and 3–6 Kb using the BluepippinTM system. Average molecular weight and concentration of the library were validated using an Agilent 2100 Bioanalyzer. All of the libraries were sequenced on the SMRTbell 1–2 Kb and 2–3 Kb libraries and the SMRTbell 3–6 Kb library using the PacBio RSII platform according to the manufacturer's instructions.

2.3. Gene Annotation and Isoform Analysis

The raw sequencing reads of full-length cDNA libraries were classified and clustered to transcript consensus using the ToFU (transcript isoforms: Full-length and Unassembled) pipeline (GitHub version, Pacific Biosciences of California, Inc. Menlo Park, CA, USA.) supported by PacificBiosciences. Briefly, the raw reads were classified to full-length (FL) and non-full-length (nFL) reads. The FL reads were clustered to isoform-level and were then used for re-clustering with nFL reads by Quiver, which is included in the ToFU pipeline. The initial transcript consensus sequences were filtered to obtain a high-quality isoform sequence with at least 99% accuracy (Figure S1). The final transcriptome isoform sequences were filtered by removing the redundant sequences with a CD-HIT that is a software for clustering and comparing protein or nucleotide sequence. Orthologous genes of the total transcript isoforms were analyzed by clustering the final transcript consensus contigs using a threshold of 0.99 identities with a cluster software (CD-HIT-est) of the CD-HIT-package v.4.6 (Weizhong Li's lab at UCSD, La Jolla, CA, US). In this software, the gene family, including paralogous and isoforms, is clustered into the orthologous groups.

TransDecoder was used to identify candidate coding regions from the final transcriptome isoform sequence (The Broad Institute. Cambridge, MA, USA). The candidate coding regions were used for BLAST analysis against the UniProt and the NCBI non-redundant (nr) protein database (DB) to evaluate sequence similarity with genes of other species at an E-value cutoff of 1×10^{-6}. All transcriptome information including isoform sequences of *H. discus hannai* is registered in the Phyzen DB (http://www.phyzen.com/haliotis/index.php).

2.4. Gene Ontology and Kyoto Encyclopedia of Genes and Genomes Pathway Analysis

Gene Ontology (GO; biological function, cell component, molecular function) and Kyoto Encyclopedia of Genes and Genomes (KEGG) pathway analysis of the contigs were performed using the GOstats program (Roswell Park Cancer Institute, Buffalo, NY, USA) and the Fisher's Exact Test ($p < 0.05$), as implemented in the sequence annotation tool Blast2GO (BioBam Bioinformatics SL, Valencia, Spain). Three main categories for biological process, cellular component, and molecular function were obtained after comparing for similarities using default parameters. Gene annotation and GO analysis were performed at the NICEM, Seoul National University (Seoul, Korea). The assembled data were arranged based on read length, gene annotation, GenBank number, E-value, and species. In each section, the specific composition of GO terms was calculated and presented as a bar chart according to percentage.

2.5. Isoform Grouping

In this study, we discovered the isoform groups using the pipeline to isoforms of full-length cDNA sequences (PIS) system that was developed with open software to assembly and blast using full-length cDNA sequence with no genome sequence DB. In the system, the total transcript isoform consensus sequences were aligned and mapped to the longest putative orthologous consensus to then search for the isoform level sequence using the ToFU pipeline and GMAP software (a genomic mapping and alignment program for mRNA and expressed sequence tags (EST)) [20] (Figure 1). Briefly, the total non-redundant high quality transcript isoform sequences were clustered using BLASTCLUST (NCBI, Bethesda, MD, USA). The longest reads were used as reference sequences to test for alternatively spliced isoform sequences; this was achieved by aligning them with an isoform-level consensus sequence. The reference sequences were re-updated to perform a three times replicate analysis and then re-clustered with an amino acid peptide sequence. The final alternative spliced isoform sequences were tested by alignment with transcript consensus sequences using GMAP and by filtering of redundant transcripts using ToFU pipelines.

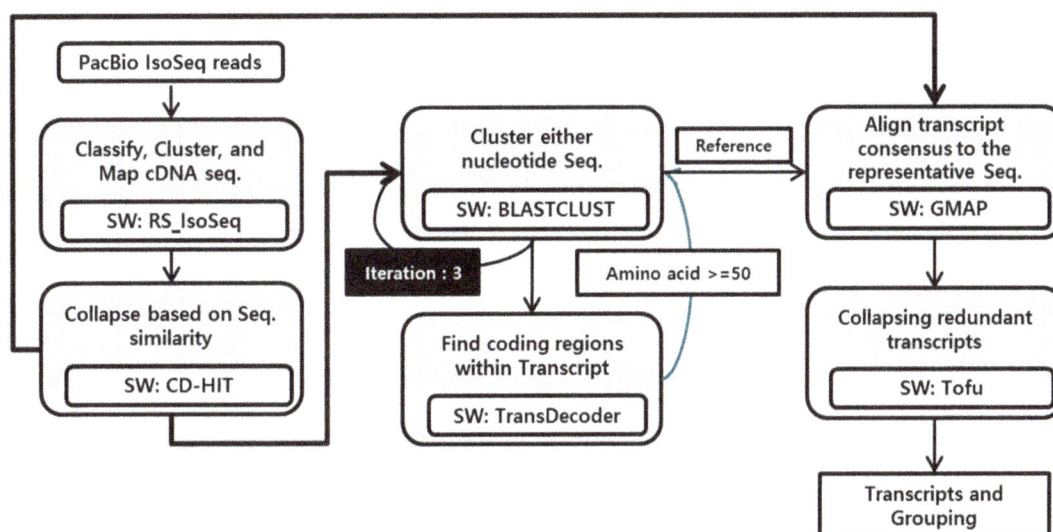

Figure 1. Schematic diagram of pipeline to isoforms of full-length complementary DNA (cDNA) sequence (PIS system).

2.6. Quantitative Real-Time RT-PCR

In order to test sex-preferential expression from each transcriptome, quantitative real-time RT-PCR (qPCR) was performed using six sex-preferential genes: *vitellogenin (VTG)*, *forkhead box protein L2 (FOXL2)*, *condensin-2*, *sperm-associated antigen 6 (SPAG6)*, *protein fem-1 homolog C-like (FEM1-like)*, and *tektin-1*, which are relatively well-characterized in marine invertebrates [9,21–24]. Based on the Minimum Information for Publication of Quantitative Real-Time PCR Experiments (MIQE) guidelines [25], transcriptional levels of each sex-preferential gene were validated using qPCR. Total RNA was extracted from the testes and ovaries of sexually mature abalone (four individuals from each sex) using the TriPure Isolation Reagent (Roche, Pleasanton, CA, USA). The maturity of gonads were examined by histological examination and revealed as being at the ripe and partial spent stages. After digestion of genomic DNA, 1 µg of total RNA was reverse transcribed using the PrimeScript RT reagent Kit (Takara Bio Inc., Shiga, Japan) with a gDNA eraser (Takara Bio Inc., Shiga, Japan). The resulting cDNAs were diluted, and an amount equivalent to 10 ng of starting RNA was assayed for mRNA expression analysis using *ribosomal protein 5 (RPL5)* as the reference gene. SYBR-based qPCR reactions (SYPR Premix Ex Taq II, Takara, Japan) were performed on an Applied Biosystems 7500 Real-Time PCR System (Applied Biosystems, Foster City, CA, USA) using the following reaction conditions: 50 °C for 2 m, 95 °C for 10 m, followed by 40 cycles of 95 °C for 15 s, 60 °C for 1 m. The primer sets used in this study are listed in Table S1. A melting curve was generated at the end of the reaction to confirm an accurate amplification of the target amplicon. PCR efficiencies of the target and reference genes were verified. The relative mRNA expression was calculated according to the formula: $2^{-(\text{Ct target gene} - \text{Ct reference gene})}$. All results are expressed as the mean $\pm$ S.E.M (standard error of mean).

2.7. Statistical Analysis

The SPSS software package (ver. 17.0; SPSS Inc., Chicago, IL, USA) was used for statistical analysis. Data are expressed as the mean $\pm$ S.E.M. Significant differences between female and male qPCR data were analyzed by the Student's t test (two-tailed). $p < 0.05$ was considered to be significant.

3. Results and Discussion

3.1. Transcriptome Assembly and Gene Annotation

To establish a sex-specific transcriptome DB of the Pacific abalone *H. discus hannai*, we sequenced the mRNA of each female and male and abalone using pooled tissues (i.e., ganglia, gills, intestine, hepatopancreas, muscle, and gonads). After trimming and assembly, a total of 81 Mb nucleotide information (including 36,273 full-length cDNAs) of female and 60 Mb (including 29,275 full-length cDNAs) of male *H. discus hannai* were obtained using the PacBio RSII (Pacific Bio-science Inc.) sequencing platform with PIS system (Table S2). The lengths of female full-length cDNA ranged from 308 to 8377 bp with an average length of 2220 bp, whereas male full-length cDNAs showed an average length of 2059 bp and ranged from 308 to 8058 bp. The N50 values of those cDNAs were 2741 bp in female and 2477 bp in male. Subsequently, 18,692 and 15,271 high quality full-length consensuses were filtered from 22,494 (female) and 18,981 (male) full-length cDNAs, respectively, using CD-HIT platform in the absence of a redundant sequence. Finally, a series of bioinformatics, which comprised five platforms, defined 15,110 and 12,145 isoforms in female and male transcriptomes, respectively (Table 1; Figure S2).

Table 1. Platforms for establishing gene sets of female and male *Haliotis discus hannai*.

Step	Data	Platform	Female	Male
1	High quality consensus sequence	RS_IsoSeq	22,494	18,981
2	Non-redundant representative sequence	CD-HIT	18,692	15,271
3	Reference isoforms	BLASTCLUST and TransDecoder	15,363	12,409
4	Final isoform transcriptome by combine representative sequence	GMAP and ToFU	15,792	12,718
5	Final gene set with representative isoforms	TransDecoder	15,110	12,145

Gene annotation of the whole transcripts was performed by BLASTx analysis using NCBI nr protein DB. The result showed that 15,110 female isoforms and 12,145 male isoforms had at least one positive BLAST hit (E-value $< 1 \times 10^{-4}$), representing 8274 and 6579 annotated genes, respectively (Tables S3 and S4). All data have been deposited in the Phyzen DB.

Distribution analysis showed that 17 species had a BLAST hit with more than 50 transcripts in both sexes. Specifically, the owl limpet *Lottia gigantea* (Mollusca, Gastropoda) showed the highest similarity, with 4542 reads for the female abalone and 4759 reads for the male abalone (Figure 2A). Among the top hit species, 83% of female contigs (9478) and 69% of male contigs (7204) matched to the phylum Mollusca with a high similarity value (Figure 2B). In addition, 65% of female contigs (7375) and 69% of male contigs (7204) showed homology to Gastropoda at class level (Figure 2C). Thus, overall gene annotation results of *H. discus hannai* showed good agreement with their phylogenetic relationships.

(A) Species

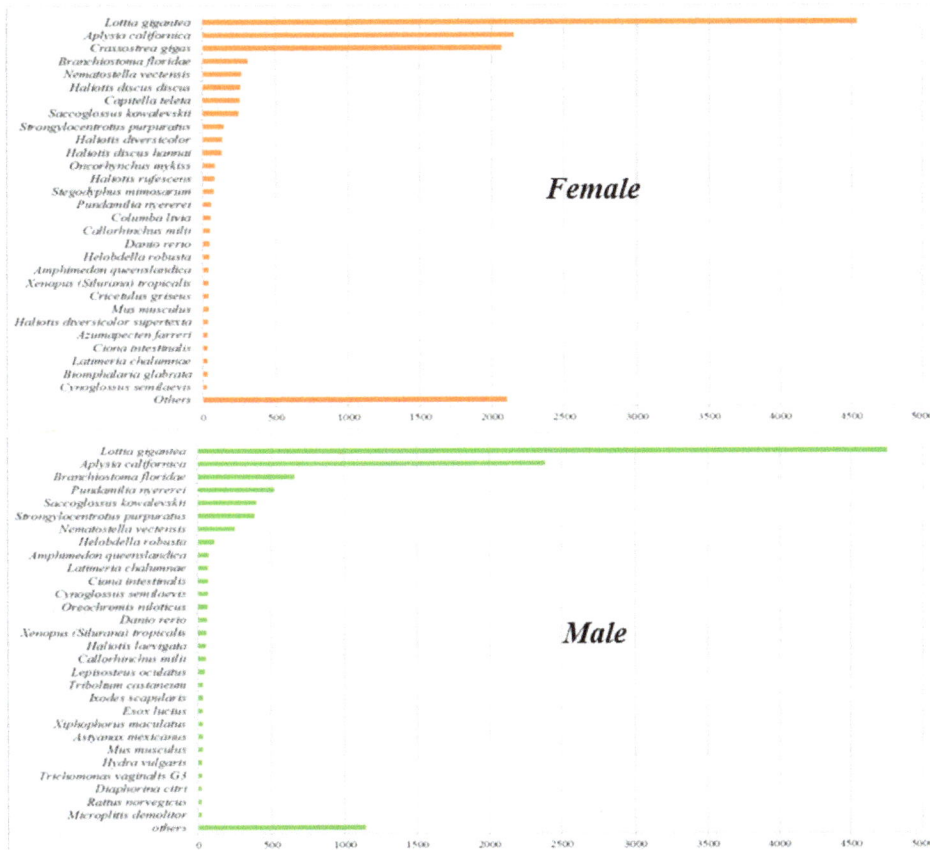

(B) Phylum

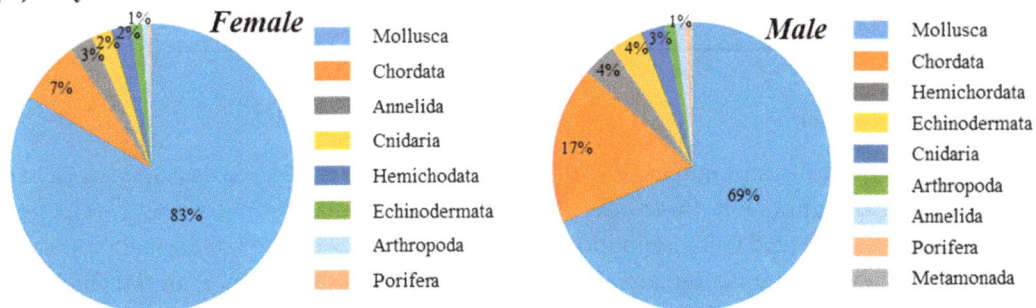

(C) Class

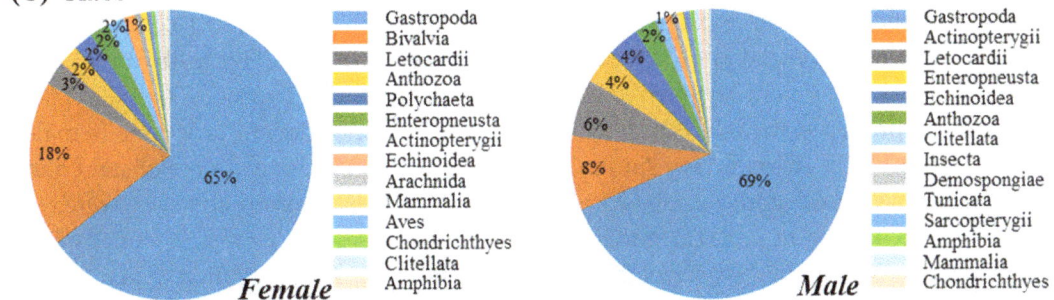

Figure 2. Homology searches of the female and male *Haliotis discus hannai* transcript contigs. (**A**) Number of BLAST hits; (**B**) top-hit phylum distribution; (**C**) top-hit class distribution.

Single copy orthologs were analyzed between female and male abalone using the reciprocal BLAST best-hit method (E-value $< 1 \times 10^{-10}$) with 18,692 female and 15,271 male nr high quality full-length cDNA sequences (Figure 3). Subsequently, 13,057 orthologs that could commonly serve in basic metabolism were determined between sexes of this species. We found that 85% female contigs had matches in the assembled contigs, whereas 91% matched with the male contigs (Table 2).

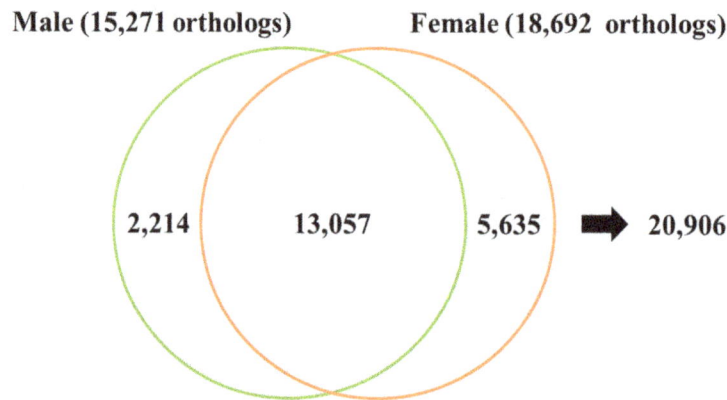

Male (15,271 orthologs)   **Female (18,692 orthologs)**

2,214   13,057   5,635   ➡ 20,906

Figure 3. Gene discovery rate of each transcriptome database. Venn diagram to compare ortholog numbers annotated in the female and male *Haliotis discus hannai*.

Table 2. Ortholog statistics between female and male *Haliotis discus hannai*.

Number of Total Orthologs	13,057
Average identity (%)	99.33
Average coverage of female (%)	85.23
Average coverage of male (%)	90.63
Number of 100% coverage orthologs	502
Number of 100% identity orthologs	597

3.2. Functional Annotation

To compare the transcriptome information of female and male *H. discus hannai*, a GO analysis in terms of cellular component, biological process, and molecular function was conducted using Blast2GO (Figure 4). Detailed GO distributions in three GO categories (i.e., biological process, cellular component, and molecular function) are incorporated in the Supplementary material (Tables S5–S10). The vast majority of genes were involved in binding (41% in both female and male) and catalytic activity functions (34% in both female and male) in the molecular function category (Figure 4A). In the cellular component class, most of the genes were related to cell (18% each), cell part (18% each), organelle (13% each), and membrane (12% each) in both female and male (Figure 4B). In the biological process class, most genes were categorized as being related to cellular processes (female: 14%; male: 14%), metabolic processes (female: 12%; male: 12%), and single-organism processes (female: 12%; male: 11%) (Figure 4C). Overall, very similar compositions (in percentage) were observed for both sexes in the three major categories using GO assignments. Although composition rates were relatively low in both sexes, there were some categories of biological processes that were differentially expressed between female and male abalones, including GO terms linked to the regulation of gene expression, positive regulation of transcription, reproduction, rhythmic process, and reproductive process. This finding indicates that some of the genes may be related to sex-specific roles.

(A) Molecular function

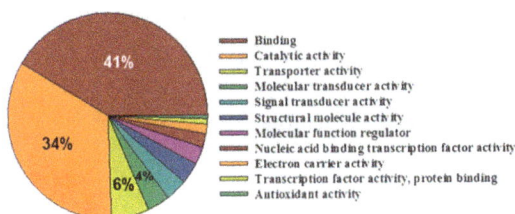

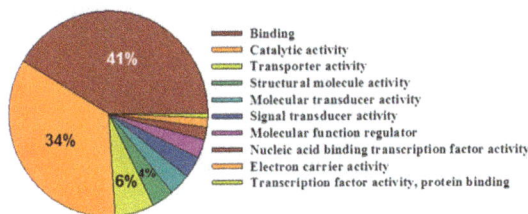

Binding
Catalytic activity
Transporter activity
Molecular transducer activity
Signal transducer activity
Structural molecule activity
Molecular function regulator
Nucleic acid binding transcription factor activity
Electron carrier activity
Transcription factor activity, protein binding
Antioxidant activity

Binding
Catalytic activity
Transporter activity
Structural molecule activity
Molecular transducer activity
Signal transducer activity
Molecular function regulator
Nucleic acid binding transcription factor activity
Electron carrier activity
Transcription factor activity, protein binding

(B) Cellular component

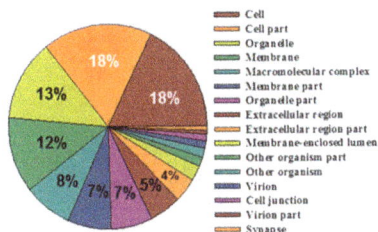

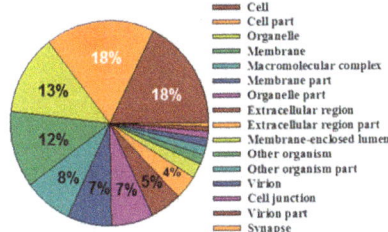

Cell
Cell part
Organelle
Membrane
Macromolecular complex
Membrane part
Organelle part
Extracellular region
Extracellular region part
Membrane-enclosed lumen
Other organism part
Other organism
Virion
Cell junction
Virion part
Synapse

Cell
Cell part
Organelle
Membrane
Macromolecular complex
Membrane part
Organelle part
Extracellular region
Extracellular region part
Membrane-enclosed lumen
Other organism
Other organism part
Virion
Cell junction
Virion part
Synapse

(C) Biological process

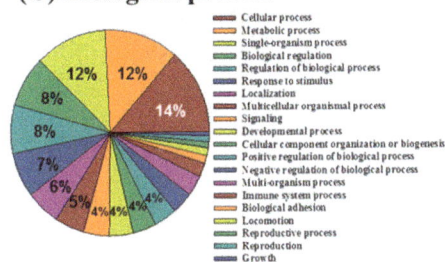

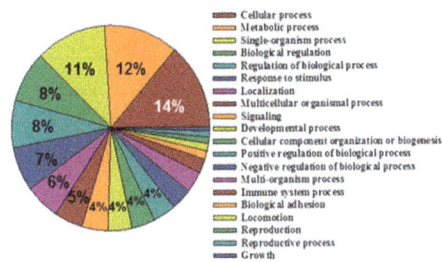

Cellular process
Metabolic process
Single-organism process
Biological regulation
Regulation of biological process
Response to stimulus
Localization
Multicellular organismal process
Signaling
Developmental process
Cellular component organization or biogenesis
Positive regulation of biological process
Negative regulation of biological process
Multi-organism process
Immune system process
Biological adhesion
Locomotion
Reproductive process
Reproduction
Growth

Cellular process
Metabolic process
Single-organism process
Biological regulation
Regulation of biological process
Response to stimulus
Localization
Multicellular organismal process
Signaling
Developmental process
Cellular component organization or biogenesis
Positive regulation of biological process
Negative regulation of biological process
Multi-organism process
Immune system process
Biological adhesion
Locomotion
Reproduction
Reproductive process
Growth

Female   **Male**

Figure 4. Gene Ontology (GO) analysis in terms of (**A**) molecular function, (**B**) cellular component, and (**C**) biological process that are enriched in the female and male *Haliotis discus hannai* transcript contigs.

To compare overall biological function between female and male transcripts, we classified all the assembled genes based on function using the KEGG pathway (Tables S11 and S12). KEGG analysis of the female and male *H. discus hannai* transcriptome revealed that the vast majority of KEGG pathways were involved in the 'purine metabolism' (female: #30; male: #30), followed by the 'starch and sucrose metabolism' (female: #25; male: #18), 'pyruvate metabolism' (female: #19) or 'pyrimidine metabolism' (male: #18), and 'glycolysis/gluconeogenesis' (female: #19; male: #17), suggesting that these metabolic pathways are actively expressed in both sexes. Similarly, the 'purine metabolism' was highly detected in the transcriptome DB of mixed organs from the South African abalone *H. midae* [26]. The 'purine metabolism' is important for providing basic components for nucleotides (i.e., DNA, RNA) and cellular energy sources like ATP. Most crucial members involved in sub-pathways (e.g., de novo biosynthesis of purine ring, salvage pathway) of the 'purine metabolism' are detected in both sexes (Figure S3). Thiamine is essential for carbohydrate metabolism and is also involved in thiazole and pyrimidine metabolism (i.e., B group of vitamin synthesis). Tissue contents of thiamin are closely associated with maximum growth of juvenile abalone [27]. The overall composition of the top ten female KEGG pathways was similar to that of its male counterpart. Thus, these results suggest the intactness of the *H. discus hannai* transcriptome, as the information dose not lack major functional GO categories or KEGG pathways compared to the transcriptomes of the genus *Haliotis*.

3.3. Isoform Analysis

Isoforms derived from alternative transcription start sites, alternative poly-adenylation, or alternative splicing were identified in *H. discus hannai* (Table 3). The numbers of genes covering at least two isoforms were as follows: 519 from 15,110 genes in female and 391 from 12,145 genes in male abalones (Tables S13 and S14). The numbers of isoforms were variable from 2 to 27 in female and from 2 to 52 in male abalones (Figure S2).

Table 3. Summary of the isoform information for *Haliotis discus hannai*.

Contig data	Female	Male
Total genes	15,110	12,145
Genes with no isoforms	14,591	11,754
Genes with at least two isoforms	519	391
Total length of genes with isoforms (bp)	1,599,611	1,166,159
Average length (bp)	3082	2982
Maximum length (bp)	8315	8058
Minimum length (bp)	741	847

Of the identified isoforms in abalone, the cubilin protein has a high number of isoforms in both sexes (Tables 4 and 5). In the GenBank DB, *cubilin* genes are annotated from 185 species; for example, there are five isoforms registered in humans. Thus, our discovery of isoforms using a PIS system could be useful for defining complex alternatively spliced patterns of certain genes. Surprisingly, the cubilin protein contains nine isoforms in both female and male abalone (Figure 5); from these, only three isoforms have the same sequence in both sexes. This is evidence that alternative splicing occurs differently in each sex. In addition, two putative *cubilin-like* genes, which are spliced from other loci by duplication, are observed to have nine and two isoforms in female and four and two isoforms in male abalones. Cubilin as an endocytic receptor is a large extracellular membrane protein (~450 kDa) [28]. In most cubilin proteins, two highly conserved domains, eight tandem epidermal growth factor domains followed by 27 tandem CUB (initially found in complement components C1r/C1s, Uegf, and bone morphogenic protein-1) domains harboring the intrinsic factor (IF)-cobalamin binding site (CUB domains 5–8), are included in extracellular modules [28,29]. Cubilin is associated with another large membrane-associated protein, megalin, to mediate the luminal uptake of a large number of proteins such as albumin. This protein complex plays a crucial role in the uptake of filtered carrier proteins such as vitamin D binding protein, retinol binding protein, transcobalamin, and transferrin [30]. Regarding the generally identifed functions of cubilin, we may suggest its putative diversified roles based on previous observations from mollusks including *Haliotis* sp.. Vitamin D metabolism is curcial in normal growth and shell mineralization of *H. discus hannai* [31,32]. Mollusks would have conserved and active retinol metabolism, as high retinyl ester storage capacity was observed with active retinoids in gastropod lineages [33,34]. Transcobalamin is a carrier protein which directly binds to cobalamin (vitamin B12). In fact, the cobalamin has been characterized as a nutrient source in most mollusks [35,36]. Tranferrin is a regulatory molecule of the innate immune system in mollusks that employs antimicrobial activity against a wide range of Gram-positive and -negative bacteria [37,38]. In *H. discus discus*, the in vivo antimicrobial property of transferrin through its Fe^{3+} binding ability was characterzied [39]. Although the functions of the cubilin protein in the renal proximal tubule of rodents and mammals are well established, information regarding the gene and its molecular mechanisms remains unclear in aquatic invertebrates. However, these results suggest that cubilin may be stronlgy involved with diverse metabolisms to maintain homeostasis and innate immunity in both female and male abalone. We assume that abalone cublin would have evolved with different transcription patterns by an alternative splicing mechanism as functional diversification. Our results also suggest that a series of bioinformatics platforms can be successfully applied to establish isoform information in the abalone.

Table 4. List of top-ranked genes containing over five isoforms in the female abalone.

Cluster ID	Length (bp)	#Isoform	Description	Matched Species	GenBank No.
F_Cluster00018	6565	27	deleted in malignant brain tumors one protein	*Columba livia*	ZDB-GENE-060228-6
F_Cluster11205	2162	11	-	-	-
F_Cluster00024	6380	9	PREDICTED: cubilin	*Haplochromis burtoni*	ZDB-GENE-060228-6
F_Cluster00089	4372	9	PREDICTED: cubilin-like	*Lingula anatina*	ZDB-GENE-060228-6
F_Cluster13261	1675	9	-	-	-
F_Cluster00812	3837	8	PREDICTED: cyclin-L1-like	*Crassostrea gigas*	H2U6Q2
F_Cluster00002	8315	7	PREDICTED: LOW QUALITY PROTEIN: sushi, von Willebrand factor type A, epidermal growth factor (EGF) and pentraxin domain-containing protein 1	*Callorhinchus milii*	F1MNH3
F_Cluster00829	3833	6	-	-	-
F_Cluster03162	3343	6	PREDICTED: serine/arginine-rich splicing factor 6-like isoform X1	*Crassostrea gigas*	A0A0D9SEM4
F_Cluster05356	3088	6	heterogeneous nuclear ribonucleoprotein L, partial	*Aplysia californica*	R4GHI6
F_Cluster11593	2114	6	-	-	-
F_Cluster00004	7437	5	hypothetical protein LOTGIDRAFT_214098	*Lottia gigantea*	NP_001116989.1
F_Cluster00011	6791	5	hypothetical protein AC249_AIPGENE2795	*Exaiptasia pallida*	F1NX90
F_Cluster10316	2268	5	PREDICTED: Na(+)/H(+) exchange regulatory cofactor NHE-RF1-like	*Biomphalaria glabrata*	XP_414851.3
F_Cluster12757	1916	5	-	-	-

Table 5. List of top-ranked genes containing over five isoforms in the male abalone.

Cluster ID	Length (bp)	#Isoform	Description	Matched species	GenBank No.
M_Cluster00016	6579	52	hypothetical protein cypCar_00021969, partial	*Cyprinus carpio*	ZDB-GENE-060228-6
M_Cluster00705	3598	11	PREDICTED: cyclin-L1-like	*Crassostrea gigas*	H2U6Q2
M_Cluster00017	6523	9	PREDICTED: cubilin	*Oreochromis niloticus*	ZDB-GENE-060228-6
M_Cluster01458	3359	8	PREDICTED: mesocentin-like	*Aplysia californica*	-
M_Cluster09253	1770	8	-	-	-
M_Cluster09585	1697	8	-	-	-
M_Cluster00226	3901	7	hypothetical protein LOTGIDRAFT_115468	*Lottia gigantea*	XP_002415964.1
M_Cluster00908	3513	7	serine-arginine protein 55	*Melipona quadrifasciata*	E1C270
M_Cluster00059	4569	5	hypothetical protein LOTGIDRAFT_200884	*Lottia gigantea*	NP_001040037.1
M_Cluster00931	3505	5	-	-	-
M_Cluster01425	3366	5	-	-	-
M_Cluster01740	3296	5	putative splicing factor, arginine/serine-rich 7	*Crassostrea gigas*	NP_064477.1
M_Cluster01748	3295	5	-	-	-
M_Cluster06621	2255	5	PREDICTED: tryptophan 2,3-dioxygenase-like	*Lingula anatina*	M3X838
M_Cluster09868	1649	5	-	-	-

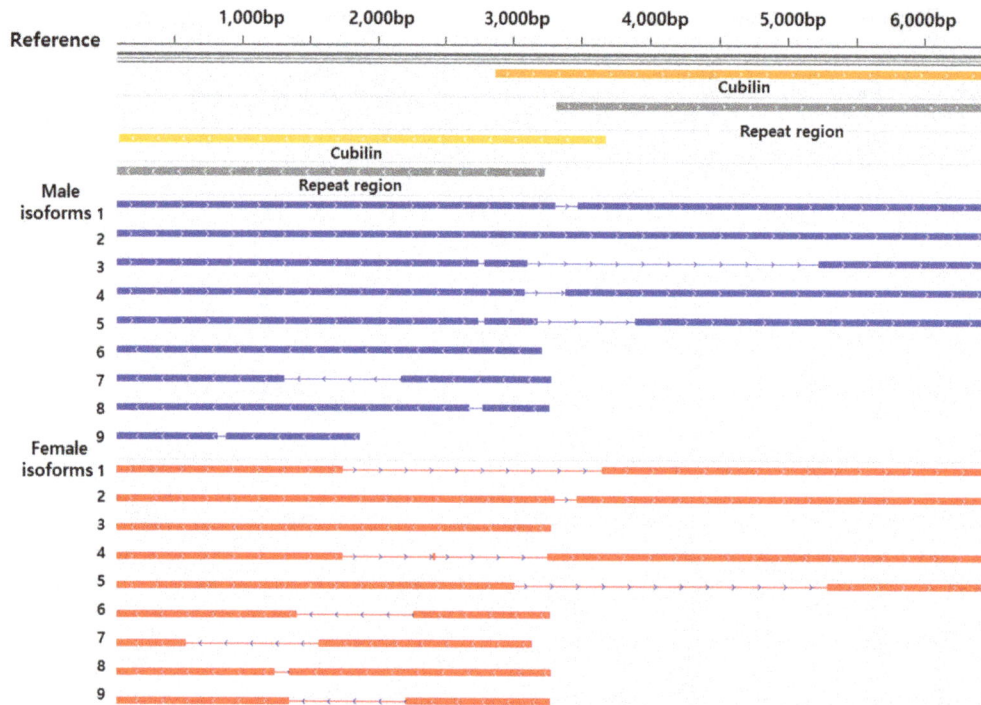

Figure 5. *Cubilin* isoforms in female and male Pacific abalone. *Cubilin* isoforms were determined by mapping consensus sequences to *cubilin* reference genes. Nine isoforms were defined in both female and male abalone, but only three isoforms showed the sequence in both sexes.

3.4. Sex-Preferential Genes Expression

To validate our sex-specific isoform DB, we analyzed the mRNA expression profiles of six sex-preferential genes, which are suggested to be sex-specific markers in mollusks. Gender-related genes such as *VTG*, *FOXL2*, *condensin-2*, *SPAG6*, *FEM1-like*, and *tektin-1* were identified in the transcriptome DB, and their transcriptional levels in the same tissues (used in de novo assembly) were measured by the qPCR method. Of three female-related genes, *VTG* and *FOXL2* were predominantly detected in mature female tissues (Figure 6A,B), whereas *condensin-2* showed no significant difference between sexes (Figure 6C). In general, the yolk protein vitelline is synthesized from its precursor VTG and is accumulated in oocytes in most oviparous animals [40]. Although VTG is known to be synthesized in extraovarian tissues such as the fat body of insects, the hepatopancreas of decapod crustaceans, and the liver of vertebrates, many reports have suggested that the ovary can be a single source of VTG protein in mollusks [41–48]. The transcription factor *FOXL2* is preferentially expressed in the ovary and plays an important role in ovarian determination and development in vertebrates [49]. Ovary-preferential expression of *FOXL2* is also observed in several mollusks, as shown in the Pacific abalone [22,50–53]. Although a microarray-based analysis using the European clam *Ruditapes decussatus* showed that *condensin-2* is overexpressed in females during gonad development [21], we were not able to detect any significant differences in the Pacific abalone.

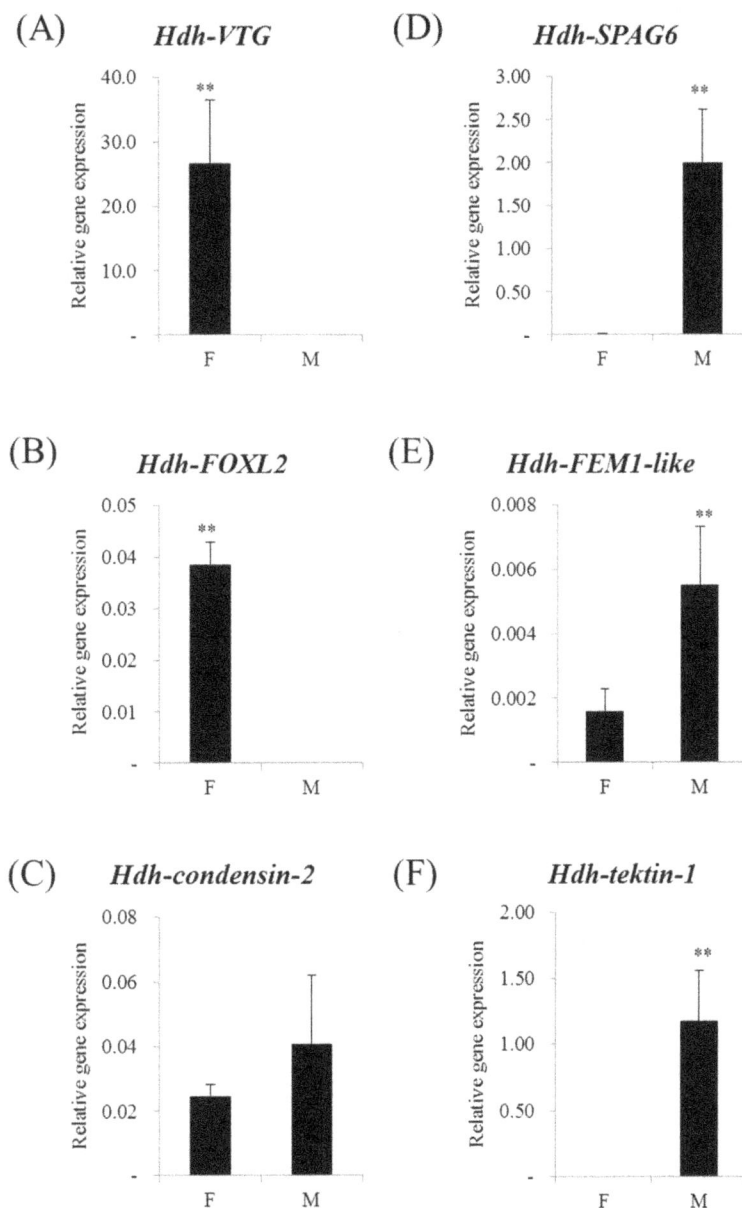

Figure 6. Relative mRNA expression profiles of the six selected genes from sexually mature *Haliotis discus hannai* by quantitative real-time RT-PCR analysis. The genes overexpressed in females were vitellogenin (*Hdh-VTG*) (**A**) and forkhead box protein L2 (*Hdh-FOXL2*) (**B**), but not condensin-2 (*Hdh-condensin-2*) (**C**). The genes overexpressed in males were sperm-associated antigen 6 (*Hdh-SPAG6*) (**D**), protein fem-1 homolog C-like (*Hdh-FEM1-like*) (**E**), and tektin-1 (*Hdh-tektin-1*) (**F**). Statistical changes were determined by the Student's *t* test (two-tailed) and are denoted as follows: ** $p < 0.01$. M and F indicate female and male, respectively.

The male-specific genes, *SPAG6*, *FEM1-like*, and *tektin-1* were highly expressed in mature male tissues (Figure 6D–F). SPAG6 is an axoneme central apparatus protein and has an essential role in the regulation of cilia/sperm flagella motility [54]. The *FEM1-like* gene is a component of the signal transduction pathway controlling sex determination in *Drosophila melanogaster* and *Caenorhabditis elegans* [55,56]. *Tektin-1*, a constitutive protein of microtubules in cilia, flagella, basal bodies, and centrioles is predominantly expressed in the testis and plays a role in spermatogenesis [57]. Male- or testis-preferential expressions of these genes has been consistently suggested in invertebrates [21,22,24,58]. Taken together, these studies strongly suggest the intactness of the sex-specific isoform DB of the Pacific abalone.

4. Conclusions

In this study, we characterized the transcriptome information for female and male *H. discus hannai*. We investigated differentially expressed isoforms involved in numerous signaling pathways and physiological metabolisms. Based on a comparative analysis of female and male abalone, the information obtained in this study represents the first significant contribution to sex-specific genomic resources, as well as isoform information. These data will provide an essential genomic reference that could be used for further diverse genetics- and physiology-based research using abalones.

Supplementary Materials: The following are available online at www.mdpi.com/2073-4425/8/3/99/s1, Figure S1: Schematic chart of raw data processing and de novo assembly of the female and male *Haliotis discus hannai* transcriptome. To construct a high-resolution assembly, the SMART portal protocol was employed using an ICE algorithm and quiver algorithm, Figure S2: Gene distribution of the isoforms in female and male abalone. A total of 519 and 391 genes possess a minimum of two isoforms in female and male abalones, respectively, Figure S3: Example of the most highly matched pathway, 'purine metabolism', in (A) female and (B) male abalones using the KEGG pathway database, Table S1: Summary of the raw read sequencing information on *Haliotis discus hannai*, Table S2: BLASTx results of female abalone transcriptome, Table S3: BLASTx results of male abalone transcriptome, Table S4: GO term analysis of female abalone transcriptome: Molecular function, Table S5: GO term analysis of female abalone transcriptome: Cellular component, Table S6: GO term analysis of female abalone transcriptome: Biological process, Table S7: GO term analysis of male abalone transcriptome: Molecular function, Table S8: GO term analysis of male abalone transcriptome: Cellular component, Table S9: GO term analysis of male abalone transcriptome: Biological process, Table S10: KEGG analysis of female abalone transcriptome, Table S11: KEGG analysis of male abalone transcriptome, Table S12: List and description of genes including Isoforms from female abalone transcriptome, Table S13: List and description of genes including Isoforms from male abalone transcriptome, Table S14: Information on oligo primer sequences used in this study.

Acknowledgments: This research was supported by the Golden Seed Project, Ministry of Agriculture, Food and Rural Affairs (MAFRA), Ministry of Oceans and Fisheries (MOF), Rural Development Administration (RDA), and Korea Forest Service (KFS), and by Basic Science Research Program through the National Research Foundation of Korea (NRF) funded by the Ministry of Education (2015R1D1A1A01060673).

Author Contributions: Mi Ae Kim, Jae-Sung Rhee, Tae Ha Kim, Jung Sick Lee, Ik-Young Choi, and Young Chang Sohn conceived and designed the experiments; Mi Ae Kim, Tae Ha Kim, Ah-Young Choi, and Beom-Soon Choi performed the experiments; Mi Ae Kim, Tae Ha Kim, Jae-Sung Rhee, Jung Sick Lee, Ik-Young Choi, and Young Chang Sohn analyzed the data; Ah-Young Choi, Beom-Soon Choi, and Ik-Young Choi contributed analysis tools; Mi Ae Kim, Jae-Sung Rhee, Jung Sick Lee, Ik-Young Choi, and Young Chang Sohn wrote the paper.

References

1.   Estes, J.A.; Lindberg, D.R.; Wray, C. Evolution of large body size in abalones (*Haliotis*): Patterns and implications. *Paleobiology* **2005**, *31*, 591–606. [CrossRef]

2.   Suleria, H.A.; Masci, P.P.; Gobe, G.C.; Osborne, S.A. Therapeutic potential of abalone and status of bioactive molecules: A comprehensive review. *Crit. Rev. Food Sci. Nutr.* **2015**, *57*, 1742–1748. [CrossRef] [PubMed]

3.   Cook, P.A.; Gordon, H.R. World abalone supply, markets, and pricing. *J. Shellfish Res.* **2010**, *29*, 569–571. [CrossRef]

4.   Cook, P.A. The worldwide abalone industry. *Mod. Econ.* **2014**, *5*, 1181–1186. [CrossRef]

5.   Sekino, M.; Hara, M. Linkage maps for the Pacific abalone (genus *Haliotis*) based on microsatellite DNA markers. *Genetics* **2007**, *175*, 945–958. [CrossRef] [PubMed]

6.   Franchini, P.; van der Merwe, M.; Roodt-Wilding, R. Transcriptome characterization of the South African abalone *Haliotis midae* using sequencing-by-synthesis. *BMC Res. Notes* **2011**, *4*, 59. [CrossRef] [PubMed]

7.   Huang, Z.X.; Chen, Z.S.; Ke, C.H.; Zhao, J.; You, W.W.; Zhang, J.; Dong, W.T.; Chen, J. Pyrosequencing of *Haliotis diversicolor* transcriptomes: Insights into early developmental molluscan gene expression. *PLoS ONE* **2012**, *7*, e51279. [CrossRef] [PubMed]

8. Palmer, M.R.; McDowall, M.H.; Stewart, L.; Ouaddi, A.; MacCoss, M.J.; Swanson, W.J. Mass spectrometry and next-generation sequencing reveal an abundant and rapidly evolving abalone sperm protein. *Mol. Reprod. Dev.* **2013**, *80*, 460–465. [CrossRef] [PubMed]

9. Mendoza-Porras, O.; Botwright, N.A.; McWilliam, S.M.; Cook, M.T.; Harris, J.O.; Wijffels, G.; Colgrave, M.L. Exploiting genomic data to identify proteins involved in abalone reproduction. *J. Proteom.* **2014**, *108*, 337–353. [CrossRef] [PubMed]

10. Shiel, B.P.; Hall, N.E.; Cooke, I.R.; Robinson, N.A.; Strugnell, J.M. De novo characterisation of the greenlip abalone transcriptome (*Haliotis laevigata*) with a focus on the heat shock protein 70 (HSP70) family. *Mar. Biotechnol.* **2015**, *17*, 23–32. [CrossRef] [PubMed]

11. Harney, E.; Dubief, B.; Boudry, P.; Basuyaux, O.; Schilhabel, M.B.; Huchette, S.; Paillard, C.; Nunes, F.L. De novo assembly and annotation of the European abalone *Haliotis tuberculata* transcriptome. *Mar. Genom.* **2016**, *28*, 11–16. [CrossRef] [PubMed]

12. Jiang, L.; You, W.; Zhang, X.; Xu, J.; Jiang, Y.; Wang, K.; Zhao, Z.; Chen, B.; Zhao, Y.; Mahboob, S.; et al. Construction of the BAC library of small abalone (*Haliotis diversicolor*) for gene screening and genome characterization. *Mar. Biotechnol.* **2016**, *18*, 49–56. [CrossRef] [PubMed]

13. Bathige, S.D.; Umasuthan, N.; Jayasinghe, J.D.; Godahewa, G.I.; Park, H.C.; Lee, J. Three novel C1q domain containing proteins from the disk abalone *Haliotis discus discus*: Genomic organization and analysis of the transcriptional changes in response to bacterial pathogens. *Fish Shellfish Immunol.* **2016**, *56*, 181–187. [CrossRef] [PubMed]

14. Wang, T.; Nuurai, P.; McDougall, C.; York, P.S.; Bose, U.; Degnan, B.M.; Cummins, S.F. Identification of a female spawn-associated Kazal-type inhibitor from the tropical abalone *Haliotis asinina*. *J. Pept. Sci.* **2016**, *22*, 461–470. [CrossRef] [PubMed]

15. Liu, X.; Liu, X.; Guo, X.; Gao, Q.; Zhao, H.; Zhang, G. A preliminary genetic linkage map of the Pacific abalone *Haliotis discus hannai* Ino. *Mar. Biotechnol.* **2006**, *8*, 386–397. [CrossRef] [PubMed]

16. Adachi, K.; Okumura, S.I. Determination of genome size of *Haliotis discus hannai* and *H. diversicolor aquatilis* (Haliotidae) and phylogenetic examination of this family. *Fish. Sci.* **2012**, *78*, 849–852. [CrossRef]

17. Nam, B.H.; Jung, M.; Subramaniyam, S.; Yoo, S.I.; Markkandan, K.; Moon, J.Y.; Kim, Y.O.; Kim, D.G.; An, C.M.; Shin, Y.; et al. Transcriptome analysis revealed changes of multiple genes involved in *Haliotis discus hannai* innate immunity during *Vibrio parahemolyticus* infection. *PLoS ONE* **2016**, *11*, e0153474. [CrossRef] [PubMed]

18. Ino, T. Biological studies on the propagation of Japanese abalone (Genus *Haliotis*). *Bulletin Tokai Reg. Fish. Res. Lab.* **1952**, *5*, 1–102.

19. Won, S.-H.; Kim, S.-K.; Kim, S.-C.; Yang, B.-K.; Lim, B.-S.; Lee, J.-H.; Lim, H.K.; Lee, J.-S.; Lee, J.-S. The morphological characteristics of four Korean abalone species in *Nordotis*. *Korean J. Malacol.* **2014**, *30*, 87–93. [CrossRef]

20. Wu, T.D.; Watanabe, C.K. GMAP: A genomic mapping and alignment program for mRNA and EST sequences. *Bioinformatics* **2005**, *21*, 1859–1875. [CrossRef] [PubMed]

21. De Sousa, J.T.; Milan, M.; Bargelloni, L.; Pauletto, M.; Matias, D.; Joaquim, S.; Matias, A.M.; Quillien, V.; Leitão, A.; Huvet, A. A microarray-based analysis of gametogenesis in two Portuguese populations of the European clam *Ruditapes decussatus*. *PLoS ONE* **2014**, *9*, e92202. [CrossRef] [PubMed]

22. Teaniniuraitemoana, V.; Huvet, A.; Levy, P.; Klopp, C.; Lhuillier, E.; Gaertner-Mazouni, N.; Gueguen, Y.; Le Moullac, G. Gonad transcriptome analysis of pearl oyster *Pinctada margaritifera*: Identification of potential sex differentiation and sex determining genes. *BMC Genom.* **2014**, *15*, 491. [CrossRef] [PubMed]

23. Teaniniuraitemoana, V.; Huvet, A.; Levy, P.; Gaertner-Mazouni, N.; Gueguen, Y.; Le Moullac, G. Molecular signatures discriminating the male and the female sexual pathways in the pearl oyster *Pinctada margaritifera*. *PLoS ONE* **2015**, *10*, e0122819. [CrossRef] [PubMed]

24. Liu, Y.; Hui, M.; Cui, Z.; Luo, D.; Song, C.; Li, Y.; Liu, L. Comparative transcriptome analysis reveals sex-biased gene expression in juvenile Chinese mitten crab *Eriocheir sinensis*. *PLoS ONE* **2015**, *10*, e0133068. [CrossRef] [PubMed]

25. Bustin, S.A.; Benes, V.; Garson, J.A.; Hellemans, J.; Huggett, J.; Kubista, M.; Mueller, R.; Nolan, T.; Pfaffl, M.W.; Shipley, G.L.; et al. The MIQE guidelines: Minimum information for publication of quantitative real-time PCR experiments. *Clin. Chem.* **2009**, *55*, 611–622. [CrossRef] [PubMed]

26. Picone, B.; Rhode, C.; Roodt-Wilding, R. Transcriptome profiles of wild and cultured South African abalone, *Haliotis midae*. *Mar. Genom.* **2015**, *20*, 3–6. [CrossRef] [PubMed]

27. Zhu, W.; Mai, K.; Wu, G. Thiamin requirement of juvenile abalone, *Haliotis discus hannai* Ino. *Aquaculture* **2002**, *207*, 331–343. [CrossRef]

28. Fyfe, J.C.; Madsen, M.; Højrup, P.; Christensen, E.I.; Tanner, S.M.; de la Chapelle, A.; He, Q.; Moestrup, S.K. The functional cobalamin (vitamin B12)-intrinsic factor receptor is a novel complex of cubilin and amnionless. *Blood* **2004**, *103*, 1573–1579. [CrossRef] [PubMed]

29. Moestrup, S.K.; Kozyraki, R.; Kristiansen, M.; Kaysen, J.H.; Rasmussen, H.H.; Brault, D.; Pontillon, F.; Goda, F.O.; Christensen, E.I.; Hammond, T.G.; et al. The intrinsic factor-vitamin B12 receptor and target of teratogenic antibodies is a megalin-binding peripheral membrane protein with homology to developmental proteins. *J. Biol. Chem.* **1998**, *273*, 5235–5242. [CrossRef] [PubMed]

30. Verroust, P.J.; Birn, H.; Nielsen, R.; Kozyraki, R.; Christensen, E.I. The tandem endocytic receptors megalin and cubilin are important proteins in renal pathology. *Kidney Int.* **2002**, *62*, 745–756. [CrossRef] [PubMed]

31. Zhang, W.; Mai, K.; Xu, W.; Ai, Q.; Tan, B.; Liufu, Z.; Ma, H. Effects of vitamin A and D on shell biomineralization of abalone *Haliotis discus hannai*, Ino. *J. Shellfish Res.* **2004**, *23*, 1065–1071.

32. Zhang, W.; Mai, K.; Xu, W.; Tan, B.; Ai, Q.; Liufu, Z.; Ma, H.; Wang, X. Interaction between vitamins A and D on growth and metabolic responses of abalone *Haliotis discus hannai*, Ino. *J. Shellfish Res.* **2007**, *26*, 51–58. [CrossRef]

33. Gesto, M.; Castro, L.F.; Reis-Henriques, M.A.; Santos, M.M. Retinol metabolism in the mollusk *Osilinus lineatus* indicates an ancient origin for retinyl ester storage capacity. *PLoS ONE* **2012**, *7*, e35138. [CrossRef] [PubMed]

34. Gesto, M.; Ruivo, R.; Páscoa, I.; André, A.; Castro, L.F.; Santos, M.M. Retinoid level dynamics during gonad recycling in the limpet *Patella vulgata*. *Gen. Comp. Endocrinol.* **2016**, *225*, 142–148. [CrossRef] [PubMed]

35. Watanabe, F.; Katsura, H.; Takenaka, S.; Enomoto, T.; Miyamoto, E.; Nakatsuka, T.; Nakano, Y. Characterization of vitamin B12 compounds from edible shellfish, clam, oyster, and mussel. *Int. J. Food Sci. Nutr.* **2001**, *52*, 263–268. [CrossRef] [PubMed]

36. Tanioka, Y.; Takenaka, S.; Furusho, T.; Yabuta, Y.; Nakano, Y.; Watanabe, F. Identification of vitamin B12 and pseudovitamin B12 from various edible shellfish using liquid chromatography-electrospray ionization/tandem mass spectrometry. *Fish. Sci.* **2014**, *80*, 1065–1071. [CrossRef]

37. Lambert, L.A.; Perri, H.; Halbrooks, P.J.; Mason, A.B. Evolution of the transferrin family: Conservation of residues associated with iron and anion binding. *Comp. Biochem. Physiol. B: Biochem. Mol. Biol.* **2005**, *142*, 129–141. [CrossRef] [PubMed]

38. Liu, J.; Zhang, S.; Li, L. A transferrin-like homolog in amphioxus *Branchiostoma belcheri*: Identification, expression and functional characterization. *Mol. Immunol.* **2009**, *46*, 3117–3124. [CrossRef] [PubMed]

39. Herath, H.M.; Elvitigala, D.A.; Godahewa, G.I.; Whang, I.; Lee, J. Molecular insights into a molluscan transferrin homolog identified from disk abalone (*Haliotis discus discus*) evidencing its detectable role in host antibacterial defense. *Dev. Comp. Immunol.* **2015**, *53*, 222–233. [CrossRef] [PubMed]

40. Matozzo, V.; Gagné, F.; Marin, M.G.; Ricciardi, F.; Blaise, C. Vitellogenin as a biomarker of exposure to estrogenic compounds in aquatic invertebrates: A review. *Environ. Int.* **2008**, *34*, 531–545. [CrossRef] [PubMed]

41. Osada, M.; Takamura, T.; Sato, H.; Mori, K. Vitellogenin synthesis in the ovary of scallop, *Patinopecten yessoensis*: Control by estradiol-17β and the central nervous system. *J. Exp. Zool. A Comp. Exp. Biol.* **2003**, *299*, 172–179. [CrossRef] [PubMed]

42. Boutet, I.; Moraga, D.; Marinovic, L.; Obreque, J.; Chavez-Crooker, P. Characterization of reproduction-specific genes in a marine bivalve mollusc: Influence of maturation stage and sex on mRNA expression. *Gene* **2008**, *407*, 130–138. [CrossRef] [PubMed]

43. Matsumoto, T.; Yamano, K.; Kitamura, M.; Hara, A. Ovarian follicle cells are the site of vitellogenin synthesis in the Pacific abalone *Haliotis discus hannai*. *Comp. Biochem. Physiol. A: Mol. Integr. Physiol.* **2008**, *149*, 293–298. [CrossRef] [PubMed]

44. Corporeau, C.; Vanderplancke, G.; Boulais, M.; Suquet, M.; Quéré, C.; Boudry, P.; Huvet, A.; Madec, S. Proteomic identification of quality factors for oocytes in the Pacific oyster *Crassostrea gigas*. *J. Proteom.* **2012**, *75*, 5554–5563. [CrossRef] [PubMed]

45. Zheng, H.; Zhang, Q.; Liu, H.; Liu, W.; Sun, Z.; Li, S.; Zhang, T. Cloning and expression of vitellogenin (*Vg*) gene and its correlations with total carotenoids content and total antioxidant capacity in noble scallop *Chlamys nobilis* (Bivalve: Pectinidae). *Aquaculture* **2012**, *366*, 46–53. [CrossRef]

46. Ni, J.; Zeng, Z.; Kong, D.; Hou, L.; Huang, H.; Ke, C. Vitellogenin of Fujian oyster, *Crassostrea angulata*: Synthesized in the ovary and controlled by estradiol-17β. *Gen. Comp. Endocrinol.* **2014**, *202*, 35–43. [CrossRef] [PubMed]

47. Wu, B.; Liu, Z.; Zhou, L.; Ji, G.; Yang, A. Molecular cloning, expression, purification and characterization of vitellogenin in scallop *Patinopecten yessoensis* with special emphasis on its antibacterial activity. *Dev. Comp. Immunol.* **2015**, *49*, 249–258. [CrossRef] [PubMed]

48. Kim, Y.-J.; Lee, N.; Woo, S.; Ryu, J.-C.; Yum, S. Transcriptomic change as evidence for cadmium-induced endocrine disruption in marine fish model of medaka, *Oryzias javanicus*. *Mol. Cell. Toxicol.* **2016**, *12*, 409–420. [CrossRef]

49. Georges, A.; Auguste, A.; Bessière, L.; Vanet, A.; Todeschini, A.L.; Veitia, R.A. FOXL2: A central transcription factor of the ovary. *J. Mol. Endocrinol.* **2013**, *52*, 17–33. [CrossRef] [PubMed]

50. Naimi, A.; Martinez, A.S.; Specq, M.L.; Diss, B.; Mathieu, M.; Sourdaine, P. Molecular cloning and gene expression of *Cg-Foxl2* during the development and the adult gametogenetic cycle in the oyster *Crassostrea gigas*. *Comp. Biochem. Physiol. B: Biochem. Mol. Biol.* **2009**, *154*, 134–142. [CrossRef] [PubMed]

51. Liu, X.L.; Zhang, Z.F.; Shao, M.Y.; Liu, J.G.; Muhammad, F. Sexually dimorphic expression of foxl2 during gametogenesis in scallop *Chlamys farreri*, conserved with vertebrates. *Dev. Genes Evol.* **2012**, *222*, 279–286. [CrossRef] [PubMed]

52. Zhang, N.; Xu, F.; Guo, X. Genomic analysis of the pacific oyster (*Crassostrea gigas*) reveals possible conservation of vertebrate sex determination in a mollusc. *G3* **2014**, *4*, 2207–2217. [CrossRef] [PubMed]

53. Tong, Y.; Zhang, Y.; Huang, J.; Xiao, S.; Zhang, Y.; Li, J.; Chen, J.; Yu, Z. Transcriptomics analysis of *Crassostrea hongkongensis* for the discovery of reproduction-related genes. *PLoS ONE* **2015**, *10*, e0134280. [CrossRef] [PubMed]

54. Sapiro, R.; Kostetskii, I.; Olds-Clarke, P.; Gerton, G.L.; Radice, G.L.; Strauss, J.F. Male infertility, impaired sperm motility, and hydrocephalus in mice deficient in sperm-associated antigen 6. *Mol. Cell. Biol.* **2002**, *22*, 6298–6305. [CrossRef] [PubMed]

55. Gaudet, J.; VanderElst, I.; Spence, A.M. Post-transcriptional regulation of sex determination in *Caenorhabditis elegans*: Widespread expression of the sex-determining gene fem-1 in both sexes. *Mol. Biol. Cell* **1996**, *7*, 1107–1121. [CrossRef] [PubMed]

56. Li, W.; Boswell, R.; Wood, W.B. *mag-1*, a homolog of *Drosophila mago nashi*, regulates hermaphrodite germ-line sex determination in *Caenorhabditis elegans*. *Dev. Biol.* **2000**, *218*, 172–182. [CrossRef] [PubMed]

57. Tanaka, H.; Iguchi, N.; Toyama, Y.; Kitamura, K.; Takahashi, T.; Kaseda, K.; Maekawa, M.; Nishimune, Y. Mice deficient in the axonemal protein Tektin-t exhibit male infertility and immotile-cilium syndrome due to impaired inner arm dynein function. *Mol. Cell. Biol.* **2004**, *24*, 7958–7964. [CrossRef] [PubMed]

58. Chen, W.; Liu, Y.X.; Jiang, G.F. De novo assembly and characterization of the testis transcriptome and development of EST-SSR markers in the cockroach *Periplaneta americana*. *Sci. Rep.* **2015**, *5*, 11144. [CrossRef] [PubMed]

4

Development of Gene-Based SSR Markers in Winged Bean (*Psophocarpus tetragonolobus* (L.) DC.) for Diversity Assessment

Quin Nee Wong [1], Alberto Stefano Tanzi [1,2], Wai Kuan Ho [1,2], Sunir Malla [3], Martin Blythe [3], Asha Karunaratne [2,4], Festo Massawe [1,2] and Sean Mayes [1,2,5,*]

[1] Biotechnology Research Centre, School of Biosciences, Faculty of Science, University of Nottingham Malaysia Campus, Jalan Broga, 43500 Semenyih, Selangor Darul Ehsan, Malaysia; khyx4asi@nottingham.edu.my (A.S.T.); waikuan@cffresearch.org (W.K.H.); festo.massawe@nottingham.edu.my (F.M.)

[2] Crops For the Future, Jalan Broga, 43500 Semenyih, Selangor Darul Ehsan, Malaysia; asha.karunaratne@cffresearch.org

[3] Deep Seq, Faculty of Medicine and Health Sciences, Queen's Medical Centre, University of Nottingham, Nottingham NG7 2UH, UK; sunir.malla@nottingham.ac.uk (S.M.); martinblythe@hotmail.com (M.B.)

[4] Department of Export Agriculture, Faculty of Agricultural Sciences, Sabaragamuwa University of Sri Lanka, Belihuloya 70140, Sri Lanka

[5] School of Biosciences, Faculty of Science, University of Nottingham Sutton Bonington Campus, Sutton Bonington, Leicestershire LE12 5RD, UK

* Correspondence: Sean.Mayes@nottingham.ac.uk

Academic Editor: J. Peter W. Young

Abstract: Winged bean (*Psophocarpus tetragonolobus*) is an herbaceous multipurpose legume grown in hot and humid countries as a pulse, vegetable (leaves and pods), or root tuber crop depending on local consumption preferences. In addition to its different nutrient-rich edible parts which could contribute to food and nutritional security, it is an efficient nitrogen fixer as a component of sustainable agricultural systems. Generating genetic resources and improved lines would help to accelerate the breeding improvement of this crop, as the lack of improved cultivars adapted to specific environments has been one of the limitations preventing wider use. A transcriptomic de novo assembly was constructed from four tissues: leaf, root, pod, and reproductive tissues from Malaysian accessions, comprising of 198,554 contigs with a N50 of 1462 bp. Of these, 138,958 (70.0%) could be annotated. Among 9682 genic simple sequence repeat (SSR) motifs identified (excluding monomer repeats), trinucleotide-repeats were the most abundant (4855), followed by di-nucleotide (4500) repeats. A total of 18 SSR markers targeting di- and tri-nucleotide repeats have been validated as polymorphic markers based on an initial assessment of nine genotypes originated from five countries. A cluster analysis revealed provisional clusters among this limited, yet diverse selection of germplasm. The developed assembly and validated genic SSRs in this study provide a foundation for a better understanding of the plant breeding system for the genetic improvement of winged bean.

Keywords: *Psophocarpus tetragonolobus* (L.) DC.; winged bean; SSR marker; transcriptome

---

1. Introduction

Winged bean (*Psophocarpus tetragonolobus* (L.) DC.) ($2n = 2x = 18$) is a tropical perennial vine species, classified in the family of Fabaceae and subfamily of Papilionoideae, that is cultivated mainly at a subsistence scale in hot and humid countries across India, Southeast Asia, and the Western Pacific islands, with a presence in a number of African countries as well [1–5]. It is grown for its green pods, tuberous roots, and mature seeds, all of which have received attention for their nutritional content in the past, as comprehensively described in 'The Winged Bean—A high-protein crop for the tropics' from the National Academy of Science in 1981 [1]. Initial interest was drawn to high crude protein levels in seeds, which are comparable to soybean [6–8]. Its vining nature and nitrogen fixation activity have seen it used as a cover crop and also incorporated into rotation or intercropping systems [9–11]. As such, winged bean could be a good candidate for diversifying diets to improve nutritional security, based on complex and more sustainable agricultural systems [12]. Despite its potential, winged bean has received limited research investment for developing molecular tools that can support breeding programmes, until recently. Recent reports include the development of inter-Simple Sequence Repeats (iSSRs) and Randomly Amplified Polymorphic DNA (RAPD) markers for genetic diversity and for clonal fidelity analyses and two small transcriptomic assemblies derived from a mix of leaf, bud, and shoot of Sri Lankan accessions and leaf tissue from a Nigerian accession, respectively [13–17]. Given that winged bean is believed to be largely self-pollinated, heterozygosity would be expected to be low, although a formal assessment is needed and the species does produce large flowers, suggesting a contribution from insect pollination, as recorded by Erskine [18]. Thus, molecular breeding will facilitate utilisation of genetic resources in winged bean breeding, especially among accessions, through combining beneficial traits. Molecular markers that are tightly linked to important agronomic traits are a precondition for undertaking molecular breeding in plants. The genetic basis of traits in winged bean remains largely unexplored, and to date there has not been any genetic linkage map reported for this crop, although controlled crosses have been reported [19–22].

In this study, we generated RNA-seq data from four tissues (leaf, root, reproductive tissues, and pod) of six locally grown accessions, followed by the identification of SSR-containing sequences and validation of a subset of genic SSR markers. To our knowledge, this is the first application of within-species genic SSR markers in winged bean accessions. The data will help to begin the development of comprehensive genetic information and tools to facilitate future breeding programmes, as well as allow the levels of natural inbreeding to be determined, to allow appropriate breeding schemes to be devised. The transcriptome will allow us to gain a better understanding of the phylogenetic relationships between winged bean and other leguminous and model plants.

2. Material and Methods

2.1. Plant Material, RNA Extraction, Complementary DNA (cDNA) Library Construction, and Sequencing

A total of six locally grown winged bean accessions (two derived from Malaysian Agricultural Research and Development Institute (MARDI) and four from local planters) were grown from August to December 2012 at Lady Bird Farm, Broga, Semenyih, Malaysia (Latitude: 2.9394 N; Longitude: 101.8971 E; Altitude: 45m asl). RNA was extracted separately from leaf, root, pod, and reproductive tissue (comprising of bud and flower) by pooling the respective tissues from all the six accessions. Extraction was performed from different tissue groups separately using TRIzol Reagent (Thermo Fisher Scientific, Waltham, MA, USA) followed by another round of purification using RNeasy MinElute Cleanup Kit (Qiagen, Hilden, Germany) before library preparation.

Total RNA was measured using the Qubit RNA BR assay kit (Thermo Fisher Scientific, Waltham, MA, USA). A total of 5 μg of RNA was used for enrichment of mRNA using NEBNext Poly(A) mRNA Magnetic Isolation Module (New England Biolabs, Beverly, MA, USA). RNA fragmentation was done using NEBNext Magnesium RNA Fragmentation Module (New England Biolabs, Beverly, MA, USA). Illumina stranded whole transcriptome sequencing libraries were prepared through a dUTP approach using NEBNext Ultra Directional RNA Library Prep Kit for Illumina (New Engladn Biolabs, Beverly, MA, USA). Libraries were gel purified using 2% E-Gel SizeSelect (Thermo Fisher Scientific, Waltham, MA, USA) and quality control was performed using bioanalyser HS kit (Agilent biotechnologies, Palo Alto, CA, USA). Quantification was done using qPCR (quantitative polymerase chain reaction) (Kapa Biosystems, Woburn, MA, USA). Equimolar amounts of barcoded libraries were mixed and subjected to 250 bp paired-end run using MiSeq V2 chemistry (Illumina, San Diego, CA, USA) on Illumina MiSeq sequencing platform according to manufacturer's instruction.

For simple sequence repeat (SSR) marker development, a total of nine plants from five accessions—each one representing a geographical origin—from the International Institute of Tropical Agriculture (IITA) genebank and MARDI, were used (as listed in Table 1 below). Two individuals were used per accession, which were collected after a cycle of single seed descent purification from January to June 2013 at the Lady Bird Farm, except for the Malaysian line.

Table 1. Winged bean accessions (two individuals per origin, except Malaysian line) used and their origins.

Individuals	Origin
Tpt53-9-8 Tpt53-9-10	Bangladesh
Tpt17-6-3 Tpt17-6-8	Indonesia
M3-3	Malaysia
Tpt10-7-5 Tpt10-7-7	Papua New Guinea
SLS319-10-3 SLS319-10-4	Sri Lanka

2.2. De Novo Transcriptome Assembly and Microsatellite Identification

Adaptors and low quality reads (below Q20) were trimmed using Scythe and Sickle [23], respectively, using default settings. Trimmed reads from all tissues were pooled to assemble a combined de novo assembly with Trinity version 2.2.0 pipeline using the strand specificity option. Trinity assembled transcripts were annotated with Trinotate software suite version 1.1 [24], with a blast e-value threshold of 1×10^{-5} from NCBI-BLAST [25], HMMER/PFAM [26], SignalP [27], EMBL eggNOG [28], and Gene Ontology (GO) [29] databases. The data is deposited in NCBI Sequence Read Archive (BioProject ID PRJNA374598) under the accession number of SRP099538; SRR5252646 (root), SRR5252647 (reproductive tissue), SRR5252648 (pod), and SRR5252649 (leaf).

This was followed by the identification of microsatellites using MIcroSAtellite (MISA) Perl script program, based on a minimum number of repeats of six for di-, five for tri-, tetra-, penta- and hexa-nucleotide repeat motif (monomer repeats were excluded), whilst the maximum number of bases interrupting two SSRs in a compound microsatellite was 100 [30].

2.3. Microsatellite Markers Development and Scoring

Primer pairs were designed from sequences harbouring a minimum of 18-bases long microsatellites (i.e., minimum 9 and 6 repetitions for di- and tri-nucleotide motifs, respectively) (Table S1). Primer3 [31] and PrimerQuest (Integrated DNA Technologies, Coralville, Iowa, USA) were used for oligo design, with the latter used whenever the first was not able to design with standard parameters for the given sequence. Where possible, two pairs of primers were designed for a single target region, so as to have an alternative if the first pair failed to amplify the target region. DNA was extracted from the leaf of nine genotypes (Table 1) using a modified cetyltrimethylammonium bromide (CTAB) method [32]. In addition, an RNase digestion step was added and 1 volume of isopropanol was used instead of 2/3 volume. Primer screening and optimisation was carried out in a three primer system for fluorescent labelling [33] using an equimolar mixture of genotypes. Each 20 μL of PCR reaction consisted of 1× Buffer S, 200 μM dNTPs, 0.02 μM forward primer, 0.18 μM M13-dye labelled primer, 0.2 μM reverse primer, 20 ng of DNA, and 1 U of Taq DNA Polymerase (Vivantis, Subang Jaya, Selangor, Malaysia). The PCR was programmed for 3 min of initial denaturation at 94 °C, followed by 35 cycles of 1 min at 94 °C, 1 min at 60 °C, and 1 min at 72 °C, with a 10 min final elongation. Initial polymorphism evaluation of primers was performed across all genotypes using a long capillary fragment analyser (Advance Analytical Technologies – AATI, Ankeny, IA, USA) with default parameters, except for using 4 μL of samples, and 4 kV of separation voltage for 180 min per run. Data was then analysed with Advance Analytical PROSize 2.0 v1.3.1.1 (Advance Analytical Technologies – AATI, Ankeny, IA, USA) to identify potential SSR markers based on presence of polymorphic amplicons across genotypes.

PCR products of single genotypes from potentially polymorphic SSR markers were then separated using an ABI Genetic Analyser ABI3730XL using Peak Scanner v2 for scoring (Applied Biosystems, Foster City, CA, USA). Validated markers were subsequently characterised using Power Marker v3.5 [34] for major allelic frequency, alleles per marker, heterozygosity, and polymorphic information content (PIC).

2.4. Cluster Analysis

A hierarchical cluster analysis was performed with the Dice (also known as Nei and Li) similarity coefficient and unweighted pair-group method with arithmetic mean (UPGMA) algorithm in Genstat 18th Edition [35].

3. Results and Discussion

3.1. Transcriptome Assembly and In Silico Identification of Microsatellites

A total of four libraries generated 12.77 million reads of cleaned 250 bp paired ends (Table 2). The de novo assembly derived from all tissues produced a total of 198,554 contigs with an average size of 798 bp and an N50 of 1462 bp.

Table 2. Summary statistics for the de novo assembled transcriptome.

Tissue	Leaf	Pod	Reproductive Tissue	Root
Number of raw read/base (bp)	3,150,356/ 1,544,004,822	3,973,092/ 1,868,456,680	3,544,968/ 1,719,303,632	3,873,893/ 1,859,527,511
Numbers of trimmed read/base (bp)	3,113,502/ 1,438,258,180	3,157,832/ 1,301,766,113	3,199,527/ 1,431,024,141	3,303,324/ 1,461,908,151
Number of contigs/base (bp)	198,554/158,382,439			
Average contig size (bp)	798			
N50	1462			

Out of 198,554 contigs, 138,958 (70.0%) could be annotated. Among them, 75,308 (54.2%), 69,172 (49.8%), 6499 (4.7%), 60,040 (43.2%), and 70,069 (50.4%) were found in NCBI-BLAST, HMMER/PFAM, SignalP, EMBL eggNOG, and GO databases, respectively, with no significant homology found from tmHMM on the prediction of transmembrane helices (Table S2). Figure 1 illustrates the abundance of transcripts classified based on gene ontology.

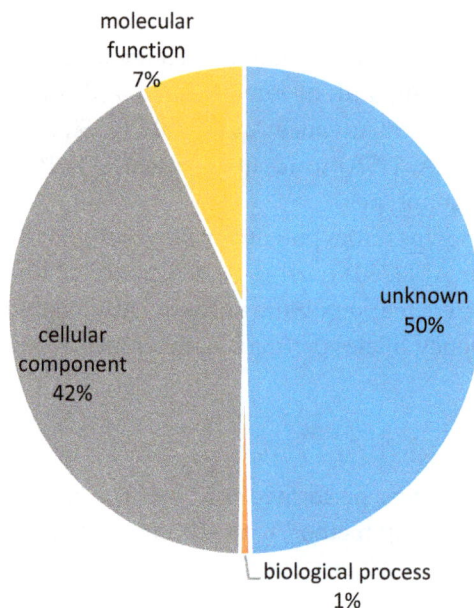

Figure 1. Distribution of first level gene ontology classification of the de novo assembly.

In this study, a total of 9682 putative SSR repeat motifs were identified from 8793 SSR containing sequences, which came from 4.4% of the total contig number in this assembly (Table S3). On average, there was one SSR locus for every 16.4 kbp of de novo assembly. After excluding mononucleotide motifs, trinucleotide repeats were the most abundant type (50.1%) (summarised in Table 3). This is consistent with Vatanparast et al.'s study [16], although hexamer motifs were not evaluated in this study. The most frequent dimer motifs were AG/GA/CT/TC type, followed by AT/TA, whereas for trimeric repeats, AAG/AGA/GAA/CTT/TCT/TTC were the most abundant (Figure 2 and Table 4). Both observations on the most common di- and tri-nucleotide repeat motif are in agreement with the winged bean transcriptome from Vatanparast et al. as well as with the soybean, medicago, and lotus Expressed Sequence Tag (EST)-SSR summarised by Jayashree et al. [16,36].

Table 3. In silico identification of microsatellites from the mixed tissue assembly. SSR, simple sequence repeat.

Total number of sequences examined	198,554
Total size of examined sequences (bp)	158,382,439
Total number of identified SSRs	9682
Number of SSR containing sequences	8793
Number of sequences containing more than one SSR	780
Number of SSRs present in compound formation	352
Number of dimer-repeat	4500
Number of trimer-repeat	4855
Number of tetramer-repeat	279
Number of pentamer-repeat	48

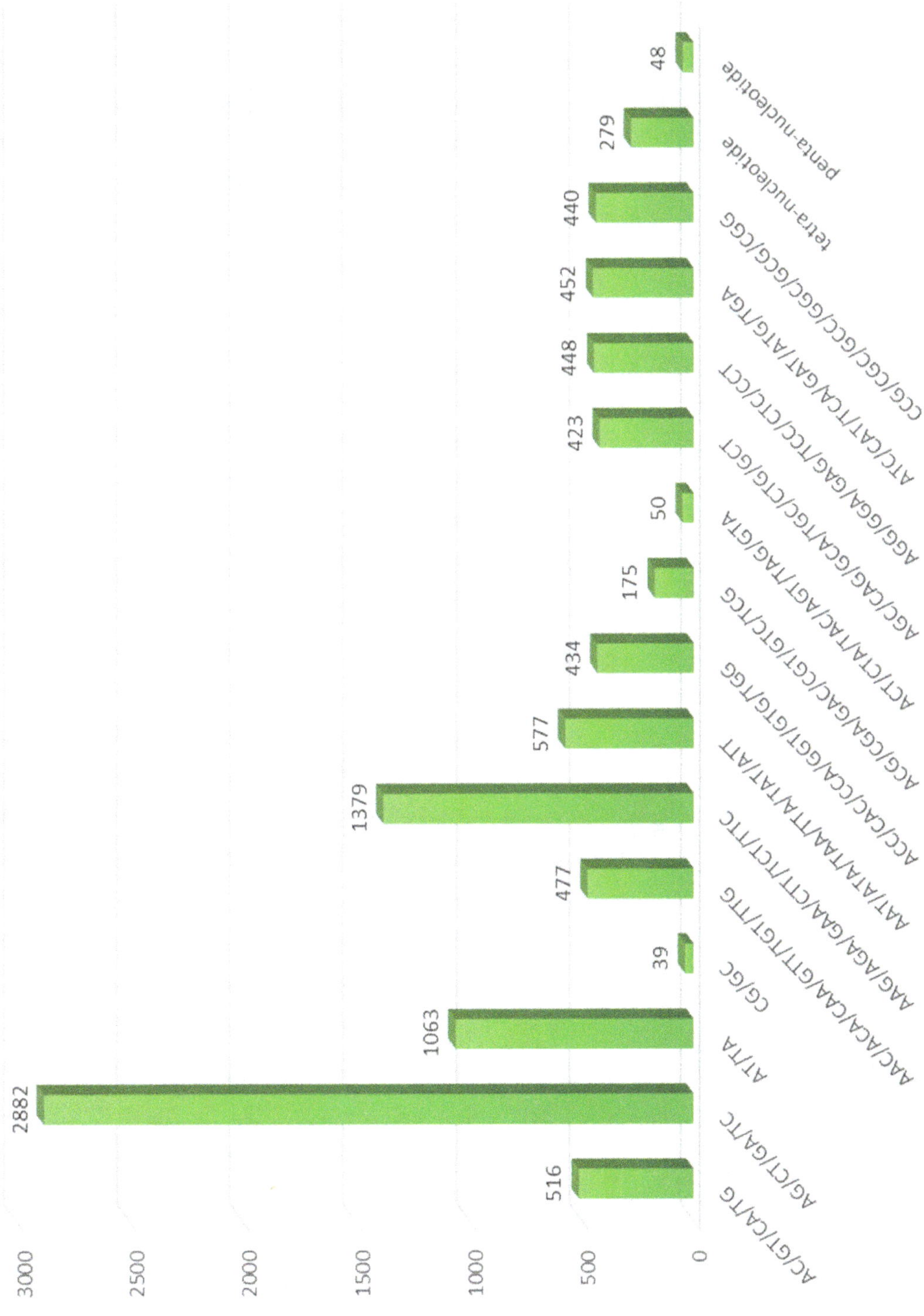

Figure 2. The number distribution of different microsatellite motif types identified. SSR, simple sequence repeat.

Table 4. Frequency distribution of di- and tri-nucleotide motif repeat in this de novo assembly.

	Number of Repeat Motif							Total	%
	5	6	7	8	9	10	>10		
Di-nucleotide									
AC/GT/CA/TG	-	256	138	63	37	12	10	516	11.5
AG/CT/GA/TC	-	995	543	330	391	434	189	2882	64.0
AT/TA	-	407	201	167	113	107	68	1063	23.6
CG/GC	-	38	1	0	0	0	0	39	0.9
Total	-	1696 (37.7%)	883 (19.6%)	560 (12.4%)	541 (12.0%)	553 (12.3%)	267 (5.9%)	4500	
Tri-nucleotide									
AAC/ACA/CAA/GTT/TGT/TTG	279	131	50	17	0	0	0	477	9.8
AAG/AGA/GAA/CTT/TCT/TTC	612	405	351	11	0	0	0	1379	28.4
AAT/ATA/TAA/TTA/TAT/ATT	305	145	112	15	0	0	0	577	11.9
ACC/CAC/CCA/GGT/GTG/TGG	307	62	55	10	0	0	0	434	8.9
ACG/CGA/GAC/CGT/GTC/TCG	90	66	12	7	0	0	0	175	3.6
ACT/CTA/TAC/AGT/TAG/GTA	36	8	3	3	0	0	0	50	1.0
AGC/CAG/GCA/TGC/CTG/GCT	271	105	38	9	0	0	0	423	8.7
AGG/GGA/GAG/TCC/CTC/CCT	247	115	75	11	0	0	0	448	9.2
ATC/CAT/TCA/GAT/ATG/TGA	311	83	24	34	0	0	0	452	9.3
CCG/CGC/GCC/GGC/GCG/CGG	247	130	55	8	0	0	0	440	9.1
Total	2705 (55.7%)	1250 (25.7%)	775 (16.0%)	125 (2.6%)	0	0	0	4855	

3.2. Development of SSR Markers and Cluster Analysis

A total of 56 (targeting 42 dimer-repeat regions) and 78 (targeting 53 trinucleotide SSR) primer pairs were designed. Subsequently, 20 dinucleotide SSR primers and 26 trinucleotide SSR primers gave good amplification products at the expected size. After polymorphism evaluation using all genotypes in this study, 18 validated SSR markers (8 for di-nucleotide and 10 for tri-nucleotide repeated motifs; Table S1) were scored and are summarised in Table 5. The low validation rate of polymorphic markers is likely to be partly due to the limited number of accessions screened, and should increase with more accessions covering a broader range of geographical origins. Residual heterozygosity could still be observed within each accession (shaded values in Table 5), even where a cycle of line purification in a controlled environment has been carried out, indicating that further cycles are needed to obtain homozygous lines, in particular for Tpt10, Tpt53, and M3. This data, along with the winged bean large flower size, also suggest that such a purification process may need to be carried out under an insect-proof enclosed environment. Using these markers, an average of 2.5 and 2.4 alleles per locus for di- and tri-nucleotide SSRs, respectively, was observed (Table 6). Individual PIC values varied from 0.16 to 0.67, which is comparable to recent legume studies in pigeonpea [37], mungbean [38], and common bean [39], although lower than in cowpea [40] and bambara groundnut [41].

The cluster analysis from the SSR scores (Figure 3) showed a few clusters with the accessions originating from Papua New Guinea closely related to the Sri Lankan accession, but sharing the least similarity with the Malaysian and Indonesian materials, comparatively. To our knowledge, the genetic relationship between germplasm from Bangladesh and Malaysia are here investigated for the first time with molecular markers, and place the Bangladesh origin closer to the Sri Lankan and Papua New Guinean germplasm. Although the number of accessions used in this study is limited, they cover a reasonable range of germplasm from different origins.

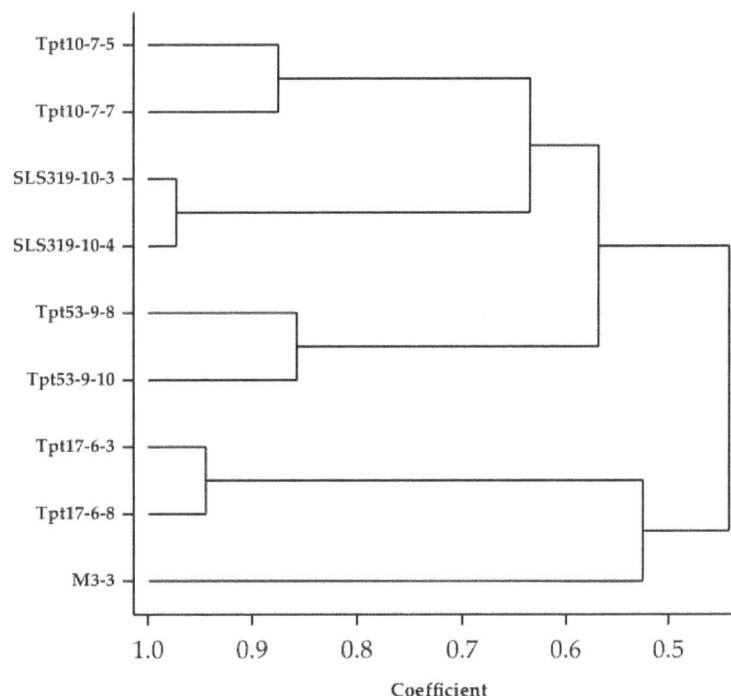

Figure 3. A dendrogram of the genetic relationship between genotypes from Papua New Guinea (Tpt10), Sri Lanka (SLS319), Bangladesh (Tpt53), Indonesia (Tpt17), and Malaysia (M3).

Table 5. Scores of 18 SSR markers from nine winged bean individuals.

Marker	Papua New Guinea		Indonesia		Bangladesh		Sri Lanka		Malaysia
	Tpt10-7-5	Tpt10-7-7	Tpt17-6-3	Tpt17-6-8	Tpt53-9-8	Tpt53-9-10	SLS319-10-3	SLS319-10-4	M3-3
P27.2	205	199/205	199	205	205	205	205	205	205
P43.2	199	199	195	195	197	199	199	199	195
Pt1.1	335	335	339	339	339	335/339	335	335	339
Pt10	226/228	228	226	226	228	228	228	228	228
Pt14	358	358	352	352	350	350	358	358	354
Pt24	219	217/219	217	217	219	219	219	219	217
Pt7.2	426/432	426	426	426	428	428	426	426	426/428
WB17	198	198	198	198	198	194/198	198	198	198
Pt53	315	309	315	315	309/315	309/315	312	312	315
Pt58	255/261	255/261	261	261	261	261	261	261	261
Pt65.1	273	273	267	267	267	267	267	267	267/273
Pt67.1	293	293	296	296	293	293	296	296	293/296
Pt68.1	226	226	229	229	226	223/226	223	223/226	226/235
Pt76.1	203	203	203	203	209	209	209	209	209
Pt78.1	306/309	306/309	306	306	309	309	306	306	309
Pt85.1	276/279	276/279	276	276	276	276	276	276	279
Pt93.1	266	266	272	272	266/272	272	266	266	276
Pt99.2	189/195	189/195	195	195	189	189	189	189	195

Table 6. A summary of data analysis of 18 SSR markers. PIC, polymorphic information content.

Marker	SSR Motif	Major Allele Frequency	No. of Alleles	Heterozygosity	PIC
P27.2	TA	0.83	2	0.11	0.24
P43.2	TA	0.56	3	0	0.49
Pt1.1	CT	0.5	2	0.11	0.38
Pt10	TC	0.72	2	0.11	0.32
Pt14	TG	0.44	4	0	0.64
Pt24	GT	0.61	2	0.11	0.36
Pt7.2	TC	0.67	3	0.22	0.4
WB17	GA	0.94	2	0.11	0.1
Average dimer SSR markers		0.66	2.5	0.1	0.37
Pt53	CGC	0.56	3	0.22	0.53
Pt58	TAG	0.89	2	0.22	0.18
Pt65.1	CAG	0.72	2	0.11	0.32
Pt67.1	AGA	0.5	2	0.11	0.38
Pt68.1	AAC	0.5	4	0.33	0.59
Pt76.1	CGC	0.56	2	0	0.37
Pt78.1	AAC	0.56	2	0.22	0.37
Pt85.1	GCG	0.78	2	0.22	0.29
Pt93.1	TGT	0.5	3	0.11	0.5
Pt99.2	TTC	0.56	2	0.22	0.37
Average trimer SSR marker		0.61	2.4	0.18	0.39

4. Conclusions

A set of validated functional winged bean genic-SSR markers is reported here for the first time, to our best knowledge. The reported residual heterozygosity across screened genotypes has suggested that further investigation needs to be carried out on the rate of natural outcrossing in winged bean, in order to understand how genetic materials should be maintained, improved, and introduced into breeding programmes. The cluster analysis provides an initial insight into the potential for these markers to be used on a larger number of winged bean accessions, to carry out a more comprehensive diversity analysis with the evaluation of germplasm from genebanks and from commonly cultivated lines. Finally, this set of 18 microsatellite markers could also be used to contribute to genetic linkage maps in winged bean, with the integration of single nucleotide polymorphisms (SNPs) markers for higher density. Such a map would be the first backbone for linkage analysis and the genetic dissection of traits with agronomic importance in this legume.

Supplementary Materials: The following are available online at www.mdpi.com/2073-4425/8/3/100/s1, Table S1: the sequences of forward and reverse SSR-primers used in this study; Table S2: Functional annotation of assembled transcripts; Table S3: Identified microsatellites.

Acknowledgments: This work was financially supported by Kouk Foundation Berhad, Crops for the Future (CFF), and the University of Nottingham Malaysia Campus (UNMC) through the CFF-UNMC Doctoral Training Partnership scheme. The authors would like to thank Victoria Wright from Deep Seq. Seeds were contributed by the International Institute of Tropical Agriculture (IITA) and Malaysian Agricultural Research and Development Institute (MARDI). This paper is dedicated to the memory of Quin Nee Wong, who passed away in a tragic car accident on 12 October 2014.

Author Contributions: Q.N.W. and A.S.T. carried out planting and sampling work and, together with W.K.H., SSR marker development and validation. A.S.T. and W.K.H. wrote the manuscript. S.M. performed library preparation for RNA-seq and M.B. performed bioinformatics analysis. A.K. contributed Sri Lanka genetic materials. F.M. helped to draft the manuscript and contributed to experimental design. S.Ms. conceived of the study, participated in its design and coordination, and revised the manuscript.

References

1. NAS. *The Winged Bean: High-Protein Crop for the Humid Tropics*, 2nd ed.; National Academy Press: Washington, DC, USA, 1981.

2. Klu, G.Y.P. Induced mutations for accelerated domestication-a case study of winged bean. *West Afr. J. Appl. Ecol.* **2000**, *1*, 47–52.

3. Khan, T.N. Papua New Guinea: A centre of genetic diversity in winged bean (*Psophocarpus tetragonologus* (L.) Dc.). *Euphytica* **1976**, *25*, 693–705. [CrossRef]

4. Harder, D.K. Chromosome Counts in Psophocarpus. *Kew Bull.* **1992**, *47*, 529–534. [CrossRef]

5. Harder, D.K.; Smartt, J. Further evidence on the origin of the cultivated winged bean, *Psophocarpus tetragonolobus* (L.) DC. *Econ. Bot.* **1992**, *46*, 187–191. [CrossRef]

6. Prakash, D.; Misra, P.N.; Misra, P.S. Amino acid profile of winged bean (*Psophocarpus tetragonolobus* (L.) DC.): A rich source of vegetable protein. *Plant Foods Hum. Nutr.* **1987**, *37*, 261–264. [CrossRef] [PubMed]

7. Kadam, S.S.; Salunkhe, D.K. Winged bean in human nutrition. *Crit. Rev. Food Sci. Nutr.* **1984**, *21*, 1–40. [CrossRef] [PubMed]

8. Okezie, B.O.; Martin, F.W. Chemical composition of dry seeds and fresh leaves of winged bean varieties grown in the U.S. and Puerto Rico. *J. Food Sci.* **1980**, *45*, 1045–1051. [CrossRef]

9. Anugroho, F.; Kitou, M.; Kinjo, K.; Kobashigawa, N. Growth and nutrient accumulation of winged bean and velvet bean as cover crops in a subtropical region. *Plant Prod. Sci.* **2010**, *13*, 360–366. [CrossRef]

10. Banerjee, A.; Bagchi, D.K.; Si, L.K. Studies on the potential of winged bean as a multipurpose legume cover crop in tropical regions. *Exp. Agric.* **2008**, *20*, 297–301. [CrossRef]

11. Hikam, S.; MacKown, C.T.; Poneleit, C.G.; Hildebrand, D.F. Growth and N accumulation in maize and winged bean as affected by N level and intercropping. *Ann. Bot.* **1991**, *68*, 17–22. [CrossRef]

12. FAO. *Coping with Climate Change—The Roles of Genetic Resources for Food and Agriculture*; FAO: Rome, Italy, 2015.

13. Mohanty, C.S.; Verma, S.; Singh, V.; Khan, S.; Gaur, P.; Gupta, P.; Nizar, M.A.; Dikshit, N.; Pattanayak, R.; Shukla, A.; et al. Characterization of winged bean (*Psophocarpus tetragonolobus* (L.) DC.) based on molecular, chemical and physiological parameters. *Am. J. Mol. Biol.* **2013**, *3*, 187–197. [CrossRef]

14. Koshy, E.P.P.; Alex, B.K.K.; John, P. Clonal fidelity studies on regenerants of *Psophocarpus tetragonolobus* (L.) DC. using RAPD markers. *Bioscan* **2013**, *8*, 763–766.

15. Chen, D.; Yi, X.; Yang, H.; Zhou, H.; Yu, Y.; Tian, Y.; Lu, X. Genetic diversity evaluation of winged bean (*Psophocarpus tetragonolobus* (L.) DC.) using inter-simple sequence repeat (ISSR). *Genet. Resour. Crop Evol.* **2015**, *62*, 823–828. [CrossRef]

16. Vatanparast, M.; Shetty, P.; Chopra, R.; Doyle, J.J.; Sathyanarayana, N.; Egan, A.N. Transcriptome sequencing and marker development in winged bean (*Psophocarpus tetragonolobus*; Leguminosae). *Sci. Rep.* **2016**, *6*, 29070. [CrossRef] [PubMed]

17. Chapman, M.A. Transcriptome sequencing and marker development for four underutilized legumes. *Appl. Plant Sci.* **2015**, *3*, 1400111. [CrossRef] [PubMed]

18. Erskine, W. Measurements of the cross-pollination of winged bean in Papua New Guinea. *SABRAO J. Breed. Genet.* **1980**, *12*, 11–14.

19. Eagleton, G.E. Evaluation of Genetic Resources in the Winged Bean (*Psophocarpus tetragonolobus* (L.) DC.) and Their Utilisation in the Development of Cultivars for Higher Latitudes. Ph.D. Dissertation, University of Western Australia, Perth, Australia, 1983.

20. De Silva, H.N.; Omran, A. Diallel analysis of yield and yield components of winged bean (*Psophocarpus tetragonolobus* (L.) D.C.). *J. Agric. Sci.* **1986**, *106*, 485–490. [CrossRef]

21. Erskine, W.; Khan, T.N. Inheritance of pigmentation and pod shape in winged bean. *Euphytica* **1977**, *26*, 829–831. [CrossRef]

22. Erskine, B.Y.W. Heritability and combining ability of vegetative and phenological characters of winged beans (*Psophocarpus tetragonolobus* (L.) DC.). *J. Agric. Sci.* **1981**, *96*, 503–508. [CrossRef]

23. Joshi, N.A.; Fass, J.N. *Sickle: A Sliding-Window, Adaptive, Quality-Based Trimming Tool for FastQ Files*, version 1.33; Software; 2011. Available online: https://github.com/najoshi/sickle (accessed on 13 March 2013).

24. Grabherr, M.G.; Haas, B.J.; Yassour, M.; Levin, J.Z.; Thompson, D.A.; Amit, I.; Adiconis, X.; Fan, L.; Raychowdhury, R.; Zeng, Q.; et al. Full-length transcriptome assembly from RNA-seq data without a reference genome. *Nat. Biotechnol.* **2011**, *29*, 644–652. [CrossRef] [PubMed]

25. Camacho, C.; Coulouris, G.; Avagyan, V.; Ma, N.; Papadopoulos, J.; Bealer, K.; Madden, T.L. BLAST+: Architecture and applications. *BMC Bioinform.* **2009**, *10*, 421. [CrossRef] [PubMed]

26. Finn, R.D.; Bateman, A.; Clements, J.; Coggill, P.; Eberhardt, R.Y.; Eddy, S.R.; Heger, A.; Hetherington, K.; Holm, L.; Mistry, J.; et al. Pfam: The protein families database. *Nucleic Acids Res.* **2014**, *42*, D222–D230. [CrossRef] [PubMed]

27. Petersen, T.N.; Brunak, S.; von Heijne, G.; Nielsen, H. SignalP 4.0: Discriminating signal peptides from transmembrane regions. *Nat. Methods* **2011**, *8*, 785–786. [CrossRef] [PubMed]

28. Powell, S.; Forslund, K.; Szklarczyk, D.; Trachana, K.; Roth, A.; Huerta-Cepas, J.; Gabaldón, T.; Rattei, T.; Creevey, C.; Kuhn, M.; et al. eggNOG v4.0: Nested orthology inference across 3686 organisms. *Nucleic Acids Res.* **2014**, *42*, D231–D239. [CrossRef] [PubMed]

29. Gene Ontology Consortium. The Gene Ontology (GO) database and informatics resource. *Nucleic Acids Res.* **2004**, *32* (Suppl. 1), D258–D261.

30. Thiel, T.; Michalek, W.; Varshney, R.K.; Graner, A. Exploiting EST databases for the development and characterization of gene-derived SSR-markers in barley (*Hordeum vulgare* L.). *Theor. Appl. Genet.* **2003**, *106*, 411–422. [CrossRef] [PubMed]

31. Koressaar, T.; Remm, M. Enhancements and modifications of primer design program Primer3. *Bioinformatics* **2007**, *23*, 1289–1291. [CrossRef] [PubMed]

32. Doyle, J. DNA Protocols for Plants. In *Molecular Techniques in Taxonomy*; Hewitt, G.M., Johnston, A.W.B., Young, J.P.W., Eds.; Springer: Berlin/Heidelberg, Germany, 1991; pp. 283–293.

33. Schuelke, M. An economic method for the fluorescent labeling of PCR fragments. *Nat. Biotechnol.* **2000**, *18*, 233–234. [CrossRef] [PubMed]

34. Liu, K.; Muse, S.V. PowerMarker: An integrated analysis environment for genetic marker analysis. *Bioinformatics* **2005**, *21*, 2128–2129. [CrossRef] [PubMed]

35. VSN International. *GenStat for Windows*, 18th ed.; VSN International: Hemel Hempstead, UK, 2015.

36. Jayashree, B.; Punna, R.; Prasad, P.; Bantte, K.; Hash, C.T.; Chandra, S.; Hoisington, D.A.; Varshney, R.K. A database of simple sequence repeats from cereal and legume expressed sequence tags mined in silico: Survey and evaluation. *In Silico Biol.* **2006**, *6*, 607–620. [PubMed]

37. Dutta, S.; Kumawat, G.; Singh, B.P.; Gupta, D.K.; Singh, S.; Dogra, V.; Gaikwad, K.; Sharma, T.R.; Raje, R.S.; Bandhopadhya, T.K.; et al. Development of genic-SSR markers by deep transcriptome sequencing in pigeonpea [*Cajanus cajan* (L.) Millspaugh]. *BMC Plant Biol.* **2011**, *11*, 17. [CrossRef] [PubMed]

38. Gupta, S.K.; Bansal, R.; Gopalakrishna, T. Development and characterization of genic SSR markers for mungbean (*Vigna radiata* (L.) Wilczek). *Euphytica* **2014**, *195*, 245–258. [CrossRef]

39. Blair, M.W.; Hurtado, N.; Chavarro, C.M.; Muñoz-Torres, M.C.; Giraldo, M.C.; Pedraza, F.; Tomkins, J.; Wing, R.; Varshney, R.; Graner, A.; et al. Gene-based SSR markers for common bean (*Phaseolus vulgaris* L.) derived from root and leaf tissue ESTs: An integration of the BMc series. *BMC Plant Biol.* **2011**, *11*, 50. [CrossRef] [PubMed]

40. Gupta, S.K.; Gopalakrishna, T. Development of unigene-derived SSR markers in cowpea (*Vigna unguiculata*) and their transferability to other Vigna species. *Genome* **2010**, *53*, 508–523. [PubMed]

41. Molosiwa, O.O.; Aliyu, S.; Stadler, F.; Mayes, K.; Massawe, F.; Kilian, A.; Mayes, S. SSR marker development, genetic diversity and population structure analysis of Bambara groundnut [*Vigna subterranean* (L.) Verdc.] landraces. *Genet. Resour. Crop Evol.* **2015**, *62*, 1225–1243. [CrossRef]

5

A Frameshift Mutation in *KIT* is Associated with White Spotting in the Arabian Camel

Heather Holl [1], Ramiro Isaza [2], Yasmin Mohamoud [3], Ayeda Ahmed [3], Faisal Almathen [4], Cherifi Youcef [5], Semir Gaouar [5], Douglas F. Antczak [6] and Samantha Brooks [1,*]

[1] Department of Animal Sciences, UF Genomics Institute, University of Florida, Gainesville, FL 32610, USA; heather.holl@ufl.edu
[2] Department of Small Animal Clinical Sciences, College of Veterinary Medicine, University of Florida, Gainesville, FL 32608, USA; isazar@ufl.edu
[3] Department of Genetic Medicine, Weill Cornell Medical College in Qatar, Doha, Qatar; yam2012@qatar-med.cornell.edu (Y.M.); ays2003@qatar-med.cornell.edu (A.A.)
[4] Veterinary Public Health and Animal Husbandry, College of Veterinary Medicine and Animal Resources, King Faisal University, Al-Ahsa 31982, Saudi Arabia; faisalvet@hotmail.com
[5] Department of Biology, University of Abou Bekr Belkaïd, Tlemcen 13000, Algeria; cherifi.youcef@ymail.com (C.Y.); suheilgaouar@gmail.com (S.G.)
[6] Baker Institute for Animal Health, College of Veterinary Medicine, Cornell University, Ithaca, NY 14853, USA; dfa1@cornell.edu
* Correspondence: samantha.brooks@ufl.edu

Academic Editor: Paolo Cinelli

Abstract: While the typical Arabian camel is characterized by a single colored coat, there are rare populations with white spotting patterns. White spotting coat patterns are found in virtually all domesticated species, but are rare in wild species. Theories suggest that white spotting is linked to the domestication process, and is occasionally associated with health disorders. Though mutations have been found in a diverse array of species, fewer than 30 genes have been associated with spotting patterns, thus providing a key set of candidate genes for the Arabian camel. We obtained 26 spotted camels and 24 solid controls for candidate gene analysis. One spotted and eight solid camels were whole genome sequenced as part of a separate project. The spotted camel was heterozygous for a frameshift deletion in *KIT* (c.1842delG, named *KIT*^{W1} for *White spotting 1*), whereas all other camels were wild-type (*KIT*$^+$/*KIT*$^+$). No additional mutations unique to the spotted camel were detected in the *EDNRB*, *EDN3*, *SOX10*, *KITLG*, *PDGFRA*, *MITF*, and *PAX3* candidate white spotting genes. Sanger sequencing of the study population identified an additional five *KIT*^{W1}/*KIT*$^+$ spotted camels. The frameshift results in a premature stop codon five amino acids downstream, thus terminating KIT at the tyrosine kinase domain. An additional 13 spotted camels tested *KIT*$^+$/*KIT*$^+$, but due to phenotypic differences when compared to the *KIT*^{W1}/*KIT*$^+$ camels, they likely represent an independent mutation. Our study suggests that there are at least two causes of white spotting in the Arabian camel, the newly described *KIT*^{W1} allele and an uncharacterized mutation.

Keywords: Arabian camel; dromedary camel; white spotting; *KIT*

1. Introduction

Coat color variation was valued throughout animal domestication [1]. Human selection favors unique morphological characteristics, resulting in the diverse array of coat pigmentation present in many species. Many genes are associated with white spotting patterns in domestic animals [2]. However, color variation is often associated with other behavioral or health traits, due to the phenomenon of genetic pleiotropy [3]. White spotting in particular is linked with conditions such as anemia, deafness, gastrointestinal abnormalities, skin cancers, and sterility [4]. However, while white spotting phenotypes are common among domesticated species, not all have known pleiotropy.

The Arabian camel is an important domestic species selected for draught, meat, milk, racing, and riding. Though usually with a uniform coat of varying shades of brown or black, there are rare spotted or fully white populations [5–8]. To our knowledge, there have been no published studies on the genetics of camel coat color patterns. In this work, we report on a candidate gene approach using whole genome sequencing to characterize the genetics of white spotting.

2. Materials and Methods

2.1. Sample Collection

Sample collection for this study was approved by the University of Florida Institutional Animal Care and Use Committee (IACUC) under protocol #201408506. All animals were voluntarily enrolled by private owners. Blood or hair samples were collected from 50 Arabian camels. Photographs were obtained for all but three animals, which were used to assess phenotype (Table 1). For the three animals without photos, phenotype was recorded as a written description at the time of sample collection. Individuals were identified as spotted if they exhibited a region of white fur with pink skin underneath on a background of normal pigmentation. Pedigree information was only available for five dam-offspring pairs (Table 1). Additionally, camels US12 and US15 as well as US14 and US17 were identified as possible half siblings, though the owner was not certain. DNA was extracted using either the Gentra Puregene buffy coat protocol (Qiagen, Valencia, CA, USA) or a modified Puregene protocol for hair [9].

Table 1. Origin and coat color information for the camels in this study.

ID	Origin	Coat	Eyes	Markings	Genotype
QA1	Qatar	Solid Brown	Dark	None	KIT^+/KIT^+
QA2	Qatar	Solid Brown	Dark	None	KIT^+/KIT^+
QA3	Qatar	Solid Brown	Dark	None	KIT^+/KIT^+
QA4	Qatar	Solid Brown	Dark	None	KIT^+/KIT^+
QA5	Qatar	Solid Brown	Dark	None	KIT^+/KIT^+
US1	US Farm 1	Solid Brown	Dark	None	KIT^+/KIT^+
US3	US Farm 1	Solid Brown	Dark	None	KIT^+/KIT^+
US4	US Farm 2	Spotted Brown	Blue	Mottled white throughout body/neck/head, colored topline	KIT^{W1}/KIT^+
US5	US Farm 2	Spotted Brown	Blue	Mottled white on head and lower half of body, dark brown patch on rump	KIT^{W1}/KIT^+
US6	US Farm 3	Solid Brown	Dark	None	KIT^+/KIT^+
US12 [1]	US Farm 4	Spotted Brown	Blue	Clear white legs/belly, mostly white face	KIT^+/KIT^+
US13	US Farm 4	Spotted Brown	Blue	Clear white legs/belly, half white face	KIT^+/KIT^+
US14 [2]	US Farm 4	Spotted Brown	Dark	Mottled white to knees/hocks, wide stripe on face, white tail tip	KIT^{W1}/KIT^+
US15 [1]	US Farm 4	Spotted Brown	Blue	Clear white from legs to flanks, mostly white face	KIT^+/KIT^+
US16	US Farm 4	Spotted Brown	Blue	Clear white to stifle/knee, clear white belly spot, white nose	KIT^+/KIT^+
US17 [2]	US Farm 4	Spotted Brown	Dark	Mottled white to knees/hocks, white spot on inner hindquarters, white tail tip	KIT^{W1}/KIT^+
US25	US Farm 5	Spotted Brown	Dark	Wide stripe on face, white spot on withers, white toe, white tail tip	KIT^{W1}/KIT^+
US26 [3]	US Farm 5	Solid Brown	Blue	Light roaning	KIT^+/KIT^+
US27 [3]	US Farm 5	Solid White	Part-Blue	None	KIT^+/KIT^+
US28 [4]	US Farm 5	Spotted Brown	Blue	Mottled white throughout body/neck/head, colored topline	KIT^{W1}/KIT^+
US29 [4]	US Farm 5	Solid Brown	Dark	None	KIT^+/KIT^+

Table 1. *Cont.*

ID	Origin	Coat	Eyes	Markings	Genotype
US30	US Farm 5	Spotted Brown	Dark	White lips, light roaning	KIT^+/KIT^+
US31	US Farm 6	Spotted Brown	Dark	Wide stripe on face	KIT^{W1}/KIT^+
US32	US Farm 6	Solid Brown	Dark	None	KIT^+/KIT^+
US33	US Farm 6	Spotted Brown	Dark	White to elbows, white tail tip	KIT^{W1}/KIT^+
US34	US Farm 6	Spotted Brown	Blue	Clear white to flank, half white face	KIT^+/KIT^+
US35	US Farm 6	Solid Brown	Dark	None	KIT^+/KIT^+
US36	US Farm 6	Spotted Brown	Blue	Clear white to flank, half white face	KIT^+/KIT^+
US37	US Farm 6	Solid Brown	Dark	None	KIT^+/KIT^+
US38	US Farm 6	Solid Brown	Dark	None	KIT^+/KIT^+
US39	US Farm 6	Solid Brown	Dark	None	KIT^+/KIT^+
US40	US Farm 6	Solid Brown	Dark	None	KIT^+/KIT^+
US41	US Farm 6	Solid White	Dark	None	KIT^+/KIT^+
US42 [5]	US Farm 6	Solid Brown	Dark	None	KIT^+/KIT^+
US43 [5]	US Farm 6	Spotted Brown	Blue	Clear white to knees, white nose	KIT^+/KIT^+
US44 [6]	US Farm 6	Solid Brown	Dark	None	KIT^+/KIT^+
US45 [6]	US Farm 6	Solid Brown	Dark	None	KIT^+/KIT^+
US46 [7]	US Farm 6	Solid Brown	Dark	None	KIT^+/KIT^+
US47 [7]	US Farm 6	Solid Brown	Dark	None	KIT^+/KIT^+
US48	US Farm 6	Spotted Brown	Dark	White stripe on face, white tail	KIT^{W1}/KIT^+
US49	US Farm 6	Spotted Brown	Dark	White tail tip	KIT^+/KIT^+
US50	US Farm 6	Solid Brown	Dark	None	KIT^+/KIT^+
SA1	Algeria	Spotted Brown	Blue	Clear white belly to flank, white face	KIT^+/KIT^+
SA2	Algeria	Spotted Brown	Blue	White to knees, white face	KIT^+/KIT^+
SA3	Algeria	Spotted Brown	Blue	Clear white to flanks, mostly white face	KIT^+/KIT^+
SA4	Algeria	Spotted Brown	Blue	Clear white to flanks, mostly white face	KIT^+/KIT^+
SA5	Algeria	Spotted Brown	Blue	White to knees, half white face	KIT^+/KIT^+
SA6	Algeria	Spotted Black	Blue	No photo available	KIT^+/KIT^+
SA7	Algeria	Spotted Black	Blue	No photo available	KIT^+/KIT^+
SA8	Algeria	Spotted Black	Blue	No photo available	KIT^+/KIT^+

[1] Possible half sibling pair; [2] Possible half sibling pair; [3] Dam (US26) offspring (US27) pair; [4] Dam (US28) offspring (US29) pair; [5] Dam (US42) offspring (US43) pair; [6] Dam (US44) offspring (US45) pair; [7] Dam (US46) offspring (US47) pair.

2.2. Next-Generation Sequencing

Whole genome sequencing was performed on eight camels (QA1-5, US1, US3-4, US6). Each sample was sequenced on a single lane of an Illumina HiSeq 2500 with 2×100 bp reads. We obtained an average of 184 million read pairs per lane. Raw sequencing reads were submitted to the ENA under accession number PRJEB15365. A custom reference sequence file was prepared, comprised of Arabian camel scaffolds matching known white spotting genes *EDNRB*, *EDN3*, *SOX10*, *KIT*, *KITLG*, *PDGFRA*, *MITF*, and *PAX3* [10,11] (Supplementary Table S1). Sequencing reads were aligned with BWA 0.7.12-r1039 using default parameters, and were converted to BAM format using SAMtools 0.1.19-44428cd [12,13].

Predicted dromedary camel mRNA sequences for the candidate genes were obtained from Genbank and aligned to the custom reference using BLAT 20140318 [14] (Supplementary Table S1). Alignments were visualized in IGV 2.3.55 [15]. The coding regions and immediately surrounding introns were manually inspected for variant annotation (Supplementary Table S2). At least three reads for homozygous variants and six for heterozygous variants were required to record a genotype.

2.3. Variant Validation and Genotyping

Variants observed only in the spotted camel genome were selected for further evaluation. PCR primers were designed using the reference scaffolds and Primer3 [16] (Supplementary Table S3). Additional samples were genotyped by Sanger sequencing of PCR amplicons. Chromatograms in PDF format were visually inspected to assess genotype. Sanger sequences for camel *KIT* exons 12 and 13 were submitted to Genbank and are available under accession numbers KX784929 (KIT^{W1}) and KX784928 (KIT^+).

3. Results and Discussion

The degree of white spotting of each animal varied greatly, ranging from white only on the lips, to white on approximately 80% of the body (Table 1, Figure 1). Camels with extensive markings had blue eyes. Samples US27 and US41 were recorded as solid white as there were no obvious colored patches of fur. "Mottled white" indicates that the border between the colored and white fur was jagged and uneven, and/or with regions where colored and white fur were intermingled (Figure 1C). "Clear white" indicates a well-defined border between colored and white fur, though with possible flecks of color within predominantly white patches (Figure 1D). Minimally spotted camels often had white restricted to the "points"—the face, legs, and tail. The US spotted camels were visibly smaller in overall body size than the solid camels, though no quantitative measures were assessed. Two of the owners mentioned that spotted camels are known for being roughly 30 cm shorter at maturity.

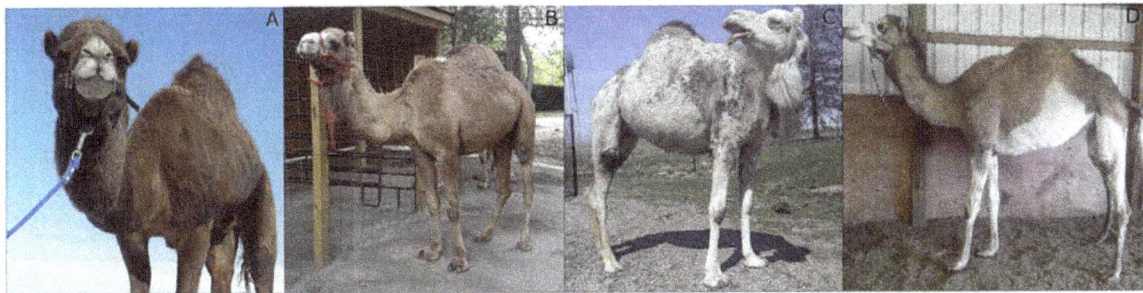

Figure 1. Range of white spotting phenotypes in this study. (**A**) A minimally spotted camel (US30), displaying white patches on both upper lips. (**B**) A minimally spotted camel (US25), showing a facial marking and white spot in front of the hump. (**C**) An extensively spotted "mottled white" camel (US28), with pigmented regions concentrated along the topline. (**D**) A moderately spotted "clear white" camel (US34), showing clear definition between pigmented and white regions.

Most US owners did not have information on the colors of the parents of the spotted camels. One owner described that many of the breeders he knew of bred white spotted bulls to non-spotted cows in order to produce color. He knew of one spotted bull (~50% of the body with white spotting) that produced all minimally spotted offspring (dark eyes with white on the face, legs, and/or tail) out of solid cows. However, the mottled white phenotype is not likely to be incompletely dominant, as one of the extensively marked camels in the study (US28) produced a completely non-spotted calf (US29). We thus assumed the mottled white trait would be dominant based on similarities to white spotting patterns in the horse.

The spotted Algerian camels, known locally as Azerghaf or Zarwala, are found in the southwest regions of the country. Some tribes prefer the spotted camels as they are born deaf, and thus are easier to raise. Interviews with local farmers indicate the color is likely autosomal recessive in inheritance. A secondary type of spotted Algerian camel (characterized by a white spot on the forehead, found in the extreme south) was not available for genotyping.

Variant screening identified one mutation in *KIT* as a likely candidate (Supplementary Table S2). There were no variants detected in *PAX3*, whereas candidate genes *SOX10, MITF,* and *KITLG* only had variants in the untranslated regions (only one of which was uniquely present in spotted camel US4). Silent variants in *KIT, EDN3, EDNRB,* and *PDGFRA* were found in both US4 and non-spotted camels (QA1-5, US1, US3, US6). There was one homozygous missense variant in *PDGFRA* present in US4, but it was not considered further as QA4 was heterozygous for this variant as well. The only coding variant unique to US4 was a heterozygous single base deletion within exon 12 of 21 in *KIT* (c.1842delG). This mutation results in a frameshift, leading to a premature stop codon five amino acids downstream (p.M614IfsX5). This mutation ultimately truncates the protein at the intracellular tyrosine kinase domain. Sanger sequencing identified eight additional heterozygotes and no homozygous mutants

(Figure 2, Table 2). Five heterozygotes were in the mottled white category, whereas the remaining four animals had too little white to define as clear or mottled (Table 1). Based on the appearance of the minimally spotted camels, we have termed the allele KIT^{W1} for *White spotting 1*.

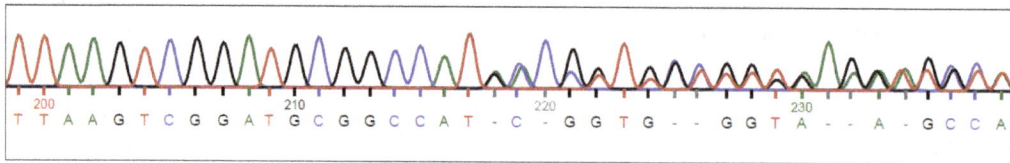

Figure 2. Sanger sequencing confirms the presence of a heterozygous deletion. The *KIT*:c.1842delG variant results in a frameshift five bases downstream, truncating the protein at the intracellular tyrosine kinase domain.

Table 2. Genotyping results for the *KIT*:c.1842delG mutation (KIT^{W1}).

Phenotype	WT/WT	WT/del	del/del	Total
Solid Brown	22	0	0	22
Solid White	2	0	0	2
Clear White	10	0	0	10
Mottled White	0	5	0	5
Other Spotted	7	4	0	11
Total	41	9	0	50

KIT, v-kit Hardy-Zuckerman 4 feline sarcoma viral oncogene homolog, is one of the most commonly implicated genes for white spotting phenotypes in animals [17–24]. There are four reported frameshift mutations in the domestic horse, all of which were associated with extensive white spotting [17,25,26]. *KIT* encodes a tyrosine kinase receptor involved in the development of erthyrocytes, melanocytes, germ cells, mast cells, and interstitial cells of Cajal [27]. As a result, loss-of-function mutations have been associated with a variety of negative pleiotropic effects, and in some cases are hypothesized to be homozygous lethal [20,28,29]. However, not all *KIT* variants have reported pleiotropy, and specific variants in cats and horses were shown to have normal hematological parameters [18,30]. All nine camels with the *KIT*:c.1842delG variant were mature without obvious health issues, although detailed histories were not available. Additionally, one of the animals had two previous successful pregnancies, and thus was fertile. The KIT^{W1} allele does not appear to be detrimental in heterozygotes, although homozygous embryos are likely not viable, similar to dogs with a frameshift mutation in *KIT* [20].

The appearance of the nine KIT^{W1}/KIT^{W+} camels was highly variable. Minimally spotted animals (white on face, legs, or tail only) had dark eyes, whereas the more extensively marked camels had blue eyes (Table 1). Blue eyes are commonly associated with white markings in a variety of species. Several of the equine *KIT* variants also show considerable variation in the degree of spotting, including horses with white found only on the head and legs [25,31,32]. Intense selection for the degree of white spotting in the mouse has demonstrated that a large number of additive and modifying loci cumulatively contribute to overall color, in addition to non-genetic factors [33]. Heterozygous *KIT* mice display the full range of color, from completely solid to completely white. It is thus likely that the more extensively marked mottled white camels possessed additional genetic variants that we did not detect.

As the remaining spotted camels were wild-type for *KIT*:c.1842delG, they likely have a different mutation responsible for white spotting. The overall appearance of the moderately marked animals was quite distinct between the "mottled white" and "clear white" patterns, with the "clear white" animals appearing similar to the "splashed white" coat color of the horse [34]. Interestingly, both the Algerian spotted camels and some splashed white horses have a congenital deafness. Splashed white is associated with mutations in *MITF* and *PAX3*, both of which are genes implicated in the human Waardenburg Syndrome (characterized by white patches of skin, blue eyes, and some degree of deafness). Thus, *MITF* and *PAX3* are attractive candidates for the "clear white" Arabian camel phenotype.

4. Conclusions

In conclusion, the c.1842delG mutation in *KIT* is likely responsible for one form of white spotting in the dromedary camel. We propose to name this allele KIT^{W1} for *White spotting 1*. However, due to the wide variation in KIT^{W1}/KIT^+ animals, and the presence of KIT^+/KIT^+ spotted camels, there are likely other mutations that we did not characterize. Further research should elucidate the genes responsible for other white phenotypes.

Supplementary Materials:
The following are available online at www.mdpi.com/2073-4425/8/3/102/s1, Table S1: NCBI accession numbers for the scaffolds and proteins used as reference sequences, Table S2: Detected variants in candidate white spotting genes, Table S3: PCR primers used for Sanger sequencing validation.

Acknowledgments: We would like to thank the owners for providing samples, as well as Dr. Fahad Alshanbari for his assistance in blood collection. This study was made possible by National Priorities Research Program (NPRP) grant 6-1303-4-023 from the Qatar National Research Fund (a member of Qatar Foundation). The statements made herein are solely the responsibility of the authors.

Author Contributions: Heather Holl and Samantha Brooks designed the study; Heather Holl, Samantha Brooks, Ramiro Isaza, Faisal Almathen, Cherifi Youcef, and Semir Gaouar collected samples; Yasmin Mohamoud and Ayeda Ahmed performed the next-generation sequencing; Heather Holl performed Sanger sequencing and data analysis; Heather Holl, Samantha Brooks, Semir Gaouar and Douglas F. Antczak drafted the manuscript. All authors read and approved the final manuscript.

References

1. Ludwig, A.; Pruvost, M.; Reissmann, M.; Benecke, N.; Brockmann, G.A.; Castaños, P.; Cieslak, M.; Lippold, S.; Llorente, L.; Malaspinas, A.S.; et al. Coat color variation at the beginning of horse domestication. *Science* **2009**. [CrossRef] [PubMed]
2. Reissmann, M.; Ludwig, A. Pleiotropic effects of coat colour-associated mutations in humans, mice, and other mammals. *Semin. Cell Dev. Biol.* **2013**. [CrossRef] [PubMed]
3. Bellone, R.R. Pleiotropic effects of pigmentation genes in horses. *Anim. Genet.* **2010**. [CrossRef] [PubMed]
4. Jackson, I. Molecular and developmental genetics of mouse coat color. *Annu. Rev. Genet.* **1994**, *28*, 189–217. [CrossRef] [PubMed]
5. Bornstein, S. The ship of the desert. The dromedary camel (*Camelus dromedarius*), a domesticated animal species well adapted to extreme conditions of aridness and heat. *Rangifer* **1990**, *10*, 231–236. [CrossRef]
6. Ishag, I.A.; Eisa, M.O.; Ahmed, M.K.A. Phenotypic characteristics of Sudanese camels (*Camelus dromedarius*). *Livestock Res. Rural Dev.* **2011**, *23*. Article #99.
7. Abdussamad, A.M.; Charruau, P.; Kalla, D.J.U.; Burger, P.A. Validating local knowledge on camels: Colour phenotypes and genetic variation of dromedaries in the Nigeria-Niger corridor. *Livest. Sci.* **2015**, *181*, 131–136. [CrossRef]
8. Cherifi, Y.A.; Gaouar, S.B.S.; Moussi, N.; Tabet, A.N.; Saïdi-Mehtar, N. Study of Camelina Biodiversity in Southwestern of Algeria. *J. Life Sci.* **2013**, *7*, 416–427.
9. Cook, D.; Gallagher, P.C.; Bailey, E. Genetics of swayback in American Saddlebred horses. *Animal Genet* **2010**, *41* (Suppl 2), 64–71. [CrossRef] [PubMed]
10. Wu, H.; Guang, X.; Al-Fageeh, M.B.; Cao, J.; Pan, S.; Zhou, H.; Zhang, L.; Abutarboush, M.H.; Xing, Y.; Xie, Z.; et al. Camelid genomes reveal evolution and adaptation to desert environments. *Nat. Commun.* **2014**. [CrossRef] [PubMed]

11. Stephenson, D.A; Mercola, M.; Anderson, E.; Wang, C.Y.; Stiles, C.D.; Bowen-Pope, D.F.; Chapman, V.M. Platelet-derived growth factor receptor alpha-subunit gene (Pdgfra) is deleted in the mouse patch (Ph) mutation. *Proc. Natl. Acad. Sci. USA.* **1991**, *88*, 6–10. [CrossRef]

12. Li, H.; Durbin, R. Fast and accurate short read alignment with Burrows-Wheeler transform. *Bioinformatics* **2009**, *25*, 1754–1760. [CrossRef] [PubMed]

13. Li, H.; Handsaker, B.; Wysoker, A.; Fennell, T.; Ruan, J.; Homer, N.; Marth, G.; Abecasis, G.; Durbin, R. 1000 Genome Project Data Processing Subgroup. The Sequence Alignment/Map format and SAMtools. *Bioinformatics* **2009**, *25*, 2078–2079. [CrossRef] [PubMed]

14. Kent, W.J. BLAT—The BLAST-like alignment tool. *Genome Res.* **2002**, *12*, 656–664. [CrossRef] [PubMed]

15. Robinson, J.T.; Thorvaldsdóttir, H.; Winckler, W.; Guttman, M.; Lander, E.S.; Getz, G.; Mesirov, J.P. Integrative Genomics Viewer. *Nat. Biotechnol.* **2011**, *29*, 24–26. [CrossRef] [PubMed]

16. Rozen, S.; Skaletsky, H. Primer3 on the WWW for general users and for biologist programmers. *Methods Mol. Biol.* **2000**, *132*, 365–386. [PubMed]

17. Haase, B.; Jagannathan, V.; Rieder, S.; Leeb, T. A novel KIT variant in an Icelandic horse with white-spotted coat color. *Anim. Genet.* **2015**. [CrossRef] [PubMed]

18. David, V.A.; Menotti-Raymond, M.; Wallace, A.C.; Roelke, M.; Kehler, J.; Leighty, R.; Eizirik, E.; Hannah, S.S.; Nelson, G.; Schäffer, A.A.; et al. Endogenous retrovirus insertion in the KIT oncogene determines white and white spotting in domestic cats. *G3 (Bethesda)* **2014**, *4*, 1881–1891. [CrossRef] [PubMed]

19. Fontanesi, L.; Vargiolu, M.; Scotti, E.; Latorre, R.; Faussone Pellegrini, M.S.; Mazzoni, M.; Asti, M.; Chiocchetti, R.; Romeo, G.; Clavenzani, P.; et al. The KIT gene is associated with the English spotting coat color locus and congenital megacolon in Checkered Giant rabbits (*Oryctolagus cuniculus*). *PLoS ONE* **2014**, *9*, e93750. [CrossRef] [PubMed]

20. Wong, A.K.; Ruhe, A.L.; Robertson, K.R.; Loew, E.R.; Williams, D.C.; Neff, M.W. A de novo mutation in KIT causes white spotting in a subpopulation of German Shepherd dogs. *Anim. Genet.* **2013**, *44*, 305–310. [CrossRef] [PubMed]

21. Haase, B.; Rieder, S.; Leeb, T. Two variants in the KIT gene as candidate causative mutations for a dominant white and a white spotting phenotype in the donkey. *Anim. Genet.* **2015**, *46*, 321–324. [CrossRef] [PubMed]

22. Johnson, J.L.; Kozysa, A.; Kharlamova, A.V.; Gulevich, R.G.; Perelman, P.L.; Fong, H.W.; Vladimirova, A.V.; Oskina, I.N.; Trut, L.N.; Kukekova, A.V. Platinum coat color in red fox (*Vulpes vulpes*) is caused by a mutation in an autosomal copy of KIT. *Anim. Genet.* **2015**, *46*, 190–199. [CrossRef] [PubMed]

23. Yan, S.Q.; Hou, J.N.; Bai, C.Y.; Jiang, Y.; Zhang, X.J.; Ren, H.L.; Sun, B.X.; Zhao, Z.H.; Sun, J.H. A base substitution in the donor site of intron 12 of KIT gene is responsible for the dominant white coat colour of blue fox (*Alopex lagopus*). *Anim. Genet.* **2014**, *45*, 293–296. [CrossRef] [PubMed]

24. Durkin, K.; Coppieters, W.; Drögemüller, C.; Ahariz, N.; Cambisano, N.; Druet, T.; Fasquelle, C.; Haile, A.; Horin, P.; Huang, L.; et al. Serial translocation by means of circular intermediates underlies colour sidedness in cattle. *Nature* **2012**, *482*, 81–84. [CrossRef] [PubMed]

25. Haase, B.; Brooks, S.A.; Tozaki, T.; Burger, D.; Poncet, P.A.; Rieder, S.; Hasegawa, T.; Penedo, C.; Leeb, T. Seven novel KIT mutations in horses with white coat colour phenotypes. *Anim. Genet.* **2009**, *40*, 623–629. [CrossRef] [PubMed]

26. Holl, H.; Brooks, S.A.; Bailey, E. De novo mutation of KIT discovered as a result of a non-hereditary white coat colour pattern. *Anim. Genet.* **2010**, *41* (Suppl 2), 196–198. [CrossRef]

27. Kitamura, Y.; Hirotab, S. Kit as a human oncogenic tyrosine kinase. *Cell Mol. Life Sci.* **2004**, *61*, 2924–2931. [CrossRef] [PubMed]

28. Bult, C.; Eppig, J.; Kadin, J.; Richardson, J.E.; Blake, J.A. Mouse Genome Database Group. The Mouse Genome Database (MGD): Mouse biology and model systems. *Nucleic Acids Res.* **2008**, *36*, D724–D728. [CrossRef] [PubMed]

29. Jackling, F.C.; Johnson, W.E.; Appleton, B.R. The genetic inheritance of the blue-eyed white phenotype in alpacas (*Vicugna pacos*). *J. Hered.* **2014**, *105*, 847–857. [CrossRef] [PubMed]

30. Haase, B.; Obexer-Ruff, G.; Dolf, G.; Rieder, S.; Burger, D.; Poncet, P.A.; Gerber, V.; Howard, J.; Leeb, T. Haematological parameters are normal in dominant white Franches-Montagnes horses carrying a KIT mutation. *Vet. J.* **2010**, *184*, 315–317. [CrossRef] [PubMed]

31. Hauswirth, R.; Jude, R.; Haase, B.; Bellone, R.R.; Archer, S.; Holl, H.; Brooks, S.A.; Tozaki, T.; Penedo, M.C.; Rieder, S.; et al. Novel variants in the KIT and PAX3 genes in horses with white-spotted coat colour phenotypes. *Anim. Genet.* **2013**, *44*, 763–765. [CrossRef] [PubMed]

32. Dürig, N.; Jude, R.; Holl, H.; Brooks, S.A.; Lafayette, C; Jagannathan, V.; Leeb, T. Whole genome sequencing reveals a novel deletion variant in the KIT gene in horses with white spotted coat colour phenotypes. *Anim. Genet.* **2017**, in press.

33. Dunn, L.C. Studies on Spotting Patterns II. Genetic Analysis of Variegated Spotting in the House Mouse. *Genetics* **1937**, *22*, 43–64. [PubMed]

34. Hauswirth, R.; Haase, B.; Blatter, M.; Brooks, S.A.; Burger, D.; Drögemüller, C.; Gerber, V.; Henke, D.; Janda, J.; Jude, R.; et al. Mutations in MITF and PAX3 cause "splashed white" and other white spotting phenotypes in horses. *PLoS Genet.* **2012**, *8*, e1002653. [CrossRef] [PubMed]

6

The Complete Chloroplast Genome Sequences of Six *Rehmannia* Species

Shuyun Zeng, Tao Zhou, Kai Han, Yanci Yang, Jianhua Zhao and Zhan-Lin Liu *

Key Laboratory of Resource Biology and Biotechnology in Western China (Ministry of Education), College of Life Science, Northwest University, Xi'an, 710069, China; zengsy.nwu@outlook.com (S.Z.); woody196@163.com (T.Z.); hank.nwu@outlook.com (K.H.); yycjyl@163.com (Y.Y.); yunjin1991@163.com (J.Z.)
* Correspondence: liuzl@nwu.edu.cn

Academic Editor: Charles Bell

Abstract: *Rehmannia* is a non-parasitic genus in Orobanchaceae including six species mainly distributed in central and north China. Its phylogenetic position and infrageneric relationships remain uncertain due to potential hybridization and polyploidization. In this study, we sequenced and compared the complete chloroplast genomes of six *Rehmannia* species using Illumina sequencing technology to elucidate the interspecific variations. *Rehmannia* plastomes exhibited typical quadripartite and circular structures with good synteny of gene order. The complete genomes ranged from 153,622 bp to 154,055 bp in length, including 133 genes encoding 88 proteins, 37 tRNAs, and 8 rRNAs. Three genes (*rpoA, rpoC2, accD*) have potentially experienced positive selection. Plastome size variation of *Rehmannia* was mainly ascribed to the expansion and contraction of the border regions between the inverted repeat (IR) region and the single-copy (SC) regions. Despite of the conserved structure in *Rehmannia* plastomes, sequence variations provide useful phylogenetic information. Phylogenetic trees of 23 Lamiales species reconstructed with the complete plastomes suggested that *Rehmannia* was monophyletic and sister to the clade of *Lindenbergia* and the parasitic taxa in Orobanchaceae. The interspecific relationships within *Rehmannia* were completely different with the previous studies. In future, population phylogenomic works based on plastomes are urgently needed to clarify the evolutionary history of *Rehmannia*.

Keywords: *Rehmannia*; chloroplast genome; repeat; positive selection; phylogeny

1. Introduction

Rehmannia Libosch. ex Fisch. et Mey. is a small genus consisting of six species, among which five (*Rehmannia chingii, Rehmannia henryi, Rehmannia elata, Rehmannia piasezkii, Rehmannia solanifolia*) are endemic to China, while *Rehmannia glutinosa*, a famous and valuable species in Chinese traditional medicine, extends its distribution range from North China to Korea and Japan [1]. The systematic position of *Rehmannia* has been debated for years. It was traditionally placed in Scrophulariaceae based on morphological traits. Recently, molecular evidence indicated that Scrophulariaceae was polyphyletic [2]. *Rehmannia* was then transferred to Plantaginaceae [3] and later placed in Orobanchaceae as the second non-parasitic branch [4,5] or treated as an independent family [6]. Besides the uncertain familial placement of *Rehmannia*, interspecific relationships within the genus are also unsolved. Despite of the differences in some flower traits, the two tetraploid species *R. glutinosa* and *R. solanifolia* share identical chloroplast and nuclear haplotypes [7,8], inferring the possibility of the symnonym of one species. Similarly, evidence from morphology, pollen, allozyme, chemical composition, and molecular data support the theory that *R. piasezkii* and *R. elata* should also be considered one species [7,9,10]. Moreover, interspecific phylogenetic relationships are incongruent when constructed by different DNA fragments. Chloroplast fragments supported *R. chingii* was the basal taxon of the genus [5] while *R. piasezkii* was confirmed as the sister group to the remaining taxa within the genus by nuclear data [7,8].

The controversy in systematic position and interspecific relationships of *Rehmannia* partly lies in the lack of sufficiently effective data. Traditional morphological classification based on limited selected characters is often deeply affected by environmental and developmental factors of samples. For example, bracteoles absence or presence, considered as the critical trait to discriminate *R. piasezkii* from *R. chingii*, are not species-specific and variable among intraspecific individuals [9]. Although molecular data such as chloroplast and/or nuclear DNA fragments provide some information for the taxonomy of *Rehmannia* [5,7,8], phylogenetic analysis based on these limited data are usually unreliable for their low resolution.

Most chloroplast (cp) genomes have a typical quadripartite structure with a pair of inverted repeats (IRs) separated by a large single-copy region (LSC) and a small single-copy region (SSC), and the genome size ranged from 120 to 160 Kb in length [11]. Previous studies indicated that the complete chloroplast genome sequences could improve the resolution at lower taxonomic level [12–14]. The Next Generation Sequencing (NGS) technique has enabled generating large amounts of sequence data at relatively low cost [15–17]. Up to now, approximately 644 plastid genomes in Viridiplantae have been sequenced and deposited in the National Center for Biotechnology Information (NCBI) Organelle Genome Resources. These massive data, together with the conservation of cp sequences, made it become a more increasingly used and effective tool for plant phylogenomic analysis than nuclear and mitochondrial genomes.

In this study, we sequenced, assembled, and characterized the plastomes of six *Rehmannia* species to verify the familial placement and evaluate the interspecific variation within the genus. These analyses will not only improve our understanding of the evolutionary mechanism of *Rehmannia* plastome and but also aid to clarify the ambiguous phylogenetic position of *Rehmannia*.

2. Materials and Methods

2.1. DNA Extraction and Sequencing

All samples used in the study were transplanted from their native habitats and cultivated in the greenhouse of Northwest University. No specific permits are required for sampling (Table 1). Healthy and fresh leaves from a single individual of the *Rehmannia* species were collected for DNA extraction.

Table 1. Sample information of six *Rehmannia* species in this study.

Species	Location	Longitude	Latitude	Clean reads	Mean coverage
R. glutinosa	Yulin, Shaanxi, China	110.57	37.77	3,721,846	1583.5×
R. henryi	Yichang, Hubei, China	10.68	31.31	31,171,142	119.6×
R. elata	Amsterdam, Holland	4.88	52.36	26,976,944	137.1×
R. piasezkii	Shiquan, Shaanxi, China	108.63	32.04	26,815,865	125.3×
R. chingii	Lishui, Zhejiang, China	120.15	28.64	25,724,095	131.5×
R. solanifolia	Chengkou, Chongqing, China	108.62	1.54	29,076,484	141.8×

The organelle-enriched DNAs of *R. glutinosa* were isolated using Percoll gradient centrifugation method [18] and CTAB extraction method. The DNA concentration was quantified using a NanoDrop spectrophotometer (Thermo Scientific, Carlsbad, CA, USA). -The DNA with concentration >30ng/μL was fragmented by mechanical interruption (ultrasonic), using PCR amplification to form a sequencing library. We sequenced the complete chloroplast genome of *R. glutinosa* with Illumina MiSeq platform at Sangon Biotech Co. (shanghai, China). A paired-end (PE) library with 265-bp insert size was constructed. Total genomic DNAs of other five *Rehmannia* species (*R. solanifolia*, *R. chingii*, *R. piasezkii*, *R. elata*, and *R. henryi*) were extracted with simplified CTAB protocol [19]. A paired-end (PE) library with 126-bp insert size was constructed using the Illumina PE DNA library kit and sequenced using an Illumina Hiseq 2500 by Biomarker technologies CO. (Beijing, China).

2.2. Chloroplast Genome Assembling and Annotation

Raw reads of *R. glutinosa* were trimmed to remove the potential low quality bases. Chloroplast genome was assembled using Velvet Assembler version 1.2.07 [20] and SPAdes [21]. Gaps and ambiguous (N) bases of the plastome were corrected using SSPACE premium version 2.2 [22]. The annotation of the plastome was performed with online tool CpGAVAS (http://www.herbalgenomics.org/cpgavas) and the gene homologies were confirmed by comparing with the NCBI's non-redundant (Nr) protein database, Cluster of Orthologous Group (COG), CDD (https://www.ncbi.nlm.nih.gov/Structure/cdd/cdd.shtml), PFAM (http://xfam.org), SWISS-PROT (http://web.expasy.org/docs/swiss-prot_guideline.html), and TREMBL (http://www.bioinfo.pte.hu/more/TrEMBL.htm) databases. The raw reads of five other *Rehmannia* species were quality-trimmed using CLC Genomics Workbench v7.5 (CLC bio, Aarhus, Denmark) with default parameters. Reference-guided assembly was then performed to reconstruct the chloroplast genomes with the program MITObim v1.7 [23] using *R. glutinosa* as the reference. The cpDNA annotation was conducted using the program GENEIOUS R8 (Biomatters Ltd., Auckland, New Zealand), and used the plastome of *R. glutinosa* as the reference, coupled with manual adjustment for start/stop codons and for intron/exon borders. Transfer RNAs (tRNAs), ribosomal RNAs (rRNAs), and coding sequences were further confirmed, and in some cases, manually adjusted after BLAST searches. The circle maps of six *Rehmannia* plastomes were obtained using Organellar Genome DRAW software (OGDRAW, http://ogdraw.mpimp-golm.mpg.de) [24]. Ambiguous (N) bases and large insertion/deletion fragment (*rpoC2*) were validated by PCR amplification and Sanger sequencing (Table S1).

2.3. Sequence Analysis and Repeat Structure

Multiple alignments of six *Rehmannia* plastomes were carried out using MAFFT version 7.017 [25]. Full alignments with annotation were visualized using the mVISTA software [26]. Genetic divergence parameter (*p*-distance) was calculated by MEGA 6.0 [27]. The percentage of variable characters for each noncoding region with an aligned length >200 bp in the genome was calculated as described in Zhang et al. [28]. Dispersed, tandem and palindromic repeats were determined by the program REPuter [29] (http://bibiserv.techfak.uni-bielefeld.de/reputer/manual.html) with a minimal size of 30 bp and >90% identity (Hamming distance equal to 3) between the two repeats. Gap size between palindromic repeats was restricted to a maximal length of 3 kb. Overlapping repeats were merged into one repeat motif whenever possible. Tandem Repeats Finder [30] (http://tandem.bu.edu/trf/trf.html) was used to identify tandem repeats in the six *Rehmannia* plastomes with default settings. A given region in the genome was designated as only one repeat type, and tandem repeat was prior to dispersed repeat if one repeat motif could be identified as both tandem and dispersed repeats.

2.4. Selective Pressure Analysis

Signals of natural selection were evaluated for all chloroplast genes located outside of IR region. Selective pressures, nonsynonymous to synonymous ratios (Ka/Ks), were computed with codeml tool of PAML package [31].

2.5. Comparative Genome Analysis

The whole plastomes of *Rehmannia* and 17 representatives of Lamiales species, including six Lamiaceae species, five Orobanchaceae species, and five species from other families (Table 2), were aligned separately by using MAUVE as implemented in Geneious with default settings [32] to test and visualize the presence of genome rearrangements and inversions

Table 2. Summary of chloroplast (cp) genomic data of all Lamiales taxa used in the study. The numbers in parenthesis indicate the genes duplicated in the inverted repeat (IR) regions.

Taxon	Species	GenBank	Length	LSC	SSC	IR	Gene	PCG	tRNA	rRNA	GC (%)
Orobanchaceae	Rehmannia glutinosa (Gaetn.) Libosch. ex Fisch. et Mey.	KX636157	153622	84605	17579	25719	133	88	37 (7)	8 (4)	38
	Rehmannia chingii Li.	KX426347	154055	84966	17675	25707	133	88	37 (7)	8 (4)	38
	Rehmannia henryi N.E. Brown	KX636158	153890	84837	17679	25687	133	88	37 (7)	8 (4)	37.9
	Rehmannia elata N.E. Brown	KX636161	153772	84788	17652	25666	133	88	37 (7)	8 (4)	38
	Rehmannia piasezkii Maxim.	KX636160	153952	84899	17674	25676	133	88	37 (7)	8 (4)	37.9
	Rehmannia solanifolia Tsoong et Chin	KX636159	153989	84839	17680	25735	133	88	37 (7)	8 (4)	37.9
	Cistanche phelypaea (L.) Coutinho	NC_025642	94380	32648	8646	26543	99	30	42 (9)	8 (4)	36.6
	Cistanche deserticola Ma	KC_128846	102657	49130	8819	22354	106	31	36 (7)	8 (4)	36.8
	Orobanche californica Cham. & Schltdl.	NC_025651	120840	62000	8516	25162	123	45	41 (6)	8 (4)	36.7
	Lindenbergia philippensis (Cham.) Benth.	NC_022859	155103	85594	17885	25812	137	85	37 (7)	8 (4)	37.8
	Schwalbea americana L.	HG_738866	160910	84756	18899	28627	128	82	37 (7)	8 (4)	38.1
Lamiaceae	Rosmarinus officinalis L.	KR_232566	152462	83355	17969	25569	134	86	37 (7)	8 (4)	38
	Salvia miltiorrhiza Bge.	NC_020098	153953	85318	17741	25447	134	86	37 (7)	8 (4)	37.9
	Origanum vulgare L.	JX_880022	151935	83135	17727	25533	134	86	37 (7)	8 (4)	37.8
	Tectona grandis L.F.	NC_020431	151328	82695	17555	25539	133	87	37 (7)	8 (4)	38
	Premna microphylla Turcz	NC_026291	155293	86078	17689	25763	133	87	37 (7)	8 (4)	37.9
	Scutellaria baicalensis Georgi	KR_233163	152731	83946	17477	25654	132	87	36 (7)	8 (4)	38.4
Scrophulariaceae	Scrophularia takesimensis Nakai	KM_590983	152425	85531	17938	23478	132	88	36 (6)	8 (4)	38.1
Gesneriaceae	Boea hygrometrica (Bunge) R. Br.	NC_016468	153493	84698	17903	25446	145	85	36 (7)	8 (4)	37.6
Acanthaceae	Andrographis paniculata (Burm. f.) Nees	NC_022451	150249	82459	17110	25340	132	87	37 (7)	8 (4)	38.3
Lentibulariaceae	Utricularia gibba L.	NC_021449	152113	81818	14187	27904	133	87	37 (6)	8 (4)	37.6
Pedaliaceae	Sesamum indicum Linn.	JN_637766	153324	85170	17872	25141	134	87	37 (7)	8 (4)	38.2
Oleaceae	Olea europaea L.	GU_931818	155889	86614	17791	25742	133	87	37 (7)	8 (4)	37.8

LSC: large single-copy; SSC: small single-copy; PCG: protein coding genes; tRNA: transfer RNA; rRNA: ribosomal RNA

2.6. Phylogenomic Analyses

The chloroplast genome sequences of six *Rehmannia* species, together with 17 Lamiales species (Table 2), were aligned with the program MAFFT version 7.017 [25] and adjusted manually when necessary. In order to test the utility of different cp regions, phylogenetic analyses were performed based on the following four datasets: (1) the complete cp DNA sequences, (2) a set of the common protein coding genes (PCGs), (3) the large single copy region, and (4) the small single copy region. Maximum likelihood (ML) analyses were implemented in RAxML version 7.2.6 [33]. RAxML searches relied on the general time reversible (GTR) model of nucleotide substitution with the gamma model of rate heterogeneity. Non-parametric bootstrapping test was implemented in the "fast bootstrap" algorithm of RAxML with 1000 replicates. Bayesian analyses were performed using the program MrBayes version 3.1.2 [34]. The best-fitting models were determined by the Akaike Information Criterion [35] as implemented in the program Modeltest 3.7 [36]. The Markov chain Monte Carlo (MCMC) algorithm was run for 200,000 generations with trees sampled every 10 generations for each data partition. The first 25% of trees from all runs were discarded as burn-in, and the remaining trees were used to construct majority-rule consensus tree. In all analyses, *Olea europaea* was set as an outgroup.

3. Results

3.1. Genome Sequencing, Assembly, and Validation

Illumina paired-end sequencing generated 1 Gb raw reads for *R. glutinosa*, accounting for 91.1% of the total reads with average length of 265 bp. The sequencing depth and coverage were approximately 6600 and 1583.5, respectively. Using the Illumina Hiseq 2500 system (Biomarker technologies CO.), five other *Rehmannia* species produced large data for each species from 25,724,095 (*R. chingii*) to 31,171,142 (*R. henryi*) clean reads (126 bp in average reads length). All paired-end reads were mapped to the reference plastome of *R. glutinosa* with the mean coverage of $119.6\times$ to $141.8\times$ (Table 1). Gaps were validated by using PCR-based sequencing with seven pairs of primers (Table S1). All six *Rehmannia* plastome sequences were deposited in GenBank (accession numbers: KX426347, KX636157- KX636161) (Table 1, Table 2).

3.2. Complete Chloroplast Genomes of Rehmannia *Species*

The six chloroplast genomes of *Rehmannia* ranged in size from 153,622 bp (*R. glutinosa*) to 154,055 bp (*R. chingii*). All of them exhibited a typical quadripartite structure consisting of a pair of IRs (25,666–25,735 bp) separated by the LSC (84,605–84,966 bp) and SSC (17,579–17,680 bp) regions (Figure 1). These six plastomes are highly conserved in gene content, gene order, and intron number. The overall GC content was about 38.0%, almost identical with each other among *Rehmannia* species (Table 2). The *Rehmannia* plastomes contained 133 genes, of which 115 occurred as a single copy and 18 were duplicated in the IR regions (Table 3). The predicted functional genes of each species were comprised of 88 protein-coding genes, 37 tRNA genes, and eight rRNA genes (Table 2). Sixteen genes (*rpl2, ndhB, petD, petB, ndhA, ndhB, rpl16, rpoC1, atpF1, rps16, trnA-UGC, trnG-GCC, trnI-GAU, trnK-UUU, trnL-UAA*, and *trnV-UAC*) had one intron, while three genes (*rps12, clpP*, and *ycf3*) contained two introns (Table 3). The *rps12* gene was a unique gene with 3′ end exon and intron located in the IR region, and the 5′ end exon in the LSC region. Unusual initiator codons were observed in *ndhD* with ATC and *rps19* with GTG in *Rehmannia* plastomes. Overlaps of adjacent genes were found in the complete genome, for example, *rps3-rpl22, atpB-atpE*, and *psbD-psbC* had a 16 bp, 4 bp, and 53 bp overlapping region, respectively. Large indels were detected in the *rpoC2* gene, which caused the gene size to vary from 2916 bp to 4185 bp among the six species (Figure S1).

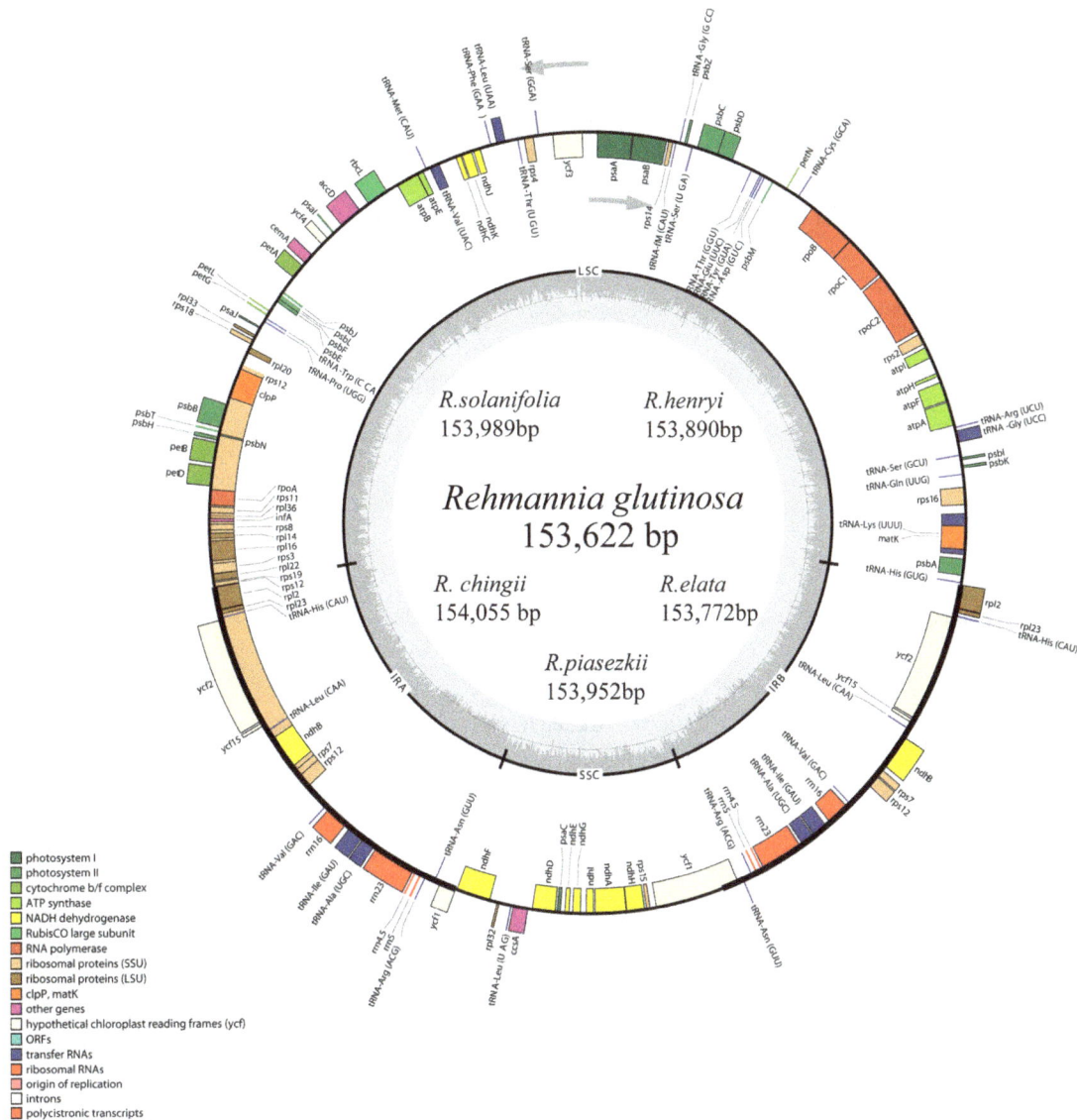

Figure 1. Gene map of *Rehmannia* chloroplast genomes. Genes shown outside the outer circle are transcribed clockwise and those inside are transcribed counterclockwise. Genes belonging to different functional groups are color coded. Dashed area in the inner circle indicates the GC content of the chloroplast genome. ORF: open reading frame.

Table 3. Gene list of plastomes of six *Rehmannia* species.

Category	Group	Name
Photosynthesis related genes	Rubisco	*rbcL*
	Photosystem I	*psaA, psaB, psaC, psaI, psaJ*
	Assembly/stability of photosystem I	** *ycf3*
	Photosystem II	*psbA, psbB, psbC, psbD, psbE, psbF, psbH, psbI, psbJ, psbK, psbL, psbM, psbN, psbT, psbZ*
	ATP synthase	*atpA, atpB, atpE, * atpF, atpH, atpI*
	cytochrome b/f complex	*petA, * petB, * petD, petG, petL, petN*
	cytochrome c synthesis	*ccsA*
	NADPH dehydrogenase	* *ndhA, *,a ndhB, ndhC, ndhD, ndhE, ndhF, ndhG, ndhH, ndhI, ndhJ, ndhK*

Table 3. Gene list of plastomes of six *Rehmannia* species.

Category	Group	Name
Transcription and translation related genes	transcription	*rpoA, rpoB, *rpoC1, rpoC2*
	ribosomal proteins	*rps2, rps3, rps4,* ᵃ *rps7, rps8, rps11,* ** *rps12, rps14, rps15, *rps16, rps18, rps19, *,ᵃrpl2, rpl14, * rpl16, rpl20, rpl22,* ᵃ *rpl23, rpl32, rpl33, rpl36*
	translation initiation factor	*infA*
RNA genes	ribosomal RNA	ᵃ*rrn5,* ᵃ *rrn4.5,* ᵃ *rrn16,* ᵃ *rrn23*
	transfer RNA	**,ᵃ trnA-UGC,* # *trnA-ACG, trnL-UAG,* ᵃ*trnA-GUU, *,ᵃtrnI-GAU,* ᵃ *trnV-(GAC),* ᵃ *trnL-CAA,* ᵃ *trnH-CAU, trnP-UGG, trnT-CCA, trnM-CAU, * trnV-UAC, trnP-GAA, * trnL-UAA, trnT-UGU, trnS-GGA, trnfM-CAU, trnG-GCC, trnS-UGA, trnT-GGU, trnG-UUC, trnT-GUA, trnA-GUC, trnC-GCA, trnA-UCU, * trnG-UCC, trnS-GCU, trnG-UUG, * trnL -UUU, trnH-GUG, trnA-GUU*
Other genes	RNA processing	*matK*
	carbon metabolism	*cemA*
	fatty acid synthesis	*accD*
	proteolysis	** *clpP*
Genes of unknown function	conserved reading frames	ᵃ *ycf1,* ᵃ *ycf2, ycf4,* ᵃ *ycf15*

* gene with one intron, ** gene with two introns, ᵃ gene with two copies.

3.3. IR Boundary Changes and Gene Rearrangement

The IR region of six *Rehmannia* chloroplast genomes was highly conserved, but structure variation was still found in the IR/SC boundary regions. To elucidate the potential contraction and expansion of IR regions, we compared the gene variation at the IR/SSC and IR/LSC boundary regions of the six plastomes. The genes *rps19-rp12-trnH* and *ycf1-ndhF* were located in the junctions of LSC/IR and SSC/IR regions. Two copies of the *ycf1* gene crossed SSC/IRa and SSC/IRb with 3 bp in the SSC region and 1083/1084 bp in the IRa region, respectively (Figure 2). Compared to the relatively fixed location of the *ycf1* and *trnH* gene in all chloroplast genomes, the LSC/IR boundary regions were more variable. The rps19 gene in *R. glutinosa*, *R. piasezkii*, and *R. chingii* crossed the LSC/IRb region with 45 bp, 3 bp, and 4 bp located at the IRb region while the intergenic spacer of *rps19-rps12* extended 3 bp, 4 bp, or 5 bp to the LSC region in *R. elata*, *R. henryi*, and *R. solanifolia*, respectively. The *rpl2* gene of *R. elata*, commonly located in the IRb region in *Rehmannia*, extended 65 bp into the LSC region and overlapped with the rps19 gene by 62 bp. To identify the potential genome rearrangements and inversions, the chloroplast genome sequences of six *Rehmannia* species, *Arabidopsis thaliana* and 16 core Lamiales taxa, were selected for synteny analyses (Table 2). No gene rearrangement and inversion events were detected in *Rehmannia*, except *Cistanche deserticola* with structure variation of a 4 kb fragment. (Supplementary Figure S2).

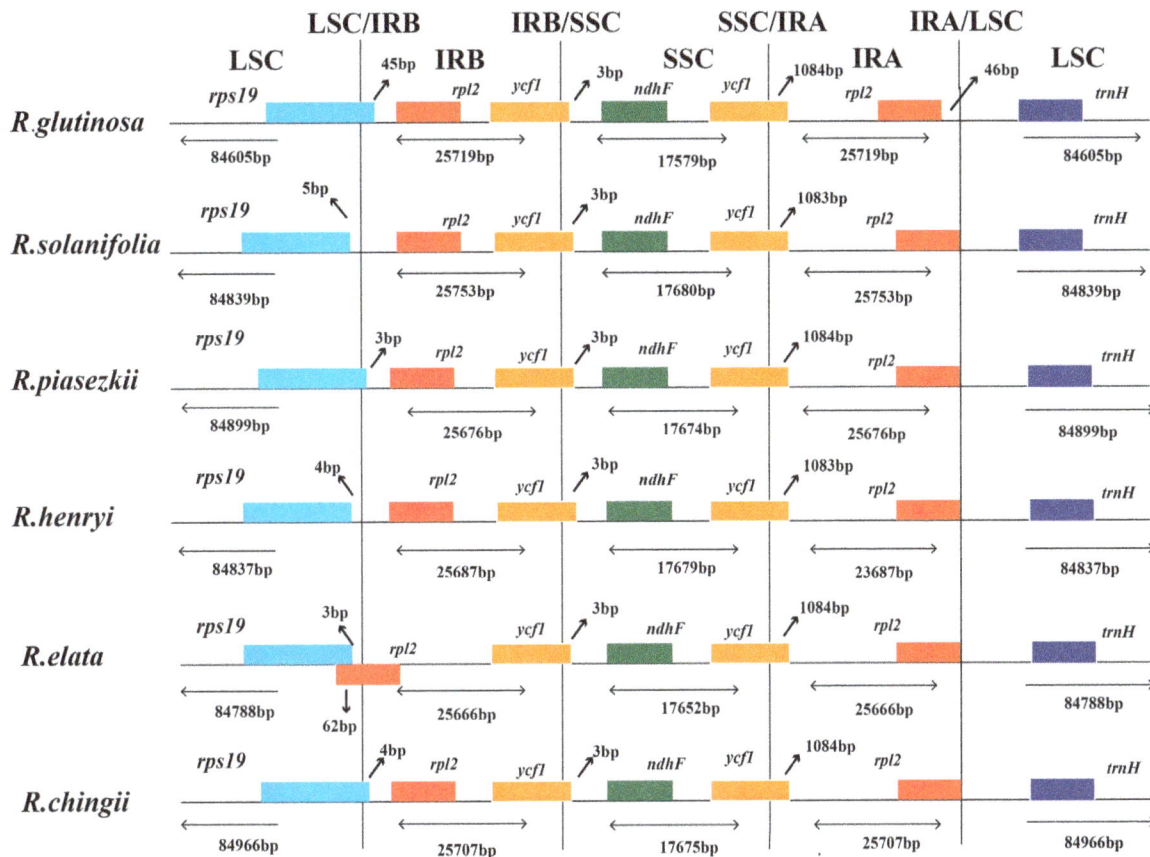

Figure 2. Comparison of the borders of large single-copy (LSC), small single-copy (SSC), and inverted repeat (IR) regions among the chloroplast genomes of six *Rehmannia* species. The location of two parts of inverted repeat region (IRA and IRB) was referred to Figure 1.

3.4. Repetitive Sequences

We classified sequence repeat motifs into three categories: dispersed, tandem, and palindromic repeats. For all repeat types, the minimal cut-off identity between two copies was set to 90%. The minimal repeat size investigated were 30 bp for dispersed, 15 bp for tandem and 20 bp for palindromic repeats, respectively. In total, 411 repeats were detected in *Rehmannia* plastomes (see Supplementary Table S2, Figure 3). Among these repeats, 24 were verified to be associated with two copies of tRNA (e.g., *trnG-UCC*) or gene duplication (e.g., *psaA/psaB*) and subsequently considered as tRNA or gene similarity repeats [37] due to their similarity in gene functions. Numbers of the three repeat types were similar among these six plastomes (Figure 3A) and their overall distribution in the plastome was highly conserved. Generally, palindromic repeats were the most common, while tandem repeats were the least in *Rehmannia* except *R. glutinosa* with dispersal repeats as the most common. The majority of repeats ranged in size from 30 bp to 44 bp (Figure 3B), even though the defined smallest size is 15 bp and 30 bp for tandem and dispersed repeats, respectively. The longest repeat is a palindromic repeat of 341 bp in *R. henryi* and *R. solanifolia*. Dispersed repeats had a wider size range (from 30 to 126 bp) than other repeat types. Palindromic repeat, accounting for 41% of total repeats, was the most common, followed by dispersed (39%), tandem (14%), and tRNA or gene similarity (6%) types (Figure 3C). A minority of repeats was found in intron (7.3%), while the majority were located in coding regions (48.9%) (such as gene *ycf2*, *rps18*, *rps11* and *rpoC2*) and intergenic spacers (43.8%) (Figure 3D).

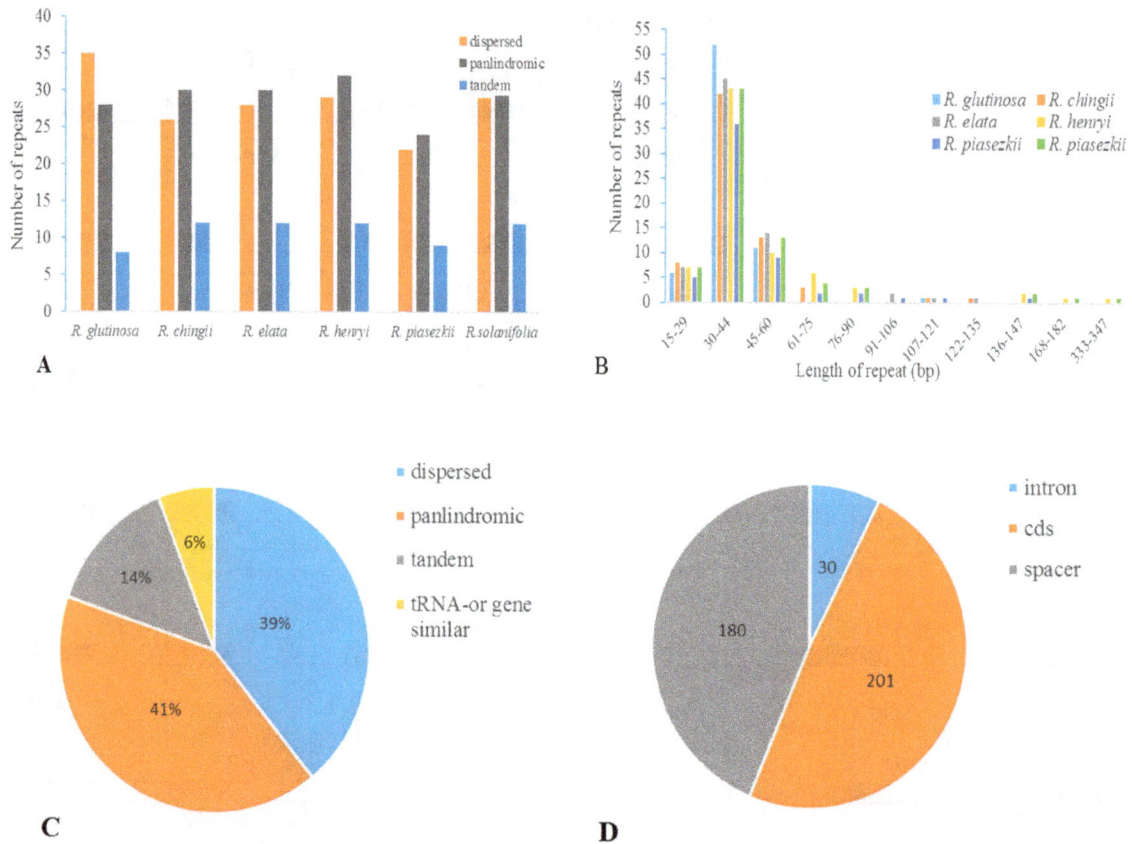

Figure 3. Analyses of repeated sequences in *Rehmannia* plastomes. (**A**) Number of three repeat types in the six chloroplast genomes; (**B**) Frequency of repeat sequences by length; (**C**) Frequency of repeat types; (**D**) Location of the all repeats from six species.

3.5. Sequence Divergence and Divergence Hotspot

To elucidate the level of the genome divergence, sequence identity among *Rehmannia* cpDNAs were plotted using the program mVISTA with *R. glutinosa* as a reference. The whole aligned sequences showed high similarities with only a few regions below 90%, suggesting that *Rehmannia* plastomes were rather conserved (Figure 4). As expected, the IRs regions were more conserved than the single-copy regions, and the coding regions were less divergent than the non-coding regions. One divergent hotspot region in LSC (*psbA-ndhJ*) region was identified (Figure 4). The complete plastome sequence divergence of six species, estimated by *p*-distance, ranged from 0.002 to 0.004 with the average value of 0.0028. We also compared the sequence divergence among the different noncoding regions. Among the 98 noncoding regions, the percentage of variation ranged from 0 to 10.41% with an average of 1.7. Nine noncoding regions had over 4% variability proportions, such as *trnH(GUG)-psbA*, *trnS(GCU)-trnG(UCC)*, *psbZ-trnG(GCC)*, *psaA-ycf3*, *trnT(UGU)-trnL(UAA)*, *cemA-petA*, *rps12-clpP*, *nhdD-psaC*, and *ndhG-ndhL* (Figure 5). These divergence hotspot regions provided abundant information for marker development in phylogenetic analyses of *Rehmannia* species.

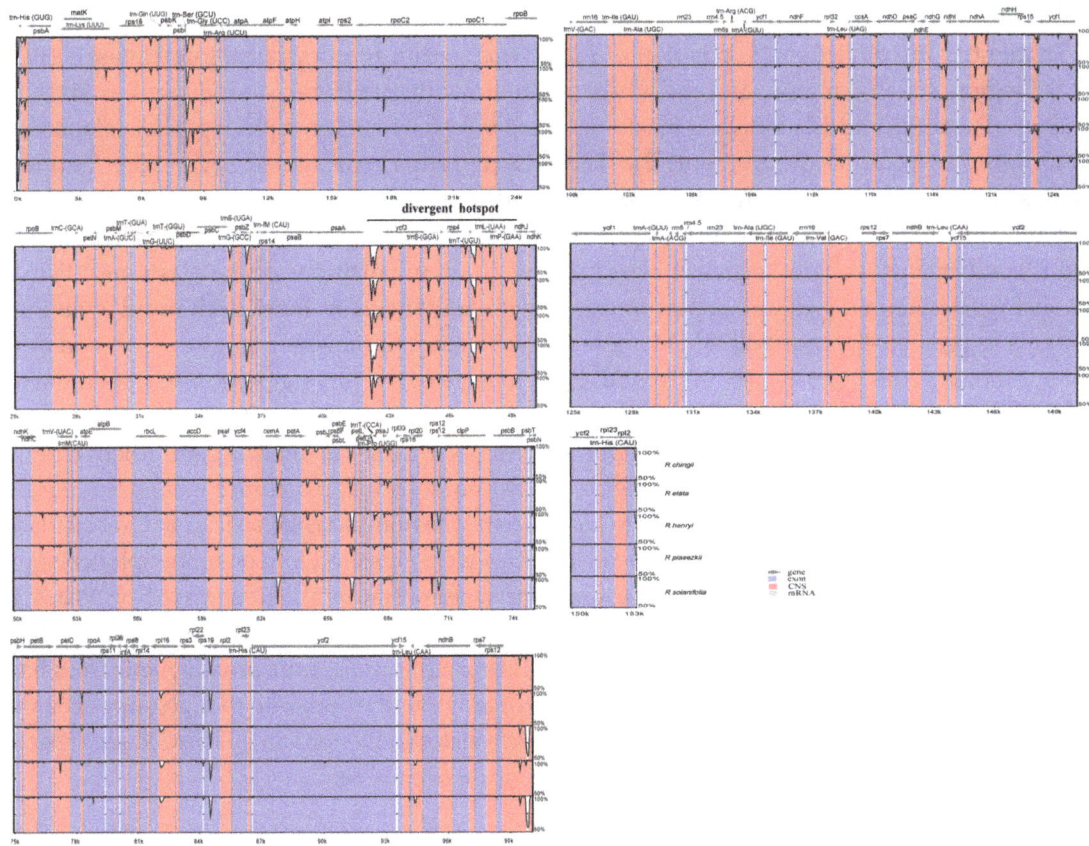

Figure 4. Visualization of alignment of the six *Rehmannia* species chloroplast genome sequences. VISTA-based identity plots showed sequence identity of six sequenced chloroplast genomes with *R. glutinosa* as a reference. The sequence similarity of the aligned regions is shown as horizontal bars indicating the average percent identity between 50% and 100% (shown on the y-axis of the graph). The x-axis represents the coordinate in the chloroplast genome. The divergent hotspot region is indicated in the chloroplast genome. Genome regions are color coded as protein coding, rRNA coding, tRNA coding or conserved noncoding sequences (CNS).

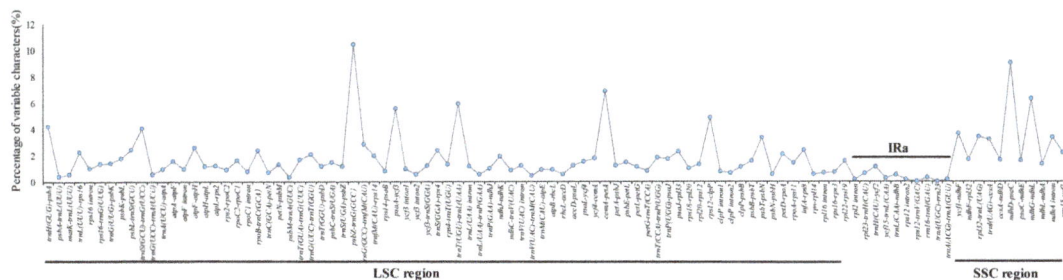

Figure 5. Percentage of variable characters in aligned noncoding regions of the six plastid genomes.

3.6. Selective Pressure Analysis

To estimate selection pressures among *Rehmannia* species, ratios of nonsynonymous (Ka) versus synonymous (Ks) substitutions were calculated for 79 protein-coding genes, generating 241 pairwise valid combinations (Table S3). The Ka/Ks ratios of the remaining comparisons were not available for Ks = 0. Three genes (*accD*, *rpoA*, *rpoC2*) located in the LSC region had Ka/Ks ratios above 1.0, which might indicate positive selection (Table S3).

3.7. Phylogenomic Analysis

To identify the phylogenetic position of *Rehmannia* within the Lamiales, four datasets (PCGs, the LSC region, the SSC region, and the whole plastome with one IR region removed) from the six *Rehmannia* plastomes and 17 published plastomes were used to reconstruct phylogenetic relationships with *O. europaea* as an outgroup (Table 2). These 22 ingroups represented seven core families in Lamiales [36]. The phylogenetic tree based on the same dataset using ML and Bayesian method had the identical topological structure with possibly different support values (Figure 6). There were no obvious conflicts between phylogenetic trees built by different partitions of the plastomes. Familial relationships based on the complete plastomes were quite identical to those with rapid evolving cp fragments as previously reported [38]. Along with the increase of sequence length, resolution power of main branches was dramatically improved. Each lineage of the phylogenetic tree with the whole plastome was well-supported with 100% bootstrap value or the Bayesian posterior probability of one (Figure 6A). Phylogenetic status of the parasitic species in Orobanchaceae was contradictive in trees based on PCGs and the LSC/SSC region (Figure 6B–D). But when all plastomes were used, parasitic species formed a monophyletic group sister to *Lindenbergia*, the non-parasitic species in Orobanchaceae (Figure 6). In all phylogenetic trees, six *Rehmannia* species were clustered into one monophyletic group sister to other Orobanchaceae taxa with high support value (Figure 6). Trees based on PCGs and the whole plastomes indicated that the Orobanchaceae clade including *Rehmannia* was sister to the Lamiaceae group. In terms of interspecific relationships of *Rehmannia*, four phylogenetic trees based on different datasets consistently showed the same topology with moderate to high support values: *R. solanifolia* and *R. henryi* were grouped into one branch sister to the remain species. *R. glutinosa* and *R. piasezkii* were successive sisters to *R. elata-R. chingii* clade (Figure 6).

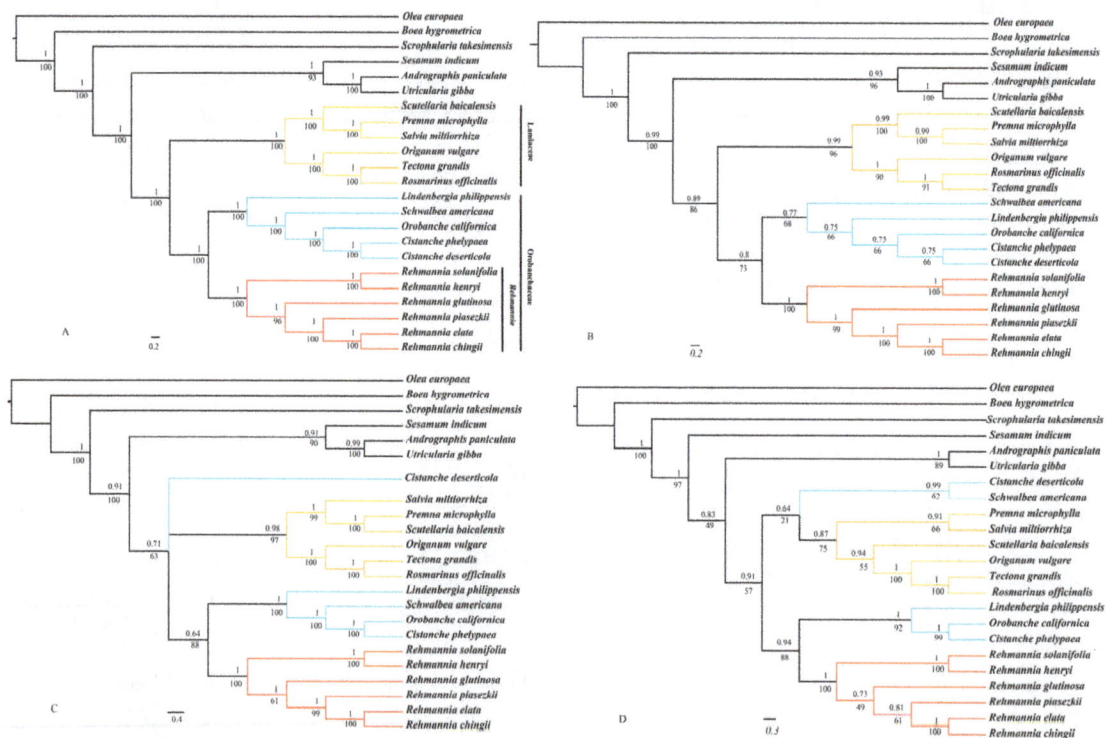

Figure 6. Phylogenetic trees of 23 species as determined from different data partitions. Support values are shown for nodes as Bayesian inference posterior probability (above branches)/maximum likelihood bootstrap (below branches). Branch lengths were calculated through Bayesian analysis, and scale bar denotes substitutions per site. (**A**) the whole chloroplast genomes; (**B**) Protein coding genes; (**C**) LSC region; (**D**) SSC region. Red represents *Rehmannia* species; Blue represents other species of Orobanchaceae; Orange represents Lamiaceae species.

4. Discussion

4.1. Genome Characteristics and Sequence Differences

Here we determined the complete plastid genome sequences from six *Rehmannia* species using Illumina sequencing technology. Although plastomes were highly conserved in terms of genomic structure and size, the IR/SC junction position variation was observed in *Rehmannia*. This may be caused by the contraction or expansion of the IR region, a common evolutionary phenomenon in plants [39–41]. As for gene contents, the same set of 88 protein-coding genes were shared by six species of *Rehmannia* species and highly conserved in aspect of gene number, gene function, gene order, and GC-content. Noncanonical start codons observed in this study could be also found in other angiosperms [42] and tree fern plants [43].

The presence of repeats in plastomes, especially in intergenic spacer regions, has been reported in all published angiosperm lineages. Compared with other angiosperm species, the number of repeats in *Rehmannia* is rather high. In all, more than 400 repeats were detected in *Rehmannia* plastomes. Previous researches have suggested that repeat sequences may play roles in rearranging sequences and producing variation through illegitimate recombination and slipped-strand mispairing [37,44,45]. However, we detected no structural rearrangements or gene loss-and-gain events in *Rehmannia* plastomes. But the regions with high divergence were generally rich in repeat units. For example, variable types of repeats could be found within the noncoding region of *psbZ-trnG* (GCC), *psaA-ycf3* and *trnS(GCU)-trnG(UCC)* gene. Lamiales has the highest diversification rates among angiosperms [46]. Thus the high repeat number in *Rehmannia* might be ascribed to the increased evolutionary rates.

Sequences of *Rehmannia* plastomes were conserved in most of regions with sequence identities above 90%. As expected, the noncoding regions exhibited higher divergent levels than the coding regions, and the single copy regions had higher variation than the IR regions. The *rpoC2* gene is an exception with lower sequence identity due to various indels, as also reported in grasses [47]. One divergent hotspot region associated with a tRNA cluster in LSC (*trnS(GGA)-trnP(GAA)*) region was identified, which is inconsonant to other herb plastomes [47,48]. We also found nine highly variable noncoding regions with variation percentages above 2%. Previous studies have also shown that noncoding regions of plastomes could be successfully used for phylogenetic studies in angiosperms [49,50]. Additional work is still necessarily needed to verify whether these highly variable regions could potentially be used as molecular targets for future phylogenetic and/or population genetics studies of *Rehmannia* species.

Our analysis indicated that three genes (*accD*, *rpoA* and *rpoC2*) were under positive selection, of which, *rpoA* and *rpoC2* were reported in Annonaceae [51]. The *accD* gene encodes a plastid-coded subunit of heteromeric acetyl-CoA carboxylase (ACCase), a key enzyme involved in fatty acid biosynthesis in plants [52]. The genes *rpoA* and *rpoC2* encode α and β'' subunit of plastid-encoded plastid RNA polymerase (PEP), respectively, a key protein responsible for most photosynthetic gene expression [53]. Field works and photosynthetic physiological study suggested that the six species of *Rehmannia* might have divergent habitats and adapt to different light intensity (unpublished data). These genes experienced natural selection might play an important role in the evolution and divergence of *Rehmannia*.

4.2. Phylogenetic Implications

Rehmannia was traditionally placed in Scrophylariaceae s.l. In our study, phylogenetic trees based on four plastome datasets generated similar topological structures except for the SSC dataset, possibly due to fewer informative sites than others. All *Rehmannia* species were clustered into a monophyletic group with high support value (BI = 1, BS = 100%), and sister to the clade of the parasitic Orobanchaceae species and *Lindenbergia philippehsis* rather than a genus of Scrophulariaceae. Chloroplast fragments analyses also suggested that *Rehmannia*, together with *Triaenophora*, represented the branch sister to Orobanchaceae s.l (including *Lindenbergia*), or a new familial clade [5]. The phylogenetic placement of

Rehmannia-Orobanchaceae clade in Lamiales remains uncertain because of the contradictive results of this clade and its related taxa of Paulowniaceae and Phrymaceae from cp fragments data [5,38]. For the lack of plastome information of Paulowniaceae and Phrymaceae species, the status of *Rehmannia*-Orobanchaceae is still unresolved in our study. But familial relationships of Lamiales taxa in this study were definite and identical to those verified by the rapid evolving cp gene fragments [38]. Therefore, analyses of the entire plastomes significantly contribute to species identification and phylogenetic studies of angiosperms [14,54,55].

Recent studies using chloroplast or nuclear gene fragments indicated that two pairs of species groups, *R. elata*-*R. piasezkii*, *R. glutinosa*-*R. solanifolia*, had completely identical sequences, respectively and could be treated as two species [7,9,10]. *R. chingii* [5] or *R. piasezkii* [7,8] were considered as the basal taxon of the genus. But our phylogenetic analyses of *Rehmannia* plastomes were inconsistent with any of these works, inferring all *Rehmannia* taxa as separate species (Figure 6). This might be ascribed to the insufficient informative characters of cp/nuclear gene fragments generating the phylogenetic trees with low support values. Of course, the entire plastomes with massive characters may also result in strong systematic biases for limited sampling [56–58]. In future, we plan to analyze plastomes at population level to elucidate the phylogenetic relationships of *Rehmannia* species. These comparative genomic analyses will not only provide insights into the chloroplast genome evolution of *Rehmannia*, but also offer valuable genetic markers for population phylogenomic study of *Rehmannia* and its close lineages.

Supplementary Materials: The following supplementary materials can be found at www.mdpi.com/2073-4425/ 8/3/103/s1, Table S1: Primers used for gap closure and *rpoC2* gene verification; Table S2: Repetitive sequence statistics; Table S3: Ka/Ks ratio between pairwise of species protein coding sequences in six *Rehmannia* species; Figure S1: Alignment of *rpoC2* fragments from all six species of *Rehmannia*. Large indels were detected in the *rpoC2* gene, which cause the gene size vary from 2916 bp to 4185 bp among the six species; Figure S2: The Mauve alignment between plastomes of *Arabidopsis thaliana* and *Rehmannia* species (A) and between those of *A. thaliana* and core Lamiales taxa (B).

Acknowledgments: This work was financially supported by the National Natural Science Foundation of China (31670219, 31370353), the Natural Science Basic Research Plan in Shaanxi Province, China (2015JM3106), and the Shaanxi Provincial Education Department (11JS093).

Author Contributions: Zhan-Lin Liu and Shuyun Zeng conceived and designed the experiments. Shuyun Zeng, Tao Zhou, Kai Han, and Yanci Yang performed the experiments and analyzed the data. Jianhua Zhao prepared the samples. Shuyun Zeng wrote the paper. Zhan-Lin Liu and Tao Zhou help to revise the paper. All authors read and approved the final manuscript.

References

1.   Rix, M. The genus *Rehmannia*. *Plantsman* **1987**, *8*, 193–195.

2.   Olmstead, R.G.; DePamphilis, C.W.; Wolfe, A.D.; Young, N.D.; Elisons, W.J.; Reeves, P.A. Disintegration of the Scrophulariaceae. *Am. J. Bot.* **2001**, *88*, 348–361. [CrossRef] [PubMed]

3.   The Angiosperm Phylogeny Group. An update of the Angiosperm Phylogeny Group classification for the orders and families of flowering plants: APG II. *Bot. J. Linn. Soc.* **2003**, *141*, 399–436.

4.   The Angiosperm Phylogeny Group. An update of the Angiosperm Phylogeny Group classification for the orders and families of flowering plants: APG IV. *Bot. J. Linn. Soc.* **2016**, *181*, 1–20.

5.   Xia, Z.; Wang, Y.-Z.; Smith, J.F. Familial placement and relations of *Rehmannia* and *Triaenophora* (Scrophulariaceae s.l.) inferred from five gene regions. *Am. J. Bot.* **2009**, *96*, 519–530. [CrossRef] [PubMed]

6.   Reveal, J.L.; Chase, M.W. APG III: Bibliographical Information and Synonymy of Magnoliidae. *Phytotaxa* **2011**, *19*, 64. [CrossRef]

7.   Huang, J.; Zeng, S.-Y.; Zhao, J.-H.; Han, K.; Li, J.; Li, Z.; Liu, Z.-L. Genetic variation and phylogenetic relationships among *Rehmannia* (Scrophulariaceae) species as revealed by a novel set of single-copy nuclear gene markers. *Biochem. Syst. Ecol.* **2016**, *66*, 43–49. [CrossRef]

8.   Liu, Z.-L.; Li, J.-F. Molecular phylogeny analysis of *Rehmannia*. *Acta Bot. Bor. Occid. Sin.* **2014**, *34*, 77–82.

9. Li, X.; Li, J. Morphology characters of leaf epidermis of the genera *Rehmannia* and *Triaenophora*. *J. Plant Sci.* **2006**, *24*, 559–564.

10. Yan, K.; Zhao, N.; Li, H.Q. Systematic relationships among *Rehmannia* (Scrophulariaceae) species. *Acta Bot. Boreal. Occident Sin.* **2007**, *27*, 1112–1120.

11. Sugiura, M. The chloroplast genome. *Plant Mol. Biol.* **1992**, *19*, 149–168. [CrossRef] [PubMed]

12. Bewick, A.J.; Chain, F.J.J.; Heled, J.; Evans, B.J. The Pipid Root. *Syst. Biol.* **2012**, *61*, 913–926. [CrossRef] [PubMed]

13. Carbonell-Caballero, J.; Alonso, R.; Ibañez, V.; Terol, J.; Talon, M.; Dopazo, J. A Phylogenetic Analysis of 34 Chloroplast Genomes Elucidates the Relationships between Wild and Domestic Species within the Genus *Citrus*. *Mol. Biol. Evol.* **2015**, *32*, 2015–2035. [CrossRef] [PubMed]

14. Jansen, R.K.; Cai, Z.; Raubeson, L.A.; Daniell, H.; dePamphilis, C.W.; Leebens-Mack, J.; Müller, K.F.; Guisinger-Bellian, M.; Haberle, R.C.; Hansen, A.K.; et al. Analysis of 81 genes from 64 plastid genomes resolves relationships in angiosperms and identifies genome-scale evolutionary patterns. *Proc. Natl. Acad. Sci. USA* **2007**, *104*, 19369–19374. [CrossRef] [PubMed]

15. Uribe-Convers, S.; Duke, J.R.; Moore, M.J.; Tank, D.C. A Long PCR–Based Approach for DNA Enrichment Prior to Next-Generation Sequencing for Systematic Studies. *Appl. Plant Sci.* **2014**, *2*. [CrossRef] [PubMed]

16. Qiao, J.; Cai, M.; Yan, G.; Wang, N.; Li, F.; Chen, B.; Gao, G.; Xu, K.; Li, J.; Wu, X. High-throughput multiplex cpDNA resequencing clarifies the genetic diversity and genetic relationships among *Brassica napus*, *Brassica rapa* and *Brassica oleracea*. *Plant Biotechnol. J.* **2016**, *14*, 409–418. [CrossRef] [PubMed]

17. Ruhsam, M.; Rai, H.S.; Mathews, S.; Ross, T.G.; Graham, S.W.; Raubeson, L.A.; Mei, W.; Thomas, P.I.; Gardner, M.F.; Ennos, R.A.; et al. Does complete plastid genome sequencing improve species discrimination and phylogenetic resolution in *Araucaria*? *Mol. Ecol. Resour.* **2015**, *15*, 1067–1078. [CrossRef] [PubMed]

18. Sandbrink, J.M.; Vellekoop, P.; Van Ham, R.; Van Brederode, J. A method for evolutionary studies on RFLP of chloroplast DNA, applicable to a range of plant species. *Biochem. Syst. Ecol.* **1989**, *17*, 45–49. [CrossRef]

19. Doyle, J.J. A rapid DNA isolation procedure for small quantities of fresh leaf tissue. *Phytochem. Bull.* **1987**, *19*, 11–15.

20. Zerbino, D.R.; Birney, E. Velvet: Algorithms for de novo short read assembly using de Bruijn graphs. *Genome Res.* **2008**, *18*, 821–829. [CrossRef] [PubMed]

21. Bankevich, A.; Nurk, S.; Antipov, D.; Gurevich, A.A.; Dvorkin, M.; Kulikov, A.S.; Lesin, V.M.; Nikolenko, S.I.; Pham, S.; Prjibelski, A.D.; et al. SPAdes: A New Genome Assembly Algorithm and Its Applications to Single-Cell Sequencing. *J. Comput. Biol.* **2012**, *19*, 455–477. [CrossRef] [PubMed]

22. Boetzer, M.; Henkel, C.V.; Jansen, H.J.; Butler, D.; Pirovano, W. Scaffolding pre-assembled contigs using SSPACE. *Bioinformatics* **2011**, *27*, 578–579. [CrossRef] [PubMed]

23. Hahn, C.; Bachmann, L.; Chevreux, B. Reconstructing mitochondrial genomes directly from genomic next-generation sequencing reads—A baiting and iterative mapping approach. *Nucleic Acids Res.* **2013**, *41*, e129. [CrossRef] [PubMed]

24. Lohse, M.; Drechsel, O.; Kahlau, S.; Bock, R. OrganellarGenomeDRAW—A suite of tools for generating physical maps of plastid and mitochondrial genomes and visualizing expression data sets. *Nucleic Acids Res.* **2013**, *41*, W575–W581. [CrossRef] [PubMed]

25. Katoh, K.; Standley, D.M. MAFFT Multiple Sequence Alignment Software Version 7: Improvements in Performance and Usability. *Mol. Biol. Evol.* **2013**, *30*, 772–780. [CrossRef] [PubMed]

26. Frazer, K.A.; Pachter, L.; Poliakov, A.; Rubin, E.M.; Dubchak, I. VISTA: Computational tools for comparative genomics. *Nucleic Acids Res.* **2004**, *32* (Suppl. 2), W273–W279. [CrossRef] [PubMed]

27. Tamura, K.; Stecher, G.; Peterson, D.; Filipski, A.; Kumar, S. MEGA6: Molecular Evolutionary Genetics Analysis Version 6.0. *Mol. Biol. Evol.* **2013**, *30*, 2725–2729. [CrossRef] [PubMed]

28. Zhang, Y.-J.; Ma, P.-F.; Li, D.-Z. High-Throughput Sequencing of Six Bamboo Chloroplast Genomes: Phylogenetic Implications for Temperate Woody Bamboos (*Poaceae: Bambusoideae*). *PLoS ONE* **2011**, *6*, e20596. [CrossRef] [PubMed]

29. Kurtz, S.; Choudhuri, J.V.; Ohlebusch, E.; Schleiermacher, C.; Stoye, J.; Giegerich, R. REPuter: The manifold applications of repeat analysis on a genomic scale. *Nucleic Acids Res.* **2001**, *29*, 4633–4642. [CrossRef] [PubMed]

30. Benson, G. Tandem repeats finder: A program to analyze DNA sequences. *Nucleic Acids Res.* **1999**, *27*, 573–580. [CrossRef] [PubMed]

31. Yang, Z. PAML 4: Phylogenetic Analysis by Maximum Likelihood. *Mol. Biol. Evol.* **2007**, *24*, 1586–1591. [CrossRef] [PubMed]

32. Darling, A.E.; Mau, B.; Perna, N.T. progressiveMauve: Multiple Genome Alignment with Gene Gain, Loss and Rearrangement. *PLOS ONE* **2010**, *5*, e11147. [CrossRef] [PubMed]

33. Stamatakis, A. RAxML-VI-HPC: Maximum likelihood-based phylogenetic analyses with thousands of taxa and mixed models. *Bioinformatics* **2006**, *22*, 2688–2690. [CrossRef] [PubMed]

34. Ronquist, F.; Huelsenbeck, J.P. MrBayes 3: Bayesian phylogenetic inference under mixed models. *Bioinformatics* **2003**, *19*, 1572–1574. [CrossRef] [PubMed]

35. Posada, D.; Buckley, T.R. Model Selection and Model Averaging in Phylogenetics: Advantages of Akaike Information Criterion and Bayesian Approaches over Likelihood Ratio Tests. *Syst. Biol.* **2004**, *53*, 793–808. [CrossRef] [PubMed]

36. Posada, D.; Crandall, K.A. Modeltest: Testing the model of DNA substitution. *Bioinformatics* **1998**, *14*, 817–818. [CrossRef] [PubMed]

37. Timme, R.E.; Kuehl, J.V.; Boore, J.L.; Jansen, R.K. A comparative analysis of the *Lactuca* and *Helianthus* (Asteraceae) plastid genomes: Identification of divergent regions and categorization of shared repeats. *Am. J. Bot.* **2007**, *94*, 302–312. [CrossRef] [PubMed]

38. Schäferhoff, B.; Fleischmann, A.; Fischer, E.; Albach, D.C.; Borsch, T.; Heubl, G.; Müller, K.F. Towards resolving Lamiales relationships: Insights from rapidly evolving chloroplast sequences. *BMC Evol. Biol.* **2010**, *10*, 352. [CrossRef] [PubMed]

39. Hansen, D.R.; Dastidar, S.G.; Cai, Z.; Penaflor, C.; Kuehl, J.V.; Boore, J.L.; Jansen, R.K. Phylogenetic and evolutionary implications of complete chloroplast genome sequences of four early-diverging angiosperms: *Buxus* (Buxaceae), *Chloranthus* (Chloranthaceae), *Dioscorea* (Dioscoreaceae), and *Illicium* (Schisandraceae). *Mol. Phylogenet. Evol.* **2007**, *45*, 547–563. [CrossRef] [PubMed]

40. Huang, H.; Shi, C.; Liu, Y.; Mao, S.-Y.; Gao, L.-Z. Thirteen *Camelliachloroplast* genome sequences determined by high-throughput sequencing: Genome structure and phylogenetic relationships. *BMC Evol. Biol.* **2014**, *14*, 151. [CrossRef] [PubMed]

41. Kim, K.J.; Lee, H.L. Complete chloroplast genome sequences from Korean ginseng (Panax schinseng Nees) and comparative analysis of sequence evolution among 17 vascular plants. *DNA Res.* **2004**, *11*, 247–261. [CrossRef] [PubMed]

42. Bortiri, E.; Coleman-Derr, D.; Lazo, G.R.; Anderson, O.D.; Gu, Y.Q. The complete chloroplast genome sequence of Brachypodium distachyon: Sequence comparison and phylogenetic analysis of eight grass plastomes. *BMC Res. Notes* **2008**, *1*, 61. [CrossRef] [PubMed]

43. Cahoon, A.B.; Sharpe, R.M.; Mysayphonh, C.; Thompson, E.J.; Ward, A.D.; Lin, A. The complete chloroplast genome of tall fescue (*Lolium arundinaceum*; Poaceae) and comparison of whole plastomes from the family Poaceae. *Am. J. Bot.* **2010**, *97*, 49–58. [CrossRef] [PubMed]

44. Asano, T.; Tsudzuki, T.; Takahashi, S.; Shimada, H.; Kadowaki, K. Complete nucleotide sequence of the sugarcane (*Saccharum officinarum*) chloroplast genome: A comparative analysis of four monocot chloroplast genomes. *DNA Res.* **2004**, *11*, 93–99. [CrossRef] [PubMed]

45. Cavalier-Smith, T. Chloroplast evolution: Secondary symbiogenesis and multiple losses. *Curr. Biol.* **2002**, *12*, R62–R64. [CrossRef]

46. Zwickl, D.J.; Hillis, D.M. Increased taxon sampling greatly reduces phylogenetic error. *Syst. Biol.* **2002**, *51*, 588–598. [CrossRef] [PubMed]

47. Diekmann, K.; Hodkinson, T.R.; Wolfe, K.H.; van den Bekerom, R.; Dix, P.J.; Barth, S. Complete Chloroplast Genome Sequence of a Major Allogamous Forage Species, Perennial Ryegrass (*Lolium perenne* L.). *DNA Res.* **2009**, *16*, 165–176. [CrossRef] [PubMed]

48. Maier, R.M.; Neckermann, K.; Igloi, G.L.; Kössel, H. Complete Sequence of the Maize Chloroplast Genome: Gene Content, Hotspots of Divergence and Fine Tuning of Genetic Information by Transcript Editing. *J. Mol. Biol.* **1995**, *251*, 614–628. [CrossRef] [PubMed]

49. Nie, X.; Lv, S.; Zhang, Y.; Du, X.; Wang, L.; Biradar, S.S.; Tan, X.; Wan, F.; Weining, S. Complete Chloroplast Genome Sequence of a Major Invasive Species, Crofton Weed (*Ageratina adenophora*). *PLoS ONE* **2012**, *7*, e36869. [CrossRef] [PubMed]

50. Wu, F.-H.; Chan, M.-T.; Liao, D.-C.; Hsu, C.-T.; Lee, Y.-W.; Daniell, H.; Duvall, M.; Lin, C.-S. Complete chloroplast genome of *Oncidium* Gower Ramsey and evaluation of molecular markers for identification and breeding in Oncidiinae. *BMC Plant Biol.* **2010**, *10*, 68. [CrossRef] [PubMed]

51. Blazier, J.C.; Ruhlman, T.A.; Weng, M.-L.; Rehman, S.K.; Sabir, J.S. M.; Jansen, R.K. Divergence of RNA polymerase α subunits in angiosperm plastid genomes is mediated by genomic rearrangement. *Sci. Rep.* **2016**, *6*, 24595. [CrossRef] [PubMed]

52. Nakkaew, A.; Chotigeat, W.; Eksomtramage, T.; Phongdara, A. Cloning and expression of a plastid-encoded subunit, beta-carboxyltransferase gene (*accD*) and a nuclear-encoded subunit, biotin carboxylase of acetyl-CoA carboxylase from oil palm (*Elaeis guineensis* Jacq.). *Plant Sci.* **2008**, *175*, 497–504. [CrossRef]

53. Hajdukiewicz, P.T.J.; Allison, L.A.; Maliga, P. The two RNA polymerases encoded by the nuclear and the plastid compartments transcribe distinct groups of genes in tobacco plastids. *EMBO J* **1997**, *16*, 4041–4048. [CrossRef] [PubMed]

54. Leebens-Mack, J.; Raubeson, L.A.; Cui, L.Y.; Kuehl, J.V.; Fourcade, M.H.; Chumley, T.W.; Boore, J.L.; Jansen, R.K.; dePamphilis, C.W. Identifying the basal angiosperm node in chloroplast genome phylogenies: Sampling one's way out of the Felsenstein zone. *Mol. Biol. Evol.* **2005**, *22*, 1948–1963. [CrossRef] [PubMed]

55. Moore, M.J.; Bell, C.D.; Soltis, P.S.; Soltis, D.E. Using plastid genome-scale data to resolve enigmatic relationships among basal angiosperms. *Proc. Natl. Acad. Sci. USA* **2007**, *104*, 19363–19368. [CrossRef] [PubMed]

56. Parks, M.; Cronn, R.; Liston, A. Increasing phylogenetic resolution at low taxonomic levels using massively parallel sequencing of chloroplast genomes. *BMC Biol.* **2009**, *7*, 84. [CrossRef] [PubMed]

57. Suzuki, Y.; Glazko, G.V.; Nei, M. Overcredibility of molecular phylogenies obtained by Bayesian phylogenetics. *Proc. Natl. Acad. Sci. USA* **2002**, *99*, 16138–16143. [CrossRef] [PubMed]

58. Wortley, A.H.; Rudall, P.J.; Harris, D.J.; Scotland, R.W. How much data are needed to resolve a difficult phylogeny? Case study in Lamiales. *Syst. Biol.* **2005**, *54*, 697–709. [CrossRef] [PubMed]

7

Development of 44 Novel Polymorphic SSR Markers for Determination of Shiitake Mushroom (*Lentinula edodes*) Cultivars

Hwa-Yong Lee [1,2,†], Suyun Moon [2,†], Donghwan Shim [3,†], Chang Pyo Hong [4], Yi Lee [5], Chang-Duck Koo [1,*], Jong-Wook Chung [5,*] and Hojin Ryu [2,*]

[1]   Department of Forest Science, Chungbuk National University, Cheongju 28644, Korea; hoasis82@chungbuk.ac.kr
[2]   Department of Biology, Chungbuk National University, Cheongju 28644, Korea; sooym21@naver.com
[3]   Department of Forest Genetic Resources, National Institute of Forest Science, Suwon 16631, Korea; shim104@korea.kr
[4]   Theragen Etex Bio Institute, Suwon 16229, Korea; changpyo.hong@theragenetex.com
[5]   Department of Industrial Plant Science and Technology, Chungbuk National University, Cheongju 28644, Korea; leeyi22@chungbuk.ac.kr
*   Correspondence: koocdm@chungbuk.ac.kr (C.-D.K.); jwchung73@chungbuk.ac.kr (J.-W.C.); hjryu96@gmail.com (H.R.)
†   These authors contributed equally to this work.

Academic Editor: Paolo Cinelli

Abstract: The shiitake mushroom (*Lentinula edodes*) is one of the most popular edible mushrooms in the world and has attracted attention for its value in medicinal and pharmacological uses. With recent advanced research and techniques, the agricultural cultivation of the shiitake mushroom has been greatly increased, especially in East Asia. Additionally, demand for the development of new cultivars with good agricultural traits has been greatly enhanced, but the development processes are complicated and more challenging than for other edible mushrooms. In this study, we developed 44 novel polymorphic simple sequence repeat (SSR) markers for the determination of shiitake mushroom cultivars based on a whole genome sequencing database of *L. edodes*. These markers were found to be polymorphic and reliable when screened in 23 shiitake mushroom cultivars. For the 44 SSR markers developed in this study, the major allele frequency ranged from 0.13 to 0.94; the number of genotypes and number of alleles were each 2–11; the observed and expected heterozygosity were 0.00–1.00 and 0.10–0.90, respectively; and the polymorphic information content value ranged from 0.10 to 0.89. These new markers can be used for molecular breeding, the determination of cultivars, and other applications.

Keywords: shiitake mushroom; simple sequence repeat (SSR) marker; whole genome sequencing

1. Introduction

The shiitake mushroom (*Lentinula edodes*) belongs to genus Lentinula, family Omphalotaceae, order Agaricales [1], and is a type of white rot fungi. This mushroom is commonly cultivated in Asian countries such as Korea, China, Japan, Taiwan, and others [2,3]. It constitutes approximately 17% of the global mushroom supply and is one of the most popular edible mushrooms in the world [4]. In addition to its value as a food, the shiitake mushroom is useful for pharmacological components such as lentinan, which shows antitumour activity [5,6].

To develop new mushroom cultivars, cross breeding, mutation breeding, transgenic breeding, and other approaches have been used [7]. The cultivar development of the shiitake mushroom is very difficult because the cultivation period is much longer than for the oyster mushroom (*Pleurotus ostreatus*), king oyster mushroom (*P. eryngii*), and winter mushroom (*Flammulina velutipes*) [8]. The traditional breeding for shiitake mushroom requires a lot of time and labor from strain selection for cultivation and identification of traits. Analysis of the genetic relationship between the relevant strains and the association of DNA markers in mushrooms can effectively increase breeding efficiency [9], because molecular markers save time in the selection process of the strain. During the last few decades, studies on the genetic diversity and population genetics in shiitake mushrooms have been conducted using various types of molecular markers, including restriction fragment length polymorphism (RFLP) [10], random amplified polymorphic DNA (RAPD) [11–13], amplified fragment length polymorphism (AFLP) [14], inter-simple sequence repeat (ISSR) [12,15,16], sequence-characterized amplified region (SCAR) [13,17,18], and sequence-related amplified polymorphism (SRAP) [12,16]. In spite of their diverse applications, the use of developed markers for the breeding and classification of shiitake mushrooms has been challengeable due to few available markers and little information regarding effectiveness for determination and specificity. Also, despite the advantages of simple sequence repeat (SSR) markers, such as co-dominant, highly polymorphic, reproducible, reliable, and distributed throughout the genome [19], the number of SSR markers available for Shiitake mushrooms are still scarce, with only a few expressed sequence tag-simple sequence repeat (EST-SSR) markers having been reported [20]. Thus, more reliable molecular markers are needed to enhance genetic analyses of the shiitake mushroom.

The traditional development of SSR markers was an experimentally long, labor-intensive, and economically costly process [21]. However, Next Generation Sequencing (NGS) technology is a powerful tool to find a large number of microsatellite loci through cost-effective and rapid identification [22]. In many recent studies, NGS-based transcriptome or genome sequencing is demonstrated to be efficient for the large-scale discovery of SSR loci in plants [21]. We have recently reported the genome sequence information of *L. edodes* [23], and have here developed higher polymorphic SSR markers based on the whole genome sequencing and determination of shiitake mushroom varieties. New SSR markers might be valuable tools to evaluate genetic variability and breeding in the shiitake mushroom.

2. Materials and Methods

2.1. Fungi Materials

To develop useful SSR markers in *L. edodes*, we selected the representative shiitake mushroom strains, which were successfully cultivated and distributed in the market in South Korea. Five strains from the National Institute of Forest Science in Korea Forest Service (http://www.forest.go.kr) and 18 strains from the Forest Mushroom Research Center (https://www.fmrc.or.kr/) were kindly provided. The list of strains are shown in Table 1. The mycelia of the strains were cultured for 10 days at 25 °C in darkness.

2.2. DNA Preparationument

For DNA extraction, the cultured mycelia were frozen in liquid nitrogen and ground into powders. DNA extraction was performed using a GenEX Plant Kit (Geneall, Seoul, Korea) following the manufacturer's instructions. The extracted DNA was stored at −80 °C.

Table 1. List of shiitake mushroom (*Lentinula edodes*) strains.

No.	Strain Name	Cultivar Name
1	KFRI 623	Baekhwahyang
2	KFRI 174	Soohyangko
3	KFRI 551	Poongnyunko
4	KFRI 2924	Sanmaru 1 h_O
5	KFRI 2925	Sanmaru 2 h_O
6	SJ101	Sanjo 101 h_O
7	SJ102	Sanjo 102 h_O
8	SJ103	Sanjo 103 h_O
9	SJ108	Sanjo 108 h_O
10	SJ109	Sanjo 109 h_O
11	SJ110	Sanjo 110 h_O
12	SJ111	Sanjo 111 h_O
13	SJ301	Sanjo 301 h_O
14	SJ501	Sanjo 501 h_O
15	SJ702	Sanjo 702 h_O
16	SJ704	Sanjo 704 h_O
17	SJ705	Sanjo 705 h_O
18	SJ706	Sanjo 706 h_O
19	SJ707	Sanjo 707 h_O
20	SJ708	Sanjo 708 h_O
21	SJ709	Sanjo 709 h_O
22	SJ710	Sanjo 710 h_O
23	SJCAR	Chamaram

No. 1~5: National Institute of Forest Science in Korea Forest Service; No. 6~23: Forest Mushroom Research Center.

2.3. Discovery of SSR Markers

Over 1000 SSR loci of the shiitake mushroom were found in whole genome sequencing performed by Shim et al. [23] using *L. edodes* monokaryon strain B17 and comparing the resequencing data of 1 strain, Chamaram. We chose 205 SSR loci to test for polymorphism among the shiitake mushroom strains, and 44 SSR markers were finally selected for proper PCR conditions fixed in 23 strains (Supplementary Materials Figure S1).

The primer design parameters were set as follows: length range, 18–23 nucleotides with 21 as the optimum; PCR product size range, 150–200 bp; optimum annealing temperature (Ta), 58 °C; and GC content 50%–61%, with 51% as the optimum.

The extracted DNA for PCR templates was diluted to 20 ng/µL after checking concentration using a K5600 micro spectrophotometer (DaAn Gene, Guangzhou, China). The PCR reaction mixture consisted of 2 µL template DNA, 1 µL each of forward and reverse primer (5 pmol), 10 µL 2× i-Taq Master Mix (Intron biotechnology, Seongnam, Korea), and 6 µL distilled water. PCR reactions were performed as follows: 95 °C for 3 min; 35 cycles of 95 °C for 30 s, 58 °C for 30 s, and 72 °C for 30 s; and finally, 72 °C for 20 min. The size of the PCR product was confirmed by fragment analyzer (Advanced Analytical Technologies, Ankeny, IA, USA).

The amplified SSR loci were scored for 23 shiitake mushroom strains. Major allele frequency (M_{AF}), number of genotypes (N_G), number of alleles (N_A), observed heterozygosity (H_O), expected heterozygosity (H_E), and polymorphic information content (PIC) values were calculated by using PowerMarker V3.25 [24].

3. Results and Discussion

The 44 SSR markers consist of di-, tri-, tetra-, and pentanucleotide DNA motifs. The SSR motifs used include 59.09% dinucleotide repeats, 31.82% trinucleotide repeats, 6.82% tetranucleotide repeats, and 2.27% pentanucleotide repeats. The SSR motifs are AG/GA, CT/TC, AT/TA, AC/CA, CG/GC, and TG dinucleotide repeats; AGG/AGA/GGA, CAG/CGA/GCA, AGA, GAT, GCT, GTT, TCA, and TCG trinucleotide repeats; TACT/TATC, and CTTT tetranucleotide repeats; and CTTCC pentanucleotide repeats (Table 2).

Table 2. Characteristics of the 44 simple sequence repeat (SSR) markers for shiitake mushroom (*Lentinula edodes*). Ta, annealing temperature.

Marker	Primer Sequences (5'-3')	Expected Size	Motif	GenBank Accession No.	Ta (°C)	Description
RL-LE-017	F: GTGCACTGTGCGATTGTTC R: CAGCAAGGATGACTCTTGGA	199	CA	NM-0418-000001	59	Subtilase family, Pro-kumamolisin, activation domain
RL-LE-018	F: CCCACAGGTTTACAGAGTTCCT R: GTGGACATCCACCTTTTGTC	152	TA	NM-0418-000002	59	-
RL-LE-019	F: TACTTTCGAAGCCAGCCA R: GTAGCTCTTTAGGTCTGCTTGG	191	CTTCC	NM-0418-000003	58	-
RL-LE-020	F: GACGGAGTTGTCAAGATCTACC R: ACCTAGGCTTTGCTCTACACAG	173	AT	NM-0418-000004	58	-
RL-LE-021	F: GCTTGAAGAGCGAGTTTGAG R: CAAGACACGCTTCGTAGTCA	200	AG	NM-0418-000005	58	Uso1/p115-like vesicle tethering protein, head region
RL-LE-022	F: CAAACGAAGGAGGAGGTAGTTC R: GAGTCCATTACTCATCGTGCTG	199	GCA	NM-0418-000006	60	-
RL-LE-023	F: GAGGTAGCACCAGTTGAGGTAA R: ATAAGACTTCGTCTCGTCCTGC	150	AGA	NM-0418-000007	59	-
RL-LE-024	F: GTAAGGCTTTAGGACTCGTCG R: CCACAGATGTTCCGAGTTG	187	TC	NM-0418-000008	59	-
RL-LE-025	F: TTGGGAGATGCGGAGTAGTTC R: ATTCAGTCGCTCAGTAGGAGAC	200	AT	NM-0418-000009	58	PCI domain, 26S proteasome subunit RPN7
RL-LE-026	F: GATTTGACGCTCACATCCC R: CCCCTAAGTATGAGCTTCCGTA	197	AG	NM-0418-000010	59	-
RL-LE-027	F: GGGTCACAAGAGCAATGTAGAC R: CTGTATGGTGATCAAGGACGAG	192	CT	NM-0418-000011	59	-
RL-LE-028	F: GAGACGACACGAGGAATTTG R: GTCGTTCTCATTGGAGACTCTG	174	CA	NM-0418-000012	59	Ras family
RL-LE-029	F: CAAGATCCGTCGCCATATAC R: AACTCCACCCTCGTCTACCTCTAC	178	GGA	NM-0418-000013	58	-
RL-LE-030	F: CTTGGGAAGGAGGAATGG R: GTGGGACCAATATGAGGACAGT	164	TACT	NM-0418-000014	59	-
RL-LE-031	F: ACTTCAGTTACAGCGACTCTGC R: GTCCGGAGACTGTGCCGTTC	194	CAG	NM-0418-000015	58	PAS domain, PAS domain
RL-LE-032	F: GTAGAAGGTGCACCAGTTTCTG R: CGTCTCTTACCAGGAATCACAC	190	AGG	NM-0418-000016	59	-

Table 2. *Cont.*

Marker	Primer Sequences (5′–3′)	Expected Size	Motif	GenBank Accession No.	Ta (°C)	Description
RL-LE-033	F: GACAGAAGAAGGACTTACCAGC R: CCAGAGCCCAAGGATAACTT	197	CT	NM-0418-000017	58	-
RL-LE-034	F: AGGTGGAGTTGAGTGTTTGAGG R: AGTCTCAGGAGACCTTCACTAGC	170	TA	NM-0418-000018	59	-
RL-LE-035	F: GTCGGAAGCTTTATGACACG R: TCAACTTTCTGCTCCCTCAC	196	GAG	NM-0418-000019	58	-
RL-LE-036	F: TCTAGCTCGGTGAGCAATGT R: GAGACCTTGAGGAAGAGACTCC	181	CG	NM-0418-000020	59	-
RL-LE-037	F: CTCTCATCCTTAAGAACCTCCC R: GAGAAGCTTACATATGGTCCCG	198	CGA	NM-0418-000021	59	-
RL-LE-038	F: CGTTTGAGTGTCAACGGTCT R: CATGTCAGACTAGTCAGGGGTC	199	AT	NM-0418-000022	59	-
RL-LE-039	F: GTACGAGGACAGCAATACAGC R: GCTTCTATATCTCTCTGCCCT	200	GA	NM-0418-000023	58	-
RL-LE-040	F: GGTTCCTCTCACACCTTACCT R: GAAAATGTGCTGTAGCGGAGC	178	CT	NM-0418-000024	59	-
RL-LE-041	F: GGTGTATAAAGAGAGCCCTTGG R: CCCCTTATCCAGTCTACTGCTAC	153	AG	NM-0418-000025	59	SNF2 family N-terminal domain, Ring finger domain
RL-LE-042	F: TCCTCTGCTTCACTAAGTCTCC R: AGTACTCGCAAGGCAGGTAAG	167	TCG	NM-0418-000026	58	STAG domain
RL-LE-043	F: GTTCGTCACTCGGTACTTTCC R: AGATGCAGGAGTATGACCTGAC	177	AC	NM-0418-000027	58	-
RL-LE-044	F: GTAAGCCTAAGGAGGGTGGAG R: CACCTCCTTCATCTGGTCC	198	GGA	NM-0418-000028	59	WH1 domain, P21-Rho-binding domain
RL-LE-045	F: ACATCTGAGAGGTCGTACGCT R: GTACCGAAGCGAGCAAGTT	164	CA	NM-0418-000029	59	Cytochrome b5-like heme/steroid binding domain, Acyl-CoA dehydrogenase, C-terminal domain
RL-LE-046	F: GCACGCAGTGATGAATAGAGAG R: ACACTTACGGATTTGGCAGG	154	AG	NM-0418-000030	60	Cytochrome P450
RL-LE-047	F: CTACCACTCGTCACTCCTTAGGT R: GAAGGAGTGTGAAGCTGAAACC	194	TC	NM-0418-000031	60	-
RL-LE-048	F: GTGGTGAAGTTACCGACAGG R: AGGTGCCCAACTTCTGGT	197	GC	NM-0418-000032	58	Pectate lyase

Table 2. *Cont.*

Marker	Primer Sequences (5'-3')	Expected Size	Motif	GenBank Accession No.	Ta (°C)	Description
RL-LE-049	F: GCTACCTAGATCCTCCTAGATCG R: GACTACGTCAAGTTGAGGATGC	184	GA	NM-0418-000033	58	-
RL-LE-050	F: TACCCGAAGGAACTAAACGAGTC R: GTCGTCGTATAAACGACTCATCC	200	TG	NM-0418-000034	59	-
RL-LE-051	F: ACTCTGCTGCCACTCTTGAC R: GACCGTCTCTAGCTTCTTGATG	172	CT	NM-0418-000035	58	Short chain dehydrogenase
RL-LE-052	F: CTAAAGCAACGGTAGACGTAGG R: ACAACAAACGCTAGAGCGAG	178	GCT	NM-0418-000036	58	-
RL-LE-053	F: CTCAAGCGTCTCATTCCCTTC R: CTCGAGTTGAGGGTGAGGTTAT	179	GTT	NM-0418-000037	58	-
RL-LE-054	F: GAATCAGCTAGACCATCTCTGC R: TCTTTACCCGTCTTGTCTGC	200	GAT	NM-0418-000038	58	-
RL-LE-055	F: CTGGGGATAGTGATATCGAGAG R: GTAAACCCGCTCCTTTGTGT	165	CTTT	NM-0418-000039	58	-
RL-LE-056	F: GCGGTCCTGAGTACAAAGTAGT R: CTACGTACGGAGGAATCTAGTGC	159	TATC	NM-0418-000040	58	-
RL-LE-057	F: AGGAGAACGGAACCGAAGTTAC R: CAGTAGACGTTGCTTACTGCAC	160	AT	NM-0418-000041	59	Protein of unknown function DUF262
RL-LE-058	F: GTCGTAGAACTTGCACGAGTC R: GAAGTTCTCCGCTATCCTCTC	163	GCA	NM-0418-000042	57	-
RL-LE-059	F: CGGAGATGTACCAATTCCTG R: GCATTCGCCGTCTATACGAT	193	TG	NM-0418-000043	59	-
RL-LE-060	F: ACTCAGCGCCACATCTAGCTT R: CAGGGAGAAGAAAGTCACGA	191	TCA	NM-0418-000044	58	-

These 44 SSR markers were analyzed in 23 cultivars of shiitake mushroom. The major allele frequency (M_{AF}) ranged from 0.13 to 0.94, with an average of 0.575. The number of genotypes (N_G), ranged from 2 to 11, with an average of 5.5, and the number of alleles (N_A) ranged from 2 to 11 with an average of 4.9 alleles. The observed heterozygosity (H_O) ranged from 0.00 to 1.00, with an average of 0.309, and the expected heterozygosity (H_E), ranged from 0.10 to 0.90, with an average of 0.552. The polymorphic information content (PIC) value ranged from 0.10 to 0.89, with an average of 0.511. M_{AF}, N_G, N_A, H_O, H_E, and PIC per locus had wide ranges among the markers (Table 3).

Table 3. Diversity statistics from each primer used for screening 23 cultivars of shiitake mushroom (*Lentinula edodes*).

Marker	M_{AF}	N_G	N_A	H_O	H_E	PIC
RL-LE-017	0.13	11	11	0.00	0.9	0.89
RL-LE-018	0.29	6	6	0.08	0.77	0.74
RL-LE-019	0.41	7	5	0.87	0.68	0.63
RL-LE-020	0.67	3	3	0.00	0.5	0.45
RL-LE-021	0.52	11	10	0.43	0.68	0.66
RL-LE-022	0.70	5	4	0.04	0.48	0.45
RL-LE-023	0.63	6	5	0.09	0.55	0.51
RL-LE-024	0.80	4	4	0.13	0.33	0.31
RL-LE-025	0.43	7	6	0.64	0.72	0.68
RL-LE-026	0.34	9	6	0.68	0.74	0.69
RL-LE-027	0.50	5	5	0.26	0.64	0.58
RL-LE-028	0.37	6	4	0.35	0.73	0.68
RL-LE-029	0.81	3	3	0.00	0.32	0.29
RL-LE-030	0.65	4	3	0.26	0.51	0.46
RL-LE-031	0.46	4	6	1.00	0.68	0.63
RL-LE-032	0.87	3	4	0.22	0.24	0.22
RL-LE-033	0.46	10	9	0.48	0.72	0.69
RL-LE-034	0.50	6	6	0.22	0.68	0.64
RL-LE-035	0.65	5	5	0.17	0.52	0.47
RL-LE-036	0.72	3	3	0.04	0.42	0.35
RL-LE-037	0.50	6	4	0.39	0.62	0.55
RL-LE-038	0.39	7	9	1.00	0.72	0.68
RL-LE-039	0.50	3	3	0.13	0.56	0.46
RL-LE-040	0.52	6	5	0.52	0.66	0.61
RL-LE-041	0.43	4	5	0.09	0.65	0.58
RL-LE-042	0.80	4	3	0.04	0.33	0.3
RL-LE-043	0.87	2	2	0.00	0.23	0.2
RL-LE-044	0.76	3	3	0.42	0.37	0.32
RL-LE-045	0.39	6	5	0.35	0.73	0.69
RL-LE-046	0.43	7	6	0.09	0.71	0.66
RL-LE-047	0.48	10	6	0.35	0.69	0.65
RL-LE-048	0.43	8	5	0.39	0.7	0.66
RL-LE-049	0.41	4	5	1.00	0.66	0.6
RL-LE-050	0.94	2	2	0.00	0.1	0.1
RL-LE-051	0.28	9	6	0.65	0.79	0.76
RL-LE-052	0.70	5	5	0.09	0.47	0.43
RL-LE-053	0.39	10	10	0.61	0.79	0.77
RL-LE-054	0.76	5	3	0.09	0.39	0.36
RL-LE-055	0.72	4	4	0.13	0.44	0.38
RL-LE-056	0.59	5	4	0.68	0.58	0.53
RL-LE-057	0.87	3	3	0.09	0.23	0.22
RL-LE-058	0.93	3	3	0.04	0.12	0.12
RL-LE-059	0.35	6	5	0.30	0.74	0.69
RL-LE-060	0.91	2	2	0.17	0.16	0.15
Mean	0.575	5.5	4.9	0.309	0.552	0.511

(M_{AF}), Major allele frequency; (N_G), number of genotypes; (N_A), number of alleles; (H_O), observed heterozygosity; (H_E), expected heterozygosity; and (PIC), polymorphic information content.

SSR marker information for the determination of cultivars in other cultivated mushrooms have been released. In the SSR markers developed for the determination of black wood ear (*Auricularia auricular-judae*) cultivars, the PIC value ranged from 0.10 to 0.84, with an average of 0.47, and N_A ranged from 2–11, with an average of 4.7 (using 17 SSR markers and 16 cultivars) [25]. In the white button mushroom (*Agaricus bisporus*), the allele frequency ranged from 0.02–0.94, with an average of 0.18, and H_O ranged from 0.00 to 0.83, with an average of 0.35 (using 33 SSR markers, 6 cultivars and 17 wild types of *A. bisporus*, and 2 wild types of *A. bisporus* var. *burnettii*) [26]. In the golden needle mushroom (*F. velutipes*), the PIC value ranged from 0.13 to 0.69, with an average of 0.42 (using 55 SSR markers and 14 cultivars) [27]. In the oyster mushroom (*P. ostreatus*), the average of N_A was approximately 4.7. H_O ranged from 0.027 to 0.946, with an average of 0.398, and H_E ranged from 0.027 to 0.810, with an average of 0.549 (using 36 SSR markers, and 37 cultivars including *P. sajor-caju*, 1 *P. eryngii*, 1 *P. cornucopiae* var. *citrinopileatus*, and 1 *P. nebrodensis*) [28]. The SSR markers developed in this study showed similar diversity values with the SSR markers of other edible mushrooms. The 20 markers with PIC values above 0.6 in the 44 newly developed markers are useful for identification among strains or cultivars of shiitake mushroom. The SSR markers developed in this study were able to distinguish 23 shiitake mushroom strains, which are broadly cultivated in Korea (Supplementary Materials Figure S2).

4. Conclusions

We have developed 44 novel SSR markers for the determination of shiitake mushroom cultivars. SSR marker development was performed based on NGS-based genome sequencing data. The efficacy and availability of the developed SSR markers were evaluated by application to distinguishing 23 shiitake mushroom cultivars. These new markers can be used for molecular breeding, cultivar determination, genetic structure research, and further applications in cultivated and wild types of shiitake mushrooms.

Supplementary Materials: The following are available online at www.mdpi.com/2073-4425/8/4/109/s1, Figure S1: The development process of SSR marker developed using genome sequencing for *Lentinula edodes*, Figure S2: Distinguished *Lentinula edodes* strains using the 44 novel SSR markers developed in this study.

Acknowledgments: This study was performed with the support of the "Golden Seed Project (Center for Horticultural Seed Development, No. 213007-05-1-SBH20)".

Author Contributions: Hwa-Yong Lee, Suyun Moon, and Donghwan Shim contributed equally to this work, analyzed the data, and wrote the paper; Chang Pyo Hong and Yi Lee performed the experiments; and Chang-Duck Koo, Jong-Wook Chung, and Hojin Ryu designed the experiments. All authors read and approved the manuscript.

References

1. International Mycological Association. Mycobank Database Fungal Databases, Nomenclature and Species Banks. Available online: http://www.mycobank.org/ (accessed on 29 September 2016).
2. Bak, W.C.; Park, Y.A.; Park, J.H. *KFRI Forest Policy Issue: Present Situation and Future of Oak Mushroom Industry*; Korea Forest Research Institute: Seoul, Korea, 2013; p. 17.
3. Kim, K.H.; Ka, K.H.; Kang, J.H.; Kim, S.; Lee, J.W.; Jeon, B.K.; Yun, J.K.; Park, S.R.; Lee, H.J. Identification of Single Nucleotide Polymorphism Markers in the Laccase Gene of Shiitake Mushrooms (*Lentinula edodes*). *Mycobiology* **2015**, *43*, 75–80. [CrossRef] [PubMed]
4. Royse, D.J. A global perspective on the high five: *Agaricus, Pleurotus, Lentinula, Auricularia & Flammulina*. In Proceedings of the 8th International Conference on Mushroom Biology and Mushroom Products, New Delhi, India, 19–22 November 2014.
5. Chihara, G.; Hamuro, J.; Maeda, Y.; Arai, Y.; Fukuoka, F. Fractionation and Purification of the Polysaccharides with Marked Antitumor Activity, Especially Lentinan, from *Lentinus edodes* (Berk.) Sing, (an Edible Mushroom). *Cancer Res.* **1970**, *30*, 2776–2781. [PubMed]

6.   Bisen, P.S.; Baghel, R.K.; Sanodiya, B.S.; Thakur, G.S.; Prasad, G.B. *Lentinus edodes*: A macrofungus with pharmacological activities. *Curr. Med. Chem.* **2010**, *17*, 2419–2430. [CrossRef] [PubMed]

7.   Chakravarty, B. Trends in mushroom cultivation and breeding. *Aust. J. Agric. Eng.* **2011**, *2*, 102–109.

8.   Forest Mushroom Research Center. *Cultivation Technique of Oak Mushroom*; Forest Mushroom Research Center: Yeoju, Korea, 2015; pp. 58–60.

9.   Sonnenberg, A.S.M.; Johan, J.P.B.; Patrick, M.H.; Brian, L.; Wei, G.; Amrah, W.; Jurriaan, J.M. Breeding and strain protection in the button mushroom *Agaricus bisporus*. In Proceedings of the 7th International Conference on Mushroom Biology and Mushroom Products, Arcachon, France, 4–7 October 2011; pp. 7–15.

10.   Kulkarni, R.K. DNA Polymorphisms in *Lentinula edodes*, the Shiitake Mushroom. *App. Environ. Microbiol.* **1991**, *57*, 1735–1739.

11.   Zhang, Y.; Molina, F.I. Strain typing of *Lentinula edodes* by random amplified polymorphic DNA assay. *FEMS Microbiol. Lett.* **1995**, *131*, 17–20. [CrossRef] [PubMed]

12.   Fu, L.Z.; Zhang, H.Y.; Wu, X.Q.; Li, H.B.; Wei, H.L.; Wu, Q.Q.; Wang, L.A. Evaluation of genetic diversity in *Lentinula edodes* strains using RAPD, ISSR and SRAP markers. *World J. Microbl. Biotech.* **2010**, *26*, 709–716. [CrossRef]

13.   Wu, X.; Li, H.; Zhao, W.; Fu, L.; Peng, H.; He, L.; Cheng, J. SCAR makers and multiplex PCR-based rapid molecular typing of *Lentinula edodes* strains. *Curr. Microbiol.* **2010**, *61*, 381–389. [CrossRef] [PubMed]

14.   Terashima, K.; Matsumoto, T. Strain typing of shiitake (*Lentinula edodes*) cultivars by AFLP analysis, focusing on a heat-dried fruiting body. *Mycoscience* **2004**, *45*, 79–82. [CrossRef]

15.   Zhang, R.; Huang, C.; Zheng, S.; Zhang, J.; Ng, T.B.; Jiang, R.; Zuo, X.; Wang, H. Strain-typing of *Lentinula edodes* in China with inter simple sequence repeat markers. *Appl. Microbiol. Biotechnol.* **2007**, *74*, 140–145. [CrossRef] [PubMed]

16.   Liu, J.; Wang, Z.R.; Li, C.; Bian, Y.B.; Xiao, Y. Evaluating genetic diversity and constructing core collections of Chinese *Lentinula edodes* cultivars using ISSR and SRAP markers. *J. Basic Microbiol.* **2015**, *55*, 749–760. [CrossRef] [PubMed]

17.   Li, H.B.; Wu, X.Q.; Peng, H.Z.; Fu, L.Z.; Wei, H.L.; Wu, Q.Q.; Jin, Q.Y.; Li, N. New available SCAR markers: Potentially useful in distinguishing a commercial strain of the superior type from other strains of *Lentinula edodes* in China. *Appl. Microbiol. Biotechnol.* **2008**, *81*, 303–309. [CrossRef] [PubMed]

18.   Liu, J.Y.; Ying, Z.H.; Liu, F.; Liu, X.R.; Xie, B.G. Evaluation of the use of SCAR markers for screening genetic diversity of *Lentinula edodes* strains. *Curr. Microbiol.* **2012**, *64*, 317–325. [CrossRef] [PubMed]

19.   Miah, G.; Rafii, M.Y.; Ismail, M.R.; Puteh, A.B.; Rahim, H.A.; Islam, Kh.N.; Latif, M.A. A review of microsatellite markers and their applications in rice breeding programs to improve blast disease resistance. *Int. J. Mol. Sci.* **2013**, *14*, 22499–22528. [CrossRef] [PubMed]

20.   Xiao, Y.; Liu, W.; Dai, Y.; Fu, C.; Bian, Y. Using SSR markers to evaluate the genetic diversity of *Lentinula edodes'* natural germplasm in China. *World J. Microbiol. Biotechnol.* **2010**, *26*, 527–536. [CrossRef]

21.   Zalapa, J.E.; Cuevas, H.; Zhu, H.; Steffan, S.; Senalik, D.; Zeldin, E.; McCown, B.; Harbut, R.; Simon, P. Using next-generation sequencing approaches to isolate simple sequence repeat (SSR) loci in the plant sciences. *Am. J. Bot.* **2012**, *99*, 193–208. [CrossRef] [PubMed]

22.   Ekblom, R.; Galindo, J. Applications of next generation sequencing in molecular ecology of non-model organisms. *Heredity* **2011**, *107*, 1–15. [CrossRef] [PubMed]

23.   Shim, D.; Park, S.G.; Kim, K.; Bae, W.; Lee, G.W.; Ha, B.S.; Ro, H.S.; Kim, M.; Ryoo, R.; Rhee, S.K.; et al. Whole genome de novo sequencing and genome annotation of the world popular cultivated edible mushroom, *Lentinula edodes*. *J. Biotechnol.* **2016**, *223*, 24–25. [CrossRef] [PubMed]

24.   Liu, K.; Muse, S.V. PowerMarker: A intergrated analysis environment for genetic marker analysis. *Bioinformatics* **2005**, *21*, 2128–2129. [CrossRef] [PubMed]

25.   Zhang, R.Y.; Hu, D.D.; Gu, J.G.; Hi, Q.X.; Zuo, X.M.; Wnag, H.X. Development of SSR markers for typing cultivars in the mushroom *Auricularia auricula-judae*. *Mycol. Prog.* **2012**, *11*, 578–592. [CrossRef]

26.   Foulongne-Oriol, M.; Spataro, M.; Savoie, J.M. Novel microsatellite markers suitable for genetic studies in the white button mushroom *Agaricus bisporus*. *Appl. Microbiol. Biotechnol.* **2009**, *84*, 1125–1135. [CrossRef] [PubMed]

27. Zhang, R.; Hu, D.; Zhang, J.; Zuo, X.; Jiang, R.; Wang, H.; Ng, T.B. Development and characterization of simple sequence repeat (SSR) markers for the mushroom *Flammulina velutipes*. *J. Biosci. Bioeng.* **2010**, *110*, 273–275. [CrossRef] [PubMed]

28. Ma, K.H.; Lee, G.A.; Lee, S.Y.; Gwag, J.G.; Kim, T.S.; Kong, W.S.; Seo, K.I.; Lee, G.S.; Park, Y.J. Development and characterization of new microsatellite markers for the oyster mushroom (*Pleurotus ostreatus*). *J. Microbiol. Biotechnol.* **2009**, *19*, 851–857. [CrossRef] [PubMed]

8

Genome-Wide Analysis of the *Sucrose Synthase* Gene Family in Grape (*Vitis vinifera*): Structure, Evolution, and Expression Profiles

Xudong Zhu, Mengqi Wang, Xiaopeng Li, Songtao Jiu, Chen Wang and Jinggui Fang *

Nanjing Agricultural University, Weigang 1 hao, 210095 Nanjing, China; 2014204002@njau.edu.cn (X.Z.); 2014104015@njau.edu.cn (M.W.); 2014204003@njau.edu.cn (X.L.); 2015204003@njau.edu.cn (S.J.); wangchen@njau.edu.cn (C.W.)
* Correspondence: fanggg@njau.edu.cn

Academic Editor: Charles Bell

Abstract: Sucrose synthase (SS) is widely considered as the key enzyme involved in the plant sugar metabolism that is critical to plant growth and development, especially quality of the fruit. The members of *SS* gene family have been identified and characterized in multiple plant genomes. However, detailed information about this gene family is lacking in grapevine (*Vitis vinifera* L.). In this study, we performed a systematic analysis of the grape (*V. vinifera*) genome and reported that there are five *SS* genes (*VvSS1–5*) in the grape genome. Comparison of the structures of grape *SS* genes showed high structural conservation of grape *SS* genes, resulting from the selection pressures during the evolutionary process. The segmental duplication of grape *SS* genes contributed to this gene family expansion. The syntenic analyses between grape and soybean (*Glycine max*) demonstrated that these genes located in corresponding syntenic blocks arose before the divergence of grape and soybean. Phylogenetic analysis revealed distinct evolutionary paths for the grape *SS* genes. *VvSS1/VvSS5*, *VvSS2/VvSS3* and *VvSS4* originated from three ancient *SS* genes, which were generated by duplication events before the split of monocots and eudicots. Bioinformatics analysis of publicly available microarray data, which was validated by quantitative real-time reverse transcription PCR (qRT-PCR), revealed distinct temporal and spatial expression patterns of *VvSS* genes in various tissues, organs and developmental stages, as well as in response to biotic and abiotic stresses. Taken together, our results will be beneficial for further investigations into the functions of *SS* gene in the processes of grape resistance to environmental stresses.

Keywords: grapevine; sucrose synthase; syntenic analysis; phylogenetic tree; expression profile

1. Introduction

Sucrose is an essential element of life cycle in higher plants. It is mainly produced by photosynthesis in source tissues and is exported to sink tissues where it serves as a carbon source of energy for various metabolic pathways [1,2]. In addition, when suffering from environmental stress such as cold, drought and salt stress, the accumulation of sucrose protects the stability of both membranes and proteins in plant cells. Further, sucrose is reported to supply energy to increase metabolism when plants recover from these stresses [3,4]. In recent years, it has also been recognized that sucrose can act as important signal in plants to modulate a wide range of processes through regulating the expression level of genes encoding enzymes, storage and transporter proteins [5–8]. Moreover, sucrose also participates in several fundamental processes, such as cell division [9], flowering induction [10], vascular tissue differentiation [11], seed germination [12], and the accumulation of storage products [13]. Thus, the study of the sucrose metabolism is pivotal in understanding various sides of plant physiology.

Sucrose synthase (SS) is one of the key enzymes that regulate the sucrose metabolism in plant. SS catalyzes the reversible reaction of sucrose and uridine diphosphate (UDP) into uridine diphosphate glucose (UDP-glucose) and fructose [14,15]. SS enzymes have pivotal roles in a large range of plant metabolic processes, such as sucrose distribution between plant source and sink tissues [16–19], starch biosynthesis, cellulose synthesis in secondary cell wall [20], nitrogen fixation [21] and response to abiotic stresses [14,18,22].

The identification and characterization of the sucrose synthase genes is the first step towards detecting their specific roles involved in different types of metabolic pathways. With the whole-genome sequencing data of many plants being released in such a short period, an increasing number of *SS* gene families have been identified subsequently and characterized. The number of members of the small multigene family differs among the plant species examined. For example, three *SS* genes were identified in the maize and pea (*Pisum sativum*) genomes [23,24]; there exist six distinct *SS* gene members in Arabidopsis, rice, *Hevea brasiliensis* and *Lotus japonicus* [22,25–27]; and diploid cotton (*Gossypium arboreum* and *Gossypium Raimondii*) genomes each contain eight *SS* genes [28]. Both of the tetraploid cotton (*G. Hirsutum*) [28] and poplar (*Populus trichocarpa*) [29] genome contain fifteen *SS* members, which belonged to one of the largest family of *SS* genes identified to date [30]. In all cases, the spatio-temporal expression patterns of the different *SS* genes across tissue types within each plant species implied that each member of *SS* gene families performed specific physiological functions in a given tissue/organ. For instance, the expression profiling of Arabidopsis *SS* genes reveals partially overlapping but distinct patterns. *AtSS2* is expressed highly only in a critical period of the seeds; in contrast, *AtSS1* and *AtSS5* are much more widely expressed, in the root, stem, flower, silliques and seed. *AtSS6* is expressed mainly in the root, flower and seed [31]. *PsSS1*, *PsSS2*, and *PsSS3* genes are expressed predominately in seed, leaf, and flower of the pea, respectively [23]. Tissue/developmental-specific expression patterns of *SS* genes have also been demonstrated in many other plant species, such as *L. japonicus*, rice, and citrus [26,27,32,33]. In conclusion, extensive studies about the *SS* gene family have been carried out in various plant species such as Arabidopsis, rice, citrus, and poplar, however, few studies about the *SS* genes in grape have been done till date.

In most fruit crops, fruit quality is determined by the contents of sugars (such as sucrose), and sucrose synthase activity can significantly influence the sucrose accumulation. Thus, the study of sucrose synthase is key to the issue of fruit quality improvement. Grapevine (*V. vinifera*) is one of the most important fruit crops in the world [34] and annually production exceeding 78.6 million tons in 2014 (FAOSTAT) [34]. The quality improvement of grape berry is a very important issue. However, prior work focused on the biological functions of SS isozymes, and there is little information on the comprehensive analysis of *SS* gene family in grapevine. Shangguan et al. [35] have identified five putative grape *SS* gene, however to date the characterization of structure, expression patterns and evolution history in the grape *SS* gene family remains elusive. To fully understand the molecular biology, evolution and possible functions of the grape *SS* gene family, we must characterize the member and their evolutionary relationships.

In the present study, we employed bioinformatics methods to analyze the characters of grape *SS* genes on a genome-wide scale in *V. vinifera* based on several publicly available data. Furthermore, we investigated their expression profiles of the grape *SS* genes at different tissues and in response to various stresses. These comprehensive results will provide an insight that will assist in better understanding the potential functions of SS enzymes in sucrose transport or sugar accumulation of grape plants in further studies.

2. Methods

2.1. Mining of Grape Sucrose Synthase Genes

To verify a complete list of grapevine *Sucrose Synthase* (*SS*) genes, we downloaded the annotated grapevine proteins from three public databases: the National Centre for Biotechnology Information (NCBI. Available online: http://www.ncbi.nlm.nih.gov/), the Genoscope (Grapevine Genome Browser.

Available online: http://www.genoscope.cns.fr/externe/GenomeBrowser/Vitis/) and the Grape Genome Database (CRIBI. Available online: http://genomes.cribi.unipd.it/grape/, V2.1). Then, the Hidden Markov Model (HMM) profiles of the core sucrose synthase domain (PF00862) and Glycosyl transferases domain (PF00534) from Pfam database (Pfam. Available online: http://pfam.xfam.org/, 30.0) was downloaded and used to survey all grapevine proteins in the 12X coverage assembly of the *V. vinifera* PN40024 genome.

All putative *SS* genes were manually verified with the InterProScan program (InterProScan. Available online: http://www.ebi.ac.uk/Tools/pfa/iprscan5/) and the Conserved Domains Database (CDD. Available online: http://www.ncbi.nlm.nih.gov/cdd) to confirm their completeness and existence of the core domains. Length of sequences, molecular weights and isoelectric points of deduced polypeptides were calculated by using tools provided at the ExPasy website (ProtParam. Available online: http://web.expasy.org/protparam/). Finally, manual annotation was performed to resolve any discrepancy between incorrectly predicted genes and the actual chromosomal locations of involved genes in question. In addition, we also use "sucrose synthase" as a query keyword to search the grapevine gene at NCBI Gene database (NCBI. Available online: https://www.ncbi.nlm.nih.gov/gene/). As *sucrose phosphate synthase* genes also possess these two domains except for sucrose-phosphatase domain (PF05116) in N-terminal, the candidate sequences of grape *SS* genes obtained from these three genome databases were submitted to InterPro and CDD databases to confirm the existence of these two domains (PF00862 and PF00534) and nonexistence of the sucrose-phosphatase domain. Those protein sequences lacking the sucrose synthase domain and Glycosyl transferases domain were removed and the longest variant with alternative splice variants were selected for further analysis.

The *SS* sequences of plant species, such as Arabidopsis, rice, maize, and poplar, were collected by searching the NCBI database (NCBI. Available online: https://www.ncbi.nlm.nih.gov), the Arabidopsis Information Resource (TAIR. Available online: http://www.arabidopsis.org/, Phytozome (Phytozome. Available online: https://phytozome.jgi.doe.gov/pz/portal.html, V11), rice genome database in the rice genome annotation project (TIGR. Available online: http://rice.tigr.org), and maize genome database (MaizeGDB. Available online: http://www.maizegdb.org/), using "sucrose synthase" as a query keyword. Other plant species such as *Solanum tuberosum* [14], *Lycopersicon esculentum* [30], *Triticum aestivum* [30], *Hordeum vulgare* [30], derived from some related articles.

2.2. Chromosomal Localization and Syntenic Analysis

Grape *SS* genes were mapped to chromosomes by identifying their chromosomal locations based on information available at the Grape Genome Database (CRIBI. Available online: http://genomes.cribi.unipd.it/grape/, V2.1). The segmental and tandem duplication regions, as well as chromosomal location, were established using PLAZA (PLAZA. Available online: http://bioinformatics.psb.ugent.be/plaza/versions/plaza_v3_dicots/, v3.0 Dicots). For syntenic analysis, synteny blocks within the grape genome and between grape and soybean genomes were downloaded from the Plant Genome Duplication Database and visualized using Circos (Circos. Available online: http://circos.ca/).

2.3. Gene Structure and Phylogenetic Analysis

The intron-exon organization analysis was carried out using Gene Structure Display Server (GSDS. Available online: http://gsds.cbi.pku.edu.cn/, 2.0) by alignment of the complementary DNA (cDNA) sequences with their corresponding genomic DNA sequences. Multiple alignments of the identified grape SS amino acid sequences were performed using ClustalX [36]. The phylogenetic tree was constructed with MEGA5.1 using the Maximum Likelihood and the bootstrap test carried out with 1000 replicates [37].

2.4. Ratio of Nonsynonymous to Synonymous Substitutions (dN/dS) and Relative Evolutionary Rate Test

The coding sequences of the grape *SS* genes were aligned following the amino acid alignment by Codon Alignment (Codon Alignment. Available online: https://www.hiv.lanl.gov/content/sequence/

CodonAlign/codonalign.html, v2.1.0). The numbers of nonsynonymous nucleotide substitutions per nonsynonymous site (dN) and the numbers of synonymous nucleotide substitutions per synonymous site (dS) were estimated with the yn00 program of PAML4 [38]. Tajima relative rate tests [39] were performed with amino acid sequences for the two grape *SS* duplicate pairs using MEGA5.1 [37].

2.5. Microarray Data Analysis

Microarray gene expression profiles of the different organs of grapevine at various developmental stages were downloaded from Gene Expression Omnibus (GEO, available online: https://www.ncbi.nlm.nih.gov/geo/) under the series entry GSE36128. The expression data obtained from the *V. vinifera* cv. "Corvina" (clone 48) expression atlas were normalized based on the mean expression value of each gene in all tissues/organs analyzed.

Expression analyses in response to abiotic and biotic stresses were based on microarray data (series matrix accession numbers GSE31594, GSE31677, GSE6404, GSE12842 and GSE31660) downloaded from the NCBI GEO datasets (GEO. Available online: https://www.ncbi.nlm.nih.gov/geo/). All the expression data were calculated as log2 fold change in treated vs. untreated samples.

The heat maps for genes expression level were created by HemI (The Cuckoo Workgroup, Wuhan, China).

2.6. Plant Material and Treatments

Two-year-old *V. vinifera* cv. Chardonnay seedlings were planted in plastic pots containing peat, vermiculite, and perlite (3:1:1, *v/v*) were grown under the following conditions: 28 °C day 12 h/18 °C night 12 h, and a relative humidity ranging from 70% to 85%, at the greenhouse of the Nanjing Agriculture University (Nanjing, China). When shoots of vines were 40–50 cm in length, the third and fourth fully expanded young grapevine leaves beneath the apex were selected for treatments and control. At each time point of each treatment, six leaves from six separate plants were combined to form one sample, and all the experiments were performed in triplicate.

For salt stress treatments, two-year-old soil-grown plants were irrigated with 100 mM NaCl. Treated leaves were collected at 0, 48, and 96 h post-treatment. Plants were sprayed with sterile water for control. For cold treatment, plants were first grown at 23 °C, and then transferred to 4 °C, and treated leaves were collected at 0, 24, and 48 h post-treatment. For high temperature treatment, plants were first grown at 23 °C, and then transferred to 42 °C, and treated leaves were collected at 0, 24, and 48 h post-treatment. Plants were grown at 23 °C for control. For drought treatment, the plants in plastic pots were first kept well-watered and then water was withheld to impose a water stress. Leaf samples were harvested at 0, 10 and 20 days after initiation of the treatments. Control plants were maintained in well-watered conditions. For dark treatment, the plants were first grown at the greenhouse, and then transferred to dark environment, and treated leaves were collected at 0, 48, and 96 h post-treatment. Plants were grown in the normal conditions for control.

These various tissues/organs such as leaves, flowers and berries of grapevine at different developmental stages were derived from six-year old *V. vinifera* cv. Chardonnay grapevine grown under normal cultivation conditions under rain shelter cultivation in Jiangpu Fruit Experimental Farms of Nanjing Agricultural University, Nanjing, China.

Tissue samples were immediately frozen in liquid nitrogen and stored at −70 °C until they were subjected to RNA extraction.

2.7. RNA Extraction, cDNA Synthesis and Quantitative Real-Time PCR (qRT-PCR)

Total RNA was extracted from each organ/tissue by using EZNA Plant RNA Kit (R6827-01, Omega Biotek, Norcross, USA) according to the manufacturer's instructions. First-strand cDNA was synthesized from 1.0 μg total RNA by using M-MLV reverse transcriptase (Fermentas, Canada) according to the manufacturer's instructions. Quantitative PCR (qPCR) was carried out using TransStart Tip Green qPCR SuperMix (TransGen Biotech, Beijing, China) on an IQ5 real time PCR machine (Bio-Rad, Hercules, CA, USA) according to the manufacturer's instructions. The PCR was conducted by following procedure: predenaturation at 94 °C for 30 s, followed by 40 cycles of denaturation at 94 °C for 5 s, primer annealing at 60 °C for 15 s, and extension at 72 °C for 10 s. Optical data were acquired after the extension step, and the PCR reactions were subjected to a melting curve analysis beginning from 65 °C to 95 °C at 0.1 °C s^{-1}. The grape *UBI* gene (Gene code LOC100259511) was used as an internal control. All reactions were performed in triplicate in each experiment and three biological repeats were conducted. Primers used for qRT-PCR are listed in Supplementary Table S3. Each relative expression level was analyzed with IQ5 software using the Normalized Expression method ($2^{-\triangle\triangle CT}$ and $2^{-\triangle CT}$ method). Expressional data consist of three replicated treatments and controls, which were calculated as 2-log-based values and were divided by the control.

3. Results

3.1. Characteristics of the Grapevine Sucrose Synthase Gene Family

We finally identified five non-redundant *sucrose synthase* genes in the three grapevine genome database, the five *SS* genes (*VIT_204s0079g00230*, *VIT_205s0077g01930*, *VIT_207s0005g00750*, *VIT_211s0016g00470* and *VIT_217s0053g00700*) were found in the V2.1 grape genome database hosted at CRIBI, and the corresponding five *SS* genes (LOC100266759, LOC100243135, LOC100267606, LOC100249279 and LOC100252799) were found in the NCBI. Similarly, five *SS* genes (*GSVIVT01035210001*, *GSVIVT01035106001*, *GSVIVT01028043001*, *GSVIVT01015018001* and *GSVIVT01029388001*) were found in the 12X grapevine genome database. The results of current analysis were consistent with previous studies [35], and no more members were identified. However, some minor difference in the properties of annotated *sucrose synthase* genes, such as the length of DNA and amino acid sequences, were noted between CRIBI V2.1 genome database and two other genome databases. Therefore, we selected the latest V2.1 grape genome database hosted at CRIBI released in 16 April 2015. Currently this V2.1 grapevine genome database is the latest one with 2258 new coding genes and 3336 putative long non-coding RNAs. Moreover, several gene models were improved and alternative splicing was described for about 30% of the genes. Based on these five genes' distributions and relative linear orders on the respective chromosomes, we named them *VvSS1* to *VvSS5* according to their chromosomal locations. Characteristics of the five *SS* genes are shown in Table 1 and included the deduced protein length, the molecular weight, the isoelectric point, the aliphatic index and the grand average of hydropathicity. The size of these five *SS* genes varied from 4229 to 9214 bp, but the coding DNA sequence (CDS) sizes were all quite similar, around 2436 to 2610 bp. Molecular analysis of the full-length deduced polypeptides indicated that the putative proteins of these grape *SS* genes contain 811 to 906 amino acids (predicted 92.4 to 102.7 kDa in molecular weight) with their isoelectric point calculated ranging from 5.73 to 8.28. This range of variability implies that different *VvSS* proteins might operate in different microenvironments.

As shown in Figure 1, five *SS* genes in grape were located on five chromosomes, and each chromosome contained only one gene. In addition, we also found *VvSS2* and *VvSS3* were located in duplicate block by the search in PLAZA v3.0 Dicots database (Figure 1). Sequence comparison revealed that the genes share a high sequence homology at the nucleotide level (54.85% to 77.55% identity) within the coding region and at the amino acid level (47.70% to 79.12% identity) (Table 2). Two unique *VvSS* pairs (*VvSS1–VvSS5* and *VvSS2–VvSS3*) were found, because a higher sequence identity of nucleotide and amino acids was observed between *VvSS2* and *VvSS3*, *VvSS1* and *VvSS5*. *VvSS4* gene was very different from the other five *VvSS* genes.

Table 1. Characteristics of Grape *Sucrose Synthase* (*SS*) genes.

Name	Locus Id [a]	Genomic DNA Size (bp)	CDS Size (bp)	Number of Amino Acids	Predicted Mw (kDa)	Theoretical pI	Ai	GRAVY	Chromosome Location	Group	Functional Domains (Start-End, bp)	
											Sucrose Synthase	Glycosyl Transferase
*Vv*SS1	VIT_204s0079g00230	4513	2457	840	95.4	8.28	83.49	−0.339	chr4:10519070-10523582	II	9-557	569-745
*Vv*SS2	VIT_205s0077g01930	7914	2508	835	95.6	5.85	91.99	−0.244	chr5:1507610-1515524	I	8-553	566-765
*Vv*SS3	VIT_207s0005g00750	9214	2436	811	92.4	5.73	90.49	−0.254	chr7:3380739-3389952	I	8-555	568-741
*Vv*SS4	VIT_211s0016g00470	4229	2448	815	93.6	6.13	92.20	−0.247	chr11:489955-494383	I	9-553	565-740
*Vv*SS5	VIT_217s0053g00700	4820	2610	906	102.7	7.55	81.19	−0.359	chr17:15994677-15999496	II	12-558	568-736

[a] IDs are available in the CRIBIV2.1 database. Abbreviations: Ai, aliphatic index; GRAVY, grand average of hydropathicity; MW, molecular weight; pI, isoelectric point; CDS, Coding DNA sequence; Chr, chromosome numbers.

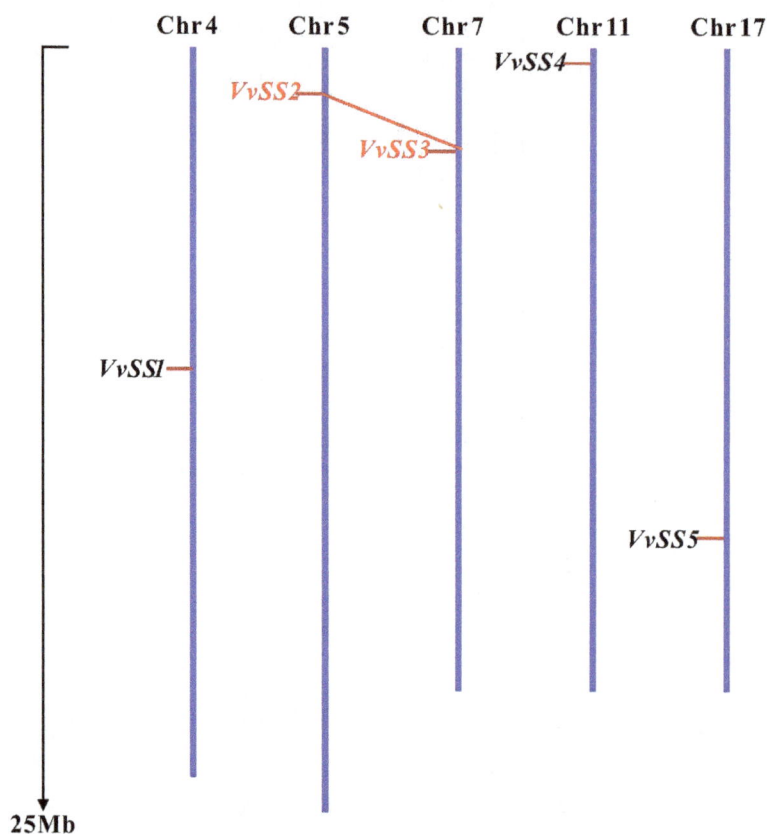

Figure 1. Chromosomal distribution of *Sucrose Synthase* (*SS*) genes in the grape genome. The chromosome number is shown at the top of each chromosome. The positions of the grape *SS* genes are marked by red lines on the chromosomes. The block duplication of *VvSS2–VvSS3* are linked by red lines.

Table 2. Coding region nucleotide (upper portion of matrix) and amino acid (bottom portion of matrix) sequence pairwise comparisons (% similarity) between grape sucrose synthase genes.

	VvSS1	*VvSS2*	*VvSS3*	*VvSS4*	*VvSS5*
VvSS1	-	51.79	53.81	53.45	70.56
VvSS2	59.18	-	79.12	66.86	47.70
VvSS3	59.93	77.55	-	71.15	50.00
VvSS4	59.60	66.36	69.47	-	59.57
VvSS5	70.50	54.85	56.60	56.85	-

3.2. Exon/intron Organization of the Grapevine Sucrose Synthase Gene Family

Analysis of the exons/introns structures of gene can provide important insights into the molecular mechanism of evolution of gene family. To investigate the exon/intron structure of the grape *SS* genes, the CDS and genomic sequences were compared. As shown in Figure 2, these genes typically consist of 13, 15, and 16 exons interrupted by 12, 14, and 15 introns of varying sizes, respectively. These exons, with lengths of 96, 174, 117, 167 and 225, appear almost at the same positions in the CDS region of most grape *SS* genes, consistent with the exons in cotton and Arabidopsis [28]. Based on the length and position of the exons, three intron/exon structure models of the *VvSS* genes were revealed and designated as I, II and III, respectively. A comparison of the intron/exon structures in the coding regions revealed significant differences among *VvSS* models (Figure 2). For each *SS* group, the gene structure also show unique features: (a) Group III members (*VvSS1* and *VvSS5*) shared the identical exon/intron pattern, such as exons block 1 (with lengths 155, 193, 119, 217 and 96) and exons block 2 (with lengths 117, 167, 225, 567 and 139) arranged in the coding region in order. (b) In Group I (*VvSS4*), the fifth position was an exon with length 336 in *VvSS4*, and in *VvSS1* of Group III the fourth and fifth exons were 119 and 217 in length. Thus, we assumed that exon with length 336 in *VvSS4* was split into two exons (exon with length 119 and length 217, respectively) in *VvSS1* through the evolutionary history. (c) In Group II (*VvSS2* and *VvSS3*), if we ignored the minor differences in length of some exons (such as the first exon with length 92 in *VvSS2* and the last exon with length 98 in *VvSS3*, the last exon with length 36 in *VvSS2* and the first exon with length 39 in *VvSS3*, etc.), and considered the arrangement order of these exons, it was a reasonable hypothesis that *VvSS2* and *VvSS3* were duplicated and retained from whole-genome duplication event (WGD).

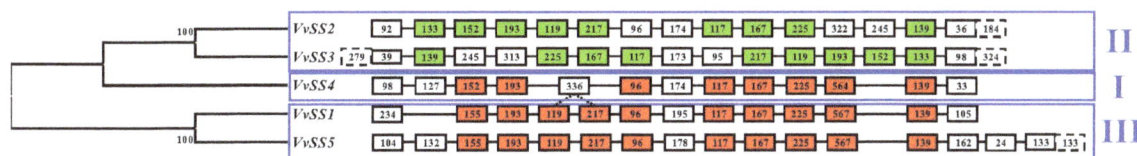

Figure 2. Phylogenetic relationships and intron-exon organization of grape *SS* genes. The unrooted phylogenetic tree was constructed using the full-length protein sequences of five grape *SS* genes by the Maximum Likelihood method with 1000 bootstrap replicates. The three groups are marked by square boxes and numbered with Roman numerals. The 5′ and 3′ untranslated regions (UTRs) are represented by a dashed box. Numbers in boxes represent the sizes (bp) of corresponding exons or UTR regions. The green and red boxes represent exons shared in Group I and II, respectively. The dashed line indicated the relationship between three exons. The solid lines represent introns whose size is not indicated.

3.3. Gene Duplication and Syntenic Analysis of the Grapevine Sucrose Synthase Genes

Gene duplication and divergence are important in gene family expansion [40] and the evolution of novel functions [41]. Grapevine has undergone whole-genome duplications during its evolutionary history [42]. To examine the impact of duplications on the *SS* gene family, we obtained tandem duplication and segmental duplication gene pairs from Plant Genome Duplication Database (PGDD. Available online: http://chibba.agtec.uga.edu/duplication/) and visualized them using Circos software (Circos. Available online: http://circos.ca/). In this study, we identified one segmental duplication pairs of grape *SS* genes (*VvSS2* and *VvSS3*) (Figure 3, Figure S1) but no tandem duplication events were detected in the grape *SS* genes. Further to explore the origin and evolution dynamics of grape *SS* genes, we investigated the syntenic relationship between grapevine and other plant species, and found that there is the syntenic relationship between grapevine and soybean (*Glycine max*) based on the results obtained from PGDD database. The syntenic analysis showed that syntenic genes included: *VvSS1*-Glyma.02G240400 (*GmSS1*)/Glyma.09G167000 (*GmSS4*)/Glyma.14G209900 (*GmSS6*), *VvSS2*-Glyma.03G216300 (*GmSS2*)/Glyma.15G151000 (*GmSS8*)/Glyma.19G212800 (*GmSS11*), *VvSS3*-Glyma.03G216300 (*GmSS2*)/Glyma.19G212800 (*GmSS11*), *VvSS4*-Glyma.13G114000 (*GmSS5*)/Glyma.17G045800 (*GmSS10*)/Glyma.15G182600 (*GmSS7*)/Glyma.09G073600 (*GmSS3*), and *VvSS5*-Glyma.09G167000 (*GmSS4*)/Glyma.16G217200 (*GmSS9*) (Figure 3, Supplementary Table S1). These results provide insights that will help infer the probable functions of grape *SS* genes.

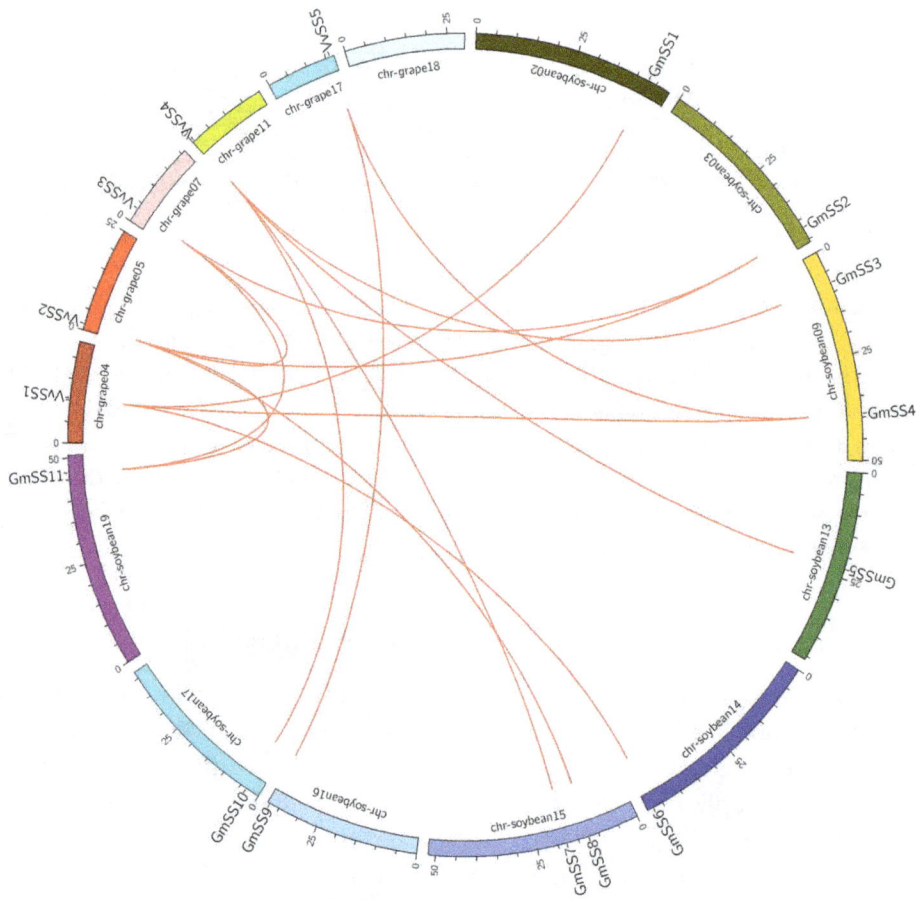

Figure 3. Segmental duplication of grape *SS* genes and syntenic analysis of grape and soybean *SS* genes. Chromosomes of *Vitis vinifera* and *Glycine max* are shown in different colors and in circular form. The approximate positions of the *GmSS* and *VvSS* genes are marked with a short black line on the circle. Colored curves denote the syntenic relationships between grape and soybean *SS* genes.

3.4. Phylogenetic Analysis of the Grapevine Sucrose Synthase Genes

To investigate the evolutionary modules and their relationship to functional ones, we constructed a maximum likelihood tree using the full-length amino acid sequences of *SS* genes from 58 dicot sequences, 19 monocot sequences and four bacteria sequences based on 1000-replicate bootstrap values (Figure 4, Supplementary Table S2). The phylogenetic analysis revealed a relatively deep evolutionary origin and relatively recent duplications. All bacterial *SS* genes clustered into the same group, whereas those from the land plants formed a monophyletic group, showing that all plant *SS* genes originated from an ancestral type [43] (Figure 4). Bacteria *SS* genes were used as the outgroup; the plant *SS* genes can be categorized into three clearly distinct subgroups based on strong statistical support. These subgroups were designated as class I, class II and class III, respectively (Figure 4). The improved resolution in the tree also enabled us to make further subdivisions within classes I, II and III. Each class was resolved into two branches, one specific for dicots and another for monocots. Thus, the phylogenetic analysis suggested that most gene duplication events that gave rise to *SS* gene classes I–III occurred before the divergence of monocot/dicot. The five grape *SS* genes were distributed in the dicot branch of classes I, II and III, *VvSS4* in the dicot subgroup of Class I, *VvSS2* and *VvSS3* in Class II, and *VvSS1* and *VvSS5* in Class III, consistent with the exon/intron organization pattern and nucleotide/amino acid sequence identity (Table 2 and Figure 2). *VvSS4* itself was clustered into a branch; *VvSS2* was clustered together with *CitSS6*; *VvSS3* was closer to *OsSS4*; *VvSS5* was closer to *CitSS4* and *MdSS6*; and *VvSS1* was closer to *CitSS5*. The apparent diversification within the grape *SS* family could imply discrete biological roles for the paralogs despite their high sequence similarity.

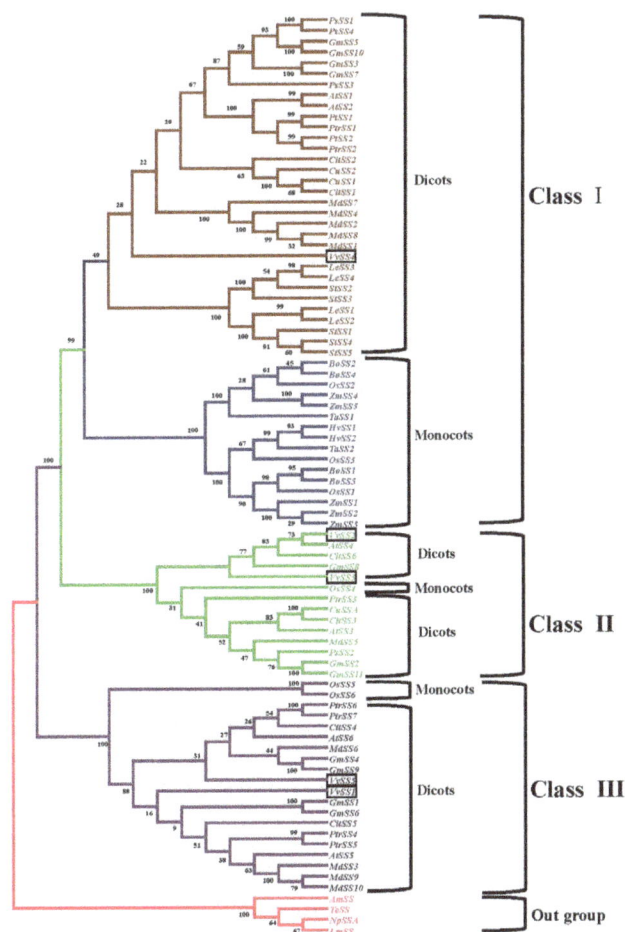

Figure 4. Phylogenetic relationships between the grape SS proteins and other SS homologous proteins in dicots and monocots (refer to Supplementary Table S2). *VvSS1* to *VvSS5* was highlighted in black boxes.

3.5. Evolution of the Coding Sequences of the Grape SS Gene Pairs

Modes of selection can be estimated by the ratio of the numbers of nonsynonymous substitutions per nonsynonymous site (dN) to the numbers of synonymous substitutions per synonymous site (dS), that is, dN/dS > 1 indicates positive selection; dN/dS < 1, purifying selection; and dN/dS = 1, neural selection [44]. The combination of phylogenetic, exon/intron structure, and paragon analyses revealed two pairs of grape *SS* gene pairs (Table 2). The nonsynonymous substitution rates of two gene pairs are markedly higher than their synonymous substitution rates and their dN/dS values are >1 (Table 3), suggesting that these duplicates likely have been subject to positive selection. Furthermore, Tajima [39] relative rate tests were conducted to investigate whether one of the *SS* gene pairs has evolved at an accelerated rate following the duplication. A statistically significant increase in evolutionary rate occurs in *VvSS2/VvSS3* duplicated pairs (Table 4).

Table 3. Divergence between paralogous *SS* gene pairs in grape.

Gene 1	Gene 2	dN	dS	dN/dS
VvSS2	*VvSS3*	0.3012	0.0957	3.1473
VvSS1	*VvSS5*	0.2759	0.1246	2.2143

dN: nonsynonymous site; dS: synonymous site.

Table 4. Tajima relative rate tests of *SS* gene pairs in grape [a].

Testing Group	Mt [b]	M1 [c]	M2 [d]	X^2	P [e]
VvSS2/VvSS3 with *GmSS2*	586	11	30	8.91	0.00450
VvSS1/VvSS5 with *GmSS4*	472	9	4	1.92	0.21690

[a] The Tajima relative rate test was used to examine the equality of evolutionary rate between grape gene pairs; [b] Mt is the sum of the identical sites and the divergent sites in all three sequences tested; [c] M1 is the number of unique differences in the first paralog; [d] M2 is the number of unique differences in the second paralog; [e] If $P < 0.05$, the test rejects the equal substitution rates between the two duplicates and infers that one of the two duplicates has an accelerated evolutionary rate.

3.6. Expression Analysis of VvSS Genes in Different Organs, Tissues and Developmental Stages

To investigate the spatial and temporal expression patterns of *SS* genes in grapevine, we retrieved the data from a global transcriptomic atlas comprising 54 tissues, organs or developmental stages [45].

The expression patterns of *VvSS* genes were analyzed in the *V. vinifera* cv. Corvina global gene expression atlas, which consists of 54 different organs/tissues at various developmental stages obtained by microarray analysis (Supplementary Table S2). All *VvSS* genes had corresponding probes on the NimbleGen array. Figure 5A presents a graphical representation of the expression pattern of each *VvSS* gene. As a whole, the expression of five *VvSS* genes showed tissue- and organ-specific pattern (Figure 5A), in which *VvSS1*, *VsSS2*, and *VvSS5* showed a relatively low expression level, while *VvSS3* and *VvSS4* remained at higher expression levels in each tested tissue. Overall, *VvSS3* is highly expressed in vegetative tissues (berry pericarp, berry flesh, and berry skin) and transport tissues (rachis, tendril, and stem). However, except for expression overlap between rachis and tendril, *VvSS4* mainly presented higher transcript levels in bud, inflorescences, and seed, which are considered as reproductive organs. The expression pattern of these genes was also evaluated by qRT-PCR in selected organs of the *V. vinifera* grapevine cultivar (Figure 5B).

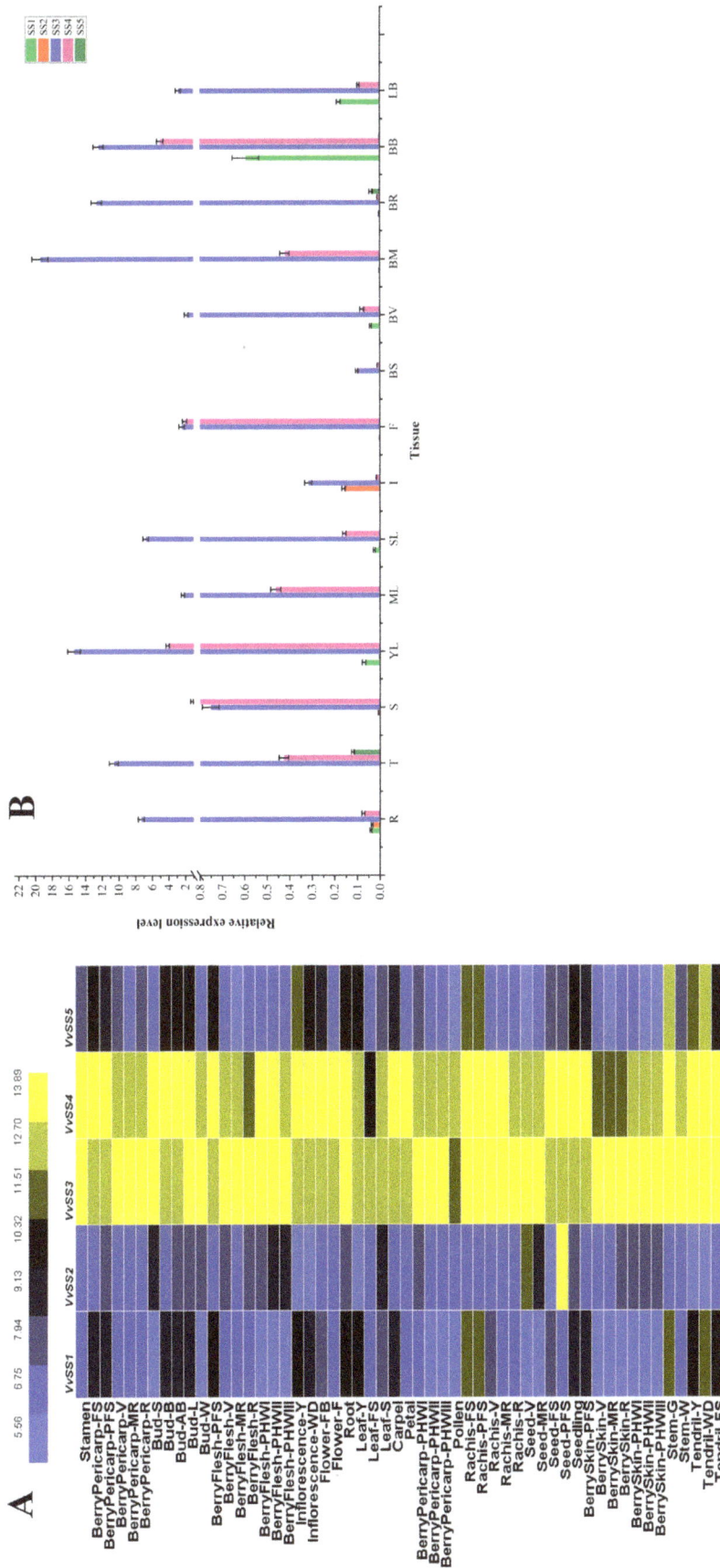

Figure 5. Expression profiles of the grapevine *VvSS* genes in different grapevine organs, tissues and developmental stages. **(A)** Expression of *VvSS* genes in the *V. vinifera* cv "Corvina" atlas (GEO Accession: GSE36128). Data were normalized based on the mean expression value of each gene in all tissues analyzed. The mean expression values were again normalized using logarithm with the base of 2 using the HemI software. Yellow and blue boxes indicate high and low expression levels, respectively, for each gene. **BerryPericarp-FS**: berry pericarp fruit set; **BerryPericarp-PFS**: berry pericarp post-fruit set; **BerryPericarp-V**: berry pericarp véraison; **BerryPericarp-MR**: berry pericarp mid-ripening; **BerryPericarp-R**: berry pericarp ripening; **Bud-S**: bud swell; **Bud-B**: bud burst (green tip); **Bud-AB**: bud after-burst (rosette of leaf tips visible); **Bud-L**: latent bud; **Bud-W**: winter bud; **BerryFlesh-PFS**: berry flesh post fruit set; **BerryFlesh-V**: berry flesh véraison; **BerryFlesh-MR**: berry flesh mid-ripening; **BerryFlesh-PHWI**: berry flesh post-harvest withering I; **BerryFlesh-PHWII**: berry flesh post-harvest berry flesh ripening; **BerryFlesh-R**: berry flesh ripening; **BerryFlesh-PHWII**: berry flesh post-harvest

withering II; **BerryFlesh-PHWIII**: berry flesh post-harvest withering III; **Inflorescence-Y**: young inflorescence (single flower in compact groups); **Inflorescence-WD**: well developed inflorescence (single flower separated); **Flower-FB**: flowering begins (10% caps off); **Flower-F**: flowering (50% caps off); **Leaf-Y**: young leaf (pool of leaves from shoot of 5 leaves); **Leaf-FS**: mature leaf (pool of leaves from shoot at fruit set); **Leaf-S**: senescencing leaf (pool of leaves at the beginning of leaf fall); **BerryPericarp-PHWI**: berry pericarp post-harvest withering I (1st month); **BerryPericarp-PHWII**: berry pericarp post-harvest withering II (2nd month); **BerryPericarp-PHWIII**: berry pericarp post-harvest withering III (3rd month); **Rachis-FS**: rachis fruit set; **Rachis-V**: rachis véraison; **Rachis-MR**: rachis mid-ripening; **Rachis-R**: rachis ripening; **Seed-V**: seed véraison; **Seed-MR**: seed mid-ripening; **Seed-FS**: seed fruit set; **Seed-PFS**: seed post fruit set; **BerrySkin-PFS**: berry skin post fruit set; **BerrySkin-V**: berry skin véraison; **BerrySkin-MR**: berry skin mid-ripening; **BerrySkin-R**: berry skin ripening; **BerrySkin-PHWI**: berry skin post-harvest withering I; **BerrySkin-PHWII**: berry skin post-harvest withering II; **BerrySkin-PHWIII**: berry skin post-harvest withering III; **Stem-G**: green stem; **Stem-W**: woody stem; **Tendril-Y**: young tendril (pool of tendrils from shoot of 7 leaves); **Tendril-WD**: well developed tendril (pool of tendrils from shoot of 12 leaves); **Tendril-FS**: mature tendril (pool of tendrils at fruit set). (**B**) qRT-PCR validation of *VvSS* expression in different tissues obtained from two-year-old grapevines of *V. vinifera* cv. Chardonnay seedlings. Transcripts were normalized to the expression of the actin gene. The mean ± s.d. of three biological replicates are presented. **R**: principal root; **T**: tendrils from shoot of 7 leaves; **S**: green stem; **YL**: young leaf, leaves from shoot of 5 leaves; **ML**: mature leaf, leaves from shoot at fruit set; **SL**: senescencing leaf, leaves at the beginning of leaf fall; **I**: inflorescence, well developed inflorescence (single flower separated); **F**: flower, flowering (50% caps off); **BS**: berry post fruit set; **BV**: berry véraison; **BM**: berry mid-ripening; **BR**: berry ripening; **BB**: bud burst (green tip); **LB**: latent bud.

In order to complement the whole-transcriptome data, steady-state mRNA levels of all *VvSS* genes were investigated by real-time PCR in various organs/tissues of the *V. vinifera* grapevine cultivar, including root, tendril, stem, leaves, flower, berry and bud at different developmental stages. As shown in Figure 5B, grape *SS* transcripts were detected in a wide range of tissues and showed distinct but partially overlapping expression patterns, suggesting that *SS* genes may be implicated in a range of physiological processes in grapevine plant. Among these, *VvSS3* and *VvSS4* expression was detected in all tissues examined. However, *VvSS3* showed strong expression in all the tissues but *VvSS4* was found to be highly expressed in stem, flower, mid-ripening berry and burst bud, and relatively lower expression was noted in root, tendril, senescencing leaf, véraison berry and latent bud. While *VvSS1* was highly expressed in leaves and buds, and found that the expression of *VvSS1* peaked at the initial stage of leaf development (young leaves), and then decreased rapidly during leaf development to a low level at the mature stage. *VvSS2* was weakly expressed in the root and inflorescence. The very low expression levels of *VvSS5* were observed in tendril and ripening berry.

3.7. Expression Patterns of VvSS Genes in Response to Biotic and Abiotic Stresses

The expression of *VvSS* genes in response to both biotic and abiotic stresses was investigated using microarray data from several previously published papers. Regarding abiotic stresses, expression datasets were obtained from two studies (GSE31594 and GSE31677) conducted on transcriptomic response in leaves of *V. vinifera* cv "Cabernet Sauvignon" to short-term salt, water and cold stress, and long-term water and salt stress. Since the Affymetrix array used in these analysis was based on the few *VvSSs* sequences (cDNA and expressed sequence tags (ESTs)) known at the time, the expression of only a limited number of *VvSS* genes (*VvSS2*, *VvSS3* and *VvSS4*) could be determined (Supplementary Table S4).

In general, *VvSS2* was down-regulated in all stress treatments, and the expression of *VvSS4* was unaffected in some cases (salt, polyethylene glycol (PEG) and cold) and down-regulated in others (water deficit and salinity), while *VvSS3* gene showed significant changes in the patterns of expression levels. We found that *VvSS3* underwent continuous and significant increase after stress treatment (Figure 6). The data presented confirmed the putative involvement of *VvSS3* in abiotic stresses responses.

To further determine whether the *SS* genes were involved in abiotic stress resistance and validate the results obtained from the analysis of the microarray data, we measured their transcript levels under drought, salt, dark, and high and low temperature using qRT-PCR. As shown in Figure 6, *VvSS3* and *VvSS5* transcript abundance increased in all five treatments, *VvSS1* showed continuously low transcript levels in all treatments. When suffering high temperature stress, the expression levels of all of the *VvSS* genes were down-regulated, except for *VvSS3* and *VvSS5*, for which the transcript levels were up-regulated significantly. *VvSS5*, the most rapidly responding gene, reached a peak of nearly five-fold at 24 h post treatment (hpt) and rapidly decreased at 48 hpt. We also measured the transcript levels of *VvSS* genes in response to low temperature (4 °C) stress. Although transcript levels varied among the family members, *VvSS* genes exhibited a similar expression pattern among high and low temperature stress. Similarly, *VvSS5* responded strongly with transcript levels increasing to 90.0-fold at 24 h cold stress and then decreased rapidly around 15-fold at 48 h treatment.

After NaCl treatment, only *VvSS2*, *VvSS3*, and *VvSS5* showed up-regulation, although the expression levels of *VvSS2* and *VvSS5* were merely slightly increased (less than twofold) under this treatment. *VvSS3* was first declined and then rapidly up-regulated, reached a peak of 26.5-fold at 96 hpt, For the dark treatment, only *VvSS3*, *VvSS4*, and *VvSS5* showed up-regulation. *VvSS3* and *VvSS4* had a peak of 2.5-, and 15.5-fold respectively at 48 hpt but then their transcript levels rapidly decreased at the 96 h. We found *VvSS5* significantly responded to dark treatment (the peak is over than 70.0-fold at 96 hpt). As for the drought treatment, *VvSS1* and *VvSS4* shared similar expression patterns, with a slow increase in transcript levels to a peak and rapidly decrease. *VvSS3* and *VvSS5* also shared similar expression patterns, with a significantly increase in transcript levels to a peak after 20-day drought treatment.

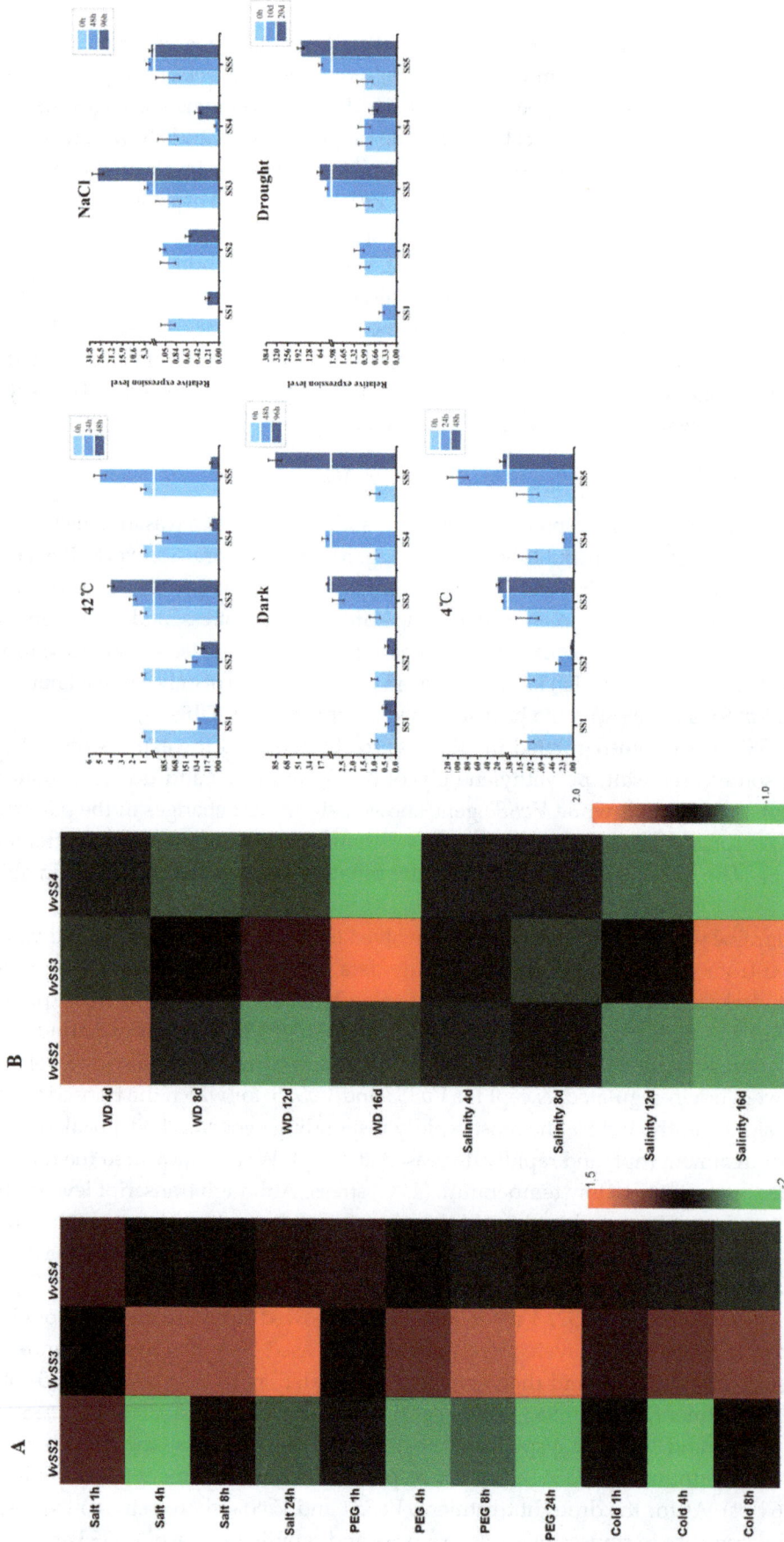

Figure 6. Expression profiles of *VvSS* genes in response to abiotic stresses. Expression of *VvSS* genes in the *V. vinifera* cv "Cabernet Sauvignon" (microarray) or two-year-old grapevines of *V. vinifera* cv. Chardonnay seedlings (qRT-PCR) in response to high temperature (42 °C), 100 mM NaCl, dark, drought (WS) and cold (4 °C). Microarray data were downloaded from the NCBI GEO datasets (GEO. Available online: https://www.ncbi.nlm.nih.gov/geo/) (GSE31594 and GSE31677), processed as log2 of the ratio between treated and untreated samples and graphically represented

with HemI software. (**A**) *V. vinifera* cv "Cabernet Sauvignon" plants grown in a hydroponic drip system were treated with 120 mM NaCl, polyethylene glycol (PEG), cold (5 °C) or left untreated. Shoots with leaves were collected at 0, 1, 4 and 8 h for all treatments, and at 24 h for all treatments except cold (GEO series GSE31594). (**B**) Potted *V. vinifera* cv "Cabernet Sauvignon" vines in the greenhouse were exposed to a water-deficit stress (WD) by withholding water or a salinity stress by watering plants with a saline solution for 16 days. Non-stressed, normally watered plants served as the control for both treatments. Shoot tips were harvested every four days (0, 4, 8, 12 and 16 days) (GEO series GSE31677). (**C**) qRT-PCR expression analysis of *VvSSs* in *V. vinifera* cv. Chardonnay seedlings subjected to stress treatments. Transcripts were normalized to the actin gene expression. The mean ± s.d. of three biological replicates are presented.

Taken together, our results showed that *VvSS* genes expression varies in response to different abiotic stresses, indicating diverse roles of the *VvSS* genes in response to abiotic stress.

In regards to *VvSS* expression in response to biotic stresses, expression datasets obtained from three different host-pathogen interaction experiments were examined, including: the inoculation of *Erysiphe necator* on leaves of the susceptible *V. vinifera* cv. Cabernet sauvignon and the tolerant *V. aestivalis* cv. Norton [46] (Supplementary Table S4); infection of *V. Vinifera* cv. Chardonnay and cv. Incrocio Manzoni infection with the *Bois Noir* phytoplasma [47] (Supplementary Table S4); and the infection of *V. vinifera* cv. Cabernet Sauvignon with grapevine leaf-roll-associated virus-3 (GLRaV-3) during véraison and ripening stages of berry development [48] (Supplementary Table S4).

The effect *E. necator* infection on *VvSS* response appeared to be much stronger in the susceptible cv. Cabernet sauvignon than in the resistant cv. Norton (Figure 7A). *VvSS2* gene was upregulated in Cabernet sauvignon after 4 h of inoculation and in Norton it was upregulated after 24 h of inoculation.

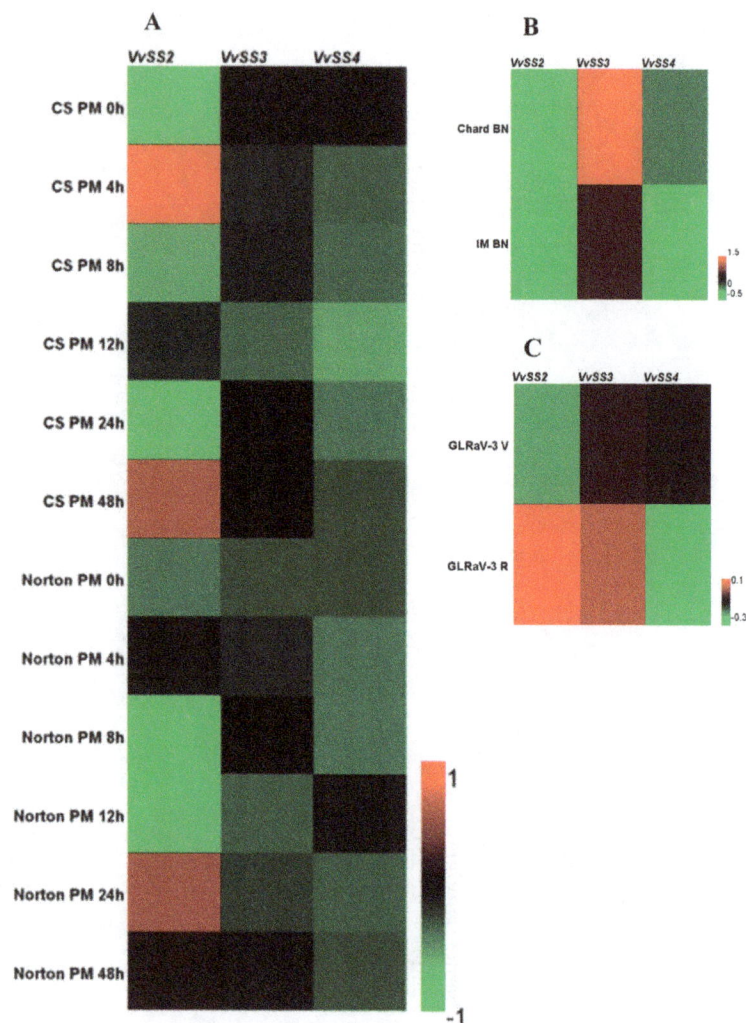

Figure 7. Expression profiles of *VvSS* genes in response to biotic stresses. (**A**) *V. vinifera* cv "Cabernet sauvignon" (CS) and *Vitis aestivalis* cv "Norton" plants were grown in an environmental chamber and inoculated with *Erysiphe necator* conidiospores (PM). Inoculated leaves were harvested at 0, 4, 8, 12, 24 and 48 h after inoculation (GEO series GSE6404). (**B**) Field-grown plants of *V. vinifera* cv "Chardonnay" (Chard) and "Incrocio Manzoni" (IM) naturally infected with *Bois Noir* phytoplasma (BN), compared to healthy samples (GEO series GSE12842). (**C**) *V. vinifera* cv "Cabernet Sauvignon" was infected with GLRaV-3 during véraison (V) and ripening (R) stages of berry development (GEO series GSE31660).

Phytoplasma infection led to an induction of *VvSS3* genes in susceptible *V. Vinifera* cv. Chardonnay compared to the tolerant cv. Incrocio Manzoni (Figure 7B), and a repression of *VvSS2* and *VvSS4* (Figure 7B). Finally, the expression profiles of *VvSS* genes in response of grapevine to GLRaV-3 infection in vèraison phases revealed a general repression of the *VvSS* family (Figure 7C). However, a significant-upregulation, limited to *VvSS2* and *VvSS3*, specifically induced during the ripening phase.

4. Discussion

Comparative genomics approaches have been used to analyze *SS* gene families in various plant species, including citrus, tobacco, poplar, and cotton [28,29,49,50]. In the present study, we verified five *SS* genes from grape (*V. Vinifera*). These grape *SS* family members shared two conserved domain; we named these as *VvSS1–5*, according to distinct molecular characteristics. Given the limited knowledge of the grape *SS* gene family prior to this study, Comprehensive understanding of the evolutionary relationships, molecular structures, and expression profiles provides the important resources for their molecular mechanisms and possible functions in grape growth and development.

4.1. Evolutionary Conservation and Divergence Among Grape SS Genes

In all plant species examined to date, the SS isoenzymes are encoded by a small, multi-gene family [8,29,40,50]. Comprehensive analysis of this multi-gene family, including its exon/intron structures, phylogeny, and syntenic analysis, makes it possible for researchers to predict the potential functions and evolutionary relationships among uncharacterized members of this gene family.

Since gene exon/intron structures are typically conserved among homologous genes of a gene family [51], analysis of exon/intron structures can provide a clue to reveal the evolutionary history of certain gene family [52]. Previous detailed studies of the *SS* genes shed light on highly conservation in the length, number and position of exon/introns in several distantly related dicot and monocot plants, bring about the speculation that the divergence of the three ancestors prior to the segregation of monocot and eudicot species [53]. In the present study, the predicted molecular features of the grape SS proteins were similar to those of previously characterized SS proteins from other plant species. We also estimated the exon/intron structures of the grape *SS* genes (Figure 2, Figure S2). The integrated gene structure model in *VvSS* was very similar to that of *SS* genes from tobacco [50], poplar [29], cotton [28] and Arabidopsis and rice [27,31]. For instance, the position of exons of five putative *SS* genes showed parallel positions, and most of the exons between Group I, II and III shared a high level of similarity but only existed difference at two ends of sequence. However, genome-wide duplication events and subsequent chromosomal rearrangements have differentially shaped the *SS* family of grape. For example, the numbers and length of introns of five *VvSS* genes were different greatly (Figure 2 and Figure S2), whereas the number and positions of exons were highly conserved among *SS* genes in citrus, tobacco, poplar, and cotton [28,29,49,50]. In addition, the arrangement and length of intron/exon of *VvSS2* and *VvSS3* showed the antiparallel relationship. We further analyzed whether these phylogenetic groups corresponded to the *VvSS* genes structure models. Indeed, clustering the intron/exon organizations of the five *SS* genes by an unrooted phylogenetic tree suggests a connection between intron/exon structures and evolutionary history.

The conservation and divergence of grape *SS* genes' structure led to the expansion of gene family members and functional conservation/differentiation of gene. In general, three pivotal mechanisms contribute to gene family evolution and expansion: exon/intron gain or loss, exonization/pseudo-exonization, and insertion/deletion [17]. For the three groups in Figure 2, according to the phylogenetic relationships of five *VvSS* genes in the present study (see the contents below) and previous work in other plants [49,50], Group III (*VvSS1/VvSS5*) was the earliest one that expanded from the evolutionary branch. As a result, *VvSS1/VvSS5* has the longest evolutionary history, leading to the complex of intron/exon structure. All members of Group I and II lack two exons and two introns at the last position, indicating that there was an intron deletion near the last exon, which increased the number of the *VvSS* family members. The divergence of intron/exon was closely related to the evolutionary history of the grape *SS* family and might result in functional diversification.

To further study the molecular evolution and explore their possible function of the *VvSS* gene family, we also investigated gene duplication events, and syntenic and phylogenetic relationships of the *VvSS* genes. Segmental duplications and tandem duplications can lead to increasing or decreasing copy numbers in gene families [54]. These processes may lead to functional redundancy, sub-functionalization and neo-functionalization. A segmental duplicated pair (*VvSS2–VvSS3*) in the grapevine genome was found, but no tandem duplicated *VvSS* genes were discovered (Figure 3, Supplementary Table S1). The duplicated gene pair (*VvSS2–VvSS3*) had high similarities in gene length, protein properties, and intron/exon structure (Tables S1 and S2, Figure 2). Furthermore, the gene pair (*VvSS2–VvSS3*) showed the tight phylogenetic relationship among plant *SS* gene families and within grape *SS* genes (Figures 2 and 4). They also showed closer phylogenetic relationships with soybean *SS* genes than that with each other, and these two genes have different syntenic genes in soybean (*G. max*) (Figure 3, Supplementary Table S1), indicating that the duplication events happened before the divergence of the grape and soybean lineages. As discussed above, *VvSS2* and *VvSS3* likely have functional redundancy.

Comparative genomics relies on the structuring of genomes into syntenic blocks that exhibit conserved features across the genomes [55]. The syntenic analysis provides evolutionary and functional characterization of *SS* gene families in grape and soybean. In this work, all five grape *SS* genes were found to have syntenic relationships with soybean genes (Figure 3). Thus, a large number of syntenic relationships indicated that some of the grape *SS* genes arose before the divergence of these two species. According to the phylogenetic tree, *VvSS* genes established closer phylogenetic relations with the corresponding *GmSS* genes (Figure 4), suggesting strongly some level of functional similarities.

Previous studies of the molecular structures and phylogenetic relationships of plant *SS* gene families divided *SS* genes into three groups (SS1, SSA, and New Group) [26,27]. This classification was corroborated a number of subsequent studies, and three groups were respectively named as the Class I, Class II, and Class III groups [2,28,56]. Later, the Class I group can be further divided into a monocot subgroup and a eudicot subgroup, as these two subgroups were obviously distinct from each other in phylogenetic trees constructed by dozens of plant *SS* gene families. The Class II and Class III groups were then classified into mix group 1 and mix group 2, as members of these groups contained monocot and eudicot plants [29]. In our study, we performed phylogenetic analysis of the *SS* genes from grape and fifteen other plant species, and found that the grape *SS* genes family had at least one member in three single groups: *VvSS4* in Class I, *VvSS2* and *VvSS3* in Class II and *VvSS1* and *VvSS5* in Class III. The five *VvSS* genes in the dicot group were divided into three subgroups, specifically, *VvSS1* and *VvSS5* clustered closely together, *VvSS2* and *VvSS3* clustered together, while *VvSS4* alone clustered into one branch. *VvSS1* and *VvSS5* were apparently apart from Arabidopsis *SS* genes. This result demonstrated that the gene duplication event that generated *VvSS1* and *VvSS5* gene occurred after monocot–eudicot separation, but before the divergence of *Vitaceae/Arabidopsis*. In addition, the generation of the *VvSS2* and *VvSS3* genes probably took place after the separation of *Vitaceae/Arabidopsis*. Grape, sweet orange, and apple are both members of Rosidae, possessed a close evolutionary relationship. There were no sweet orange and apple *SS* genes show closely relationship with *VvSS4*, suggesting that *VvSS4* should be more recent than the differentiation between grape and sweet orange, apple. Thus, *VvSS1* and *VvSS5* were older than the other grape *SS* genes, the diverged duplications of *VvSS4* have occurred more recently and may split after the divergence of the Rosidae, and grape might have lost a copy after its divergence from Arabidopsis compared with the gene copy number of Arabidopsis. We also found positive selection contributed to evolution of this gene family. Furthermore, Tajima relative rate tests identified accelerated evolutionary rates in *VvSS2/VvSS3* duplicates (Table 4). Based on the genome of *V. vinifera* [42], and the facts that grape *SS* family was distributed on five chromosomes (Figure 2), our phylogenetic analysis, the exon/intron structure, and *VvSS2* and *VvSS3* located in systemic blocks, we propose an interpretation of evolutionary history of the grape *SS* genes. Before the monocot–eudicot divergence, an early gene duplication in the ancestor gave rise to the three progenitors of the three *SS* groups with conserved exon/intron structure. Two of the three precursors evolved independently and finally retained one single gene in the Class II group (*SS2*) and Class III group (*SS1*), respectively. After the divergence of monocots and eudicots, duplication (may be whole-genome duplication event) of the *SS* precursor generated *SS3* and *SS5*, whereas *SS4* were produced after differentiation within the Rosidae (Figure 4).

4.2. VvSS Genes Involved in Grapevine Growth and Development

Functional diversity caused by the gene duplication resulted in the altered expression profiles and/or protein property, and it was a major evolutionary driver to increase the fitness to new environment of plants [57]. Analysis of gene expression profiles can be used in some level to predict the physiological processes genes involved in. To date, although the detailed expression profiles of *SS* genes have been examined in other plant species, such as Arabidopsis, rice, cotton, poplar, and rubber tree [2,22,25,27–29], there have been no detailed surveys of the expression of grape *SS* genes.

In the present study, the expression patterns of *VvSS* genes in different grapevine tissues and at different developmental stages were examined using an expression atlas of *V. vinifera* cv Corvina. The analysis revealed that the abundant expression of *VvSS3* and *VvSS4* gene in specific grapevine tissues, possibly reflected their involvement in a common metabolic and/or developmental process (Figure 5).

The expression analysis revealed that *VvSS3* were highly expressed in vegetative and transport tissues, suggesting a key role of *VvSS3* in the regulation of sugar accumulation in berry and sugar transport in stem and tendril of grapevine. *VvSS4* was found to be specifically expressed in reproductive tissues and highly accumulated during flower development, which was shown to may have a role in energy supplying to support the process from bud to flower. Similarity, the spatial-temporal expression of *SS* genes has been previously reported in many other plants. For example, at the early stage of fruit development of apple, the transcript levels of *SS* are high. As the fruit continues to grow due to cell expansion, the transcript levels of *SS* were down-regulated [58]. For tissues and organs, *CitSus1* and *CitSus2* were predominantly expressed in fruit juice sacs (JS), whereas *CitSus3* and *CitSus4* were predominantly expressed in early leaves (immature leaves), and *CitSus5* and *CitSus6* were predominantly expressed in fruit JS and in mature leaves. During fruit development, *CitSus5* transcript increased significantly and *CitSus6* transcript decreased significantly in fruit JS. In addition, in tobacco, *Sus2* (Ntab0259170) and *Sus3* (Ntab0259180) gene based on high transcript levels also play a predominant role in sucrose metabolism during leaf development [49]. Zou et al., (2013) [28] demonstrated that most *Sus* genes were differentially expressed in various tissues and *GrSUS1*, *GrSUS3*, and *GrSUS5* showed significantly higher expression levels and underwent significant changes in expression during fiber development in the three *Gossypium* species.

VvSS1, *VvSS2* and *VvSS5* genes were not expressed, or were expressed at low levels, suggesting their redundant function in the normal development and growth process of grape. However, in certain tissues, other *VvSS* genes except for *VvSS3* and 4, also plays an important role. For example, *VvSS2* was barely expressed at a high level in seed-PFS, and *VvSS1* was similarly expressed at a high level in tendril, and tendril. These results suggest that the functions of *VvSS* genes are diversified and yet partially overlap. Since sucrose affects cell division and vascular tissue differentiation in plant leaves, *VvSS4* remained at a steady high level, and *VvSS3* significantly decreased during the course of leaf development (which is same expression pattern in bud). Overall, comparison of the transcripts of all *VvSS* genes in each single tissue revealed the predominant role of *VvSS3* and *VvSS4* isoforms in the growth and development of grapevine.

4.3. Stress Induced VvSS Expression in Grapevine

Sucrose synthases have been reported to be associated with plant responds to various environmental stresses. For instance, in Arabidopsis, expression of the *AtSS1* gene could be induced by cold or drought treatment. *AtSS3* is used as a molecular marker of dehydration [25]. In addition, two barley *SS* genes (*HvSS1* and *HvSS3*) and one rubber tree *SS* gene (*HbSS5*) significantly responded to low temperature and drought stresses [22,59]. The higher expression levels of *SS* genes may result from the increased glycolytic demand under abiotic stresses [60].

In our study, the transcriptomic databases generated in previous studies of grapevine subjected to biotic and abiotic stresses, together with our qRT-PCR analysis, allowed us to identify *VvSS* genes putatively involved in stress response.

The members of grape *SS* gene family exhibited different responses of expression patterns in the mature leaf in response to different abiotic stress. In detail, under all abiotic stress, the transcription levels of *VvSS3* and *VvSS5* were continuously up-regulated, and thus the key enzymes of *VvSS3* and *VvSS5* encoded might be the predominant isoforms of leaf to respond to abiotic stress. Simultaneously, the transcription levels of the other *VvSS* genes were significantly down-regulated after abiotic stress treatment.   The similar expression response of *SS* genes was also found in rubber tree (*Hevea brasiliensis*) [22] and barley (*Hordeum vulgare*) [59]. Low temperature and drought treatments conspicuously induced *HbSus5* expression in root and leaf, suggesting a role in stress responses of rubber tree. Only *HvSs1* is up-regulated by anoxia and cold temperatures from the four *SS* genes in barley. In addition, expression of *Sus5* (*Ntab0288750*) and *Sus7* (*Ntab0234340*) were conspicuously induced by low temperature and virus treatment, indicating that these two isozymes are important in meeting the increased glycolytic demand that occurs during abiotic stress [50].

Regarding biotic stresses, to date the involvement of *VvSS* genes in biotic responses has been not previously reported. In this study, a significant induction of *VvSSs* did not occur in response to *E. necator*, *Bois Noir*, and GLRaV-3 infection. In contrast, a significant repression of *VvSSs* was observed, suggesting that these *VvSS* genes may be not involved in the pathogen response pathway. It is important to note that despite showing a high fold change of *VvSS2* in infected vs. mock-inoculated leaves in the biotic stresses, the baseline levels of *VvSS2* transcript were always lower (Supplementary Table S1).   However, *VvSS3* was significantly upregulated in response to *Bois Noir* phytoplasma infection, indicating that *VvSS3* might play an important role in plant defense against *Bois Noir*.

In addition, the duplicated gene pair, *VvSS2–VvSS3* possessed apparently unique expression profiles, further suggesting that the duplicated gene pair *VvSS2–VvSS3* might have undergone sub-functionalization. It was inconsistent with the previous research that members of the same *SS* genes orthologous group had very similar expression patterns in all three *Gossypium* species [28].

Overall, members of the grape sucrose synthase family exhibited different expression patterns in different tissues and in response to biotic and abiotic stresses. Despite many recent advances in functional studies of *SS*s in grapevine, the biological function of most *VvSS* genes in physiological and developmental processes and plant defense still needs to be elucidated. The bioinformatic analysis and expression patterns of the *VvSS* gene family conducted in the present study provide an overall picture of the composition and expression of *sucrose synthase* genes in grapevine that will facilitate selecting candidate genes for cloning and further functional characterization.

5. Conclusions

Five *sucrose synthase* genes in grapevine were bioinformatically identified and characterized. We propose that *VvSS1/VvSS5*, *VvSS2/VvSS3* and *VvSS4* originated from three ancient *SS* genes, respectively, which were generated by duplication events before the split of monocots and eudicots. The spatio-temporal expression profiles of the *VvSS* genes in various tissues at various developmental stages suggested *VvSS3* and *VvSS4* play a predominant role in sucrose metabolism during the growth and development grape. Expression of *VvSS3* was significantly induced by stress treatment, indicating that the isozyme played an important role in releasing stress.

Supplementary Materials:
The following supplementary material can be found online at www.mdpi.com/2073-4425/8/4/111/s1, Table S1 Syntenic relationships among grape and soybean sucrose synthase genes and within grape *SS* genes.

Table S2 List of sucrose synthase gene sequences used in this study. Table S3 Gene-specific primers used for quantitative RT-PCR analysis of *VvSS* gene expression. Table S4 Microarray data of *VvSS* genes downloaded from the GEO datasets.

Acknowledgments: This work was supported by grants from the Natural Science Foundation of China (NSFC) (No. 31672131); the Joint Israel-China Project of Natural Science Foundation of China (NSFC); the Israel Science Foundation (ISF) (NSFC-ISF) (No. 31361140358); and the Important National Science & Technology Specific Projects (No. 2012FY110100-3).

Author Contributions: Xudong Zhu conceived of the study and drafted the manuscript., Xudong Zhu and Mengqi Wang conducted the RT-PCR experiment, Xiaopeng Li and Songtao Jiu conducted the bioinformatics analysis, Chen Wang and Jinggui Fang revised the manuscript.

References

1. Lunn, J.E.; Furbank, R.T. Sucrose biosynthesis in C4 plants. *New Phytol.* **1999**, *143*, 221–237. [CrossRef]
2. Chen, A.; He, S.; Li, F.; Li, Z.; Ding, M.; Liu, Q.; Rong, J. Analyses of the sucrose synthase gene family in cotton: Structure, phylogeny and expression patterns. *BMC Plant Biol.* **2012**, *12*, 85. [CrossRef] [PubMed]
3. Yang, J.; Zhang, J.; Wang, Z.; Zhu, Q. Activities of starch hydrolytic enzymes and sucrose-phosphate synthase in the stems of rice subjected to water stress during grain filling. *J. Exp. Bot.* **2001**, *52*, 2169–2179. [CrossRef] [PubMed]
4. Strand, A.; Foyer, C.H.; Gustafsson, P.; Gardestrom, P.; Hurry, V. Altering flux through the sucrose biosynthesis pathway in transgenic *Arabidopsis thaliana* modifies photosynthetic acclimation at low temperatures and the development of freezing tolerance. *Plant Cell Environ.* **2003**, *26*, 523–536. [CrossRef]
5. Ciereszko, I.; Johansson, H.; Kleczkowski, L.A. Sucrose and light regulation of a cold-inducible UDP-glucose pyrophosphorylase gene via a hexokinase-independent and abscisic acid-insensitive pathway in Arabidopsis. *Biochem. J.* **2001**, *354*, 67–72. [CrossRef] [PubMed]
6. Stitt, M.; Muller, C.; Matt, P.; Gibon, Y.; Carillo, P.; Morcuende, R.; Scheible, W.R.; Krapp, A. Steps towards an integrated view of nitrogen metabolism. *J. Exp. Bot.* **2002**, *53*, 959–970. [CrossRef] [PubMed]
7. Vaughn, M.W.; Harrington, G.N.; Bush, D.R. Sucrose-mediated transcriptional regulation of sucrose symporter activity in the phloem. *Proc. Natl. Acad. Sci. USA* **2002**, *99*, 10876–10880. [PubMed]
8. Zourelidou, M.; de Torres-Zabala, M.; Smith, C.; Bevan, M.W. Storekeeper defines a new class of plant-specific DNA-binding proteins and is a putative regulator of patatin expression. *Plant J.* **2002**, *30*, 489–497. [CrossRef] [PubMed]
9. Gaudin, V.; Lunness, P.A.; Fobert, P.R.; Towers, M.; Riou-Khamlichi, C.; Murray, J.A.; Coen, E.; Doonan, J.H. The expression of *D-cyclin* genes defines distinct developmental zones in snapdragon apical meristems and is locally regulated by the *Cycloidea* gene. *Plant Physiol.* **2000**, *122*, 1137–1148. [PubMed]
10. Ohto, M.; Onai, K.; Furukawa, Y.; Aoki, E.; Araki, T.; Nakamura, K. Effects of sugar on vegetative development and floral transition in Arabidopsis. *Plant Physiol.* **2001**, *127*, 252–261. [CrossRef] [PubMed]
11. Uggla, C.; Magel, E.; Moritz, T.; Sundberg, B. Function and dynamics of auxin and carbohydrates during earlywood/latewood transition in scots pine. *Plant Physiol.* **2001**, *125*, 2029–2039. [CrossRef] [PubMed]
12. Iraqi, D.; Tremblay, F.M. Analysis of carbohydrate metabolism enzymes and cellular contents of sugars and proteins during spruce somatic embryogenesis suggests a regulatory role of exogenous sucrose in embryo development. *J. Exp. Bot.* **2001**, *52*, 2301–2311. [CrossRef] [PubMed]
13. Rook, F.; Corke, F.; Card, R.; Munz, G.; Smith, C.; Bevan, M.W. Impaired sucrose-induction mutants reveal the modulation of sugar-induced starch biosynthetic gene expression by abscisic acid signalling. *Plant J.* **2001**, *26*, 421–433. [CrossRef] [PubMed]
14. Geigenberger, P.; Stitt, M. Sucrose synthase catalyzes a readily reversible reaction in vivo in developing potato tubers and other plant tissues. *Planta* **1993**, *189*, 329–339. [CrossRef] [PubMed]
15. Koch, K. Sucrose metabolism: Regulatory mechanisms and pivotal roles in sugar sensing and plant development. *Curr. Opin. Plant Biol.* **2004**, *7*, 235–246. [PubMed]
16. Jiang, N.; Jin, L.F.; da Silva, J.A.T.; Islam, M.D.Z.; Gao, H.W.; Liu, Y.Z.; Peng, S.A. Activities of enzymes directly related with sucrose and citric acid metabolism in citrus fruit in response to soil plastic film mulch. *Sci. Hortic.* **2014**, *168*, 73–80. [CrossRef]

17. Xu, S.M.; Brill, E.; Llewellyn, D.J.; Furbank, R.T.; Ruan, Y.L. Overexpression of a potato sucrose synthase gene in cotton accelerates leaf expansion, reduces seed abortion, and enhances fiber production. *Mol. Plant* **2012**, *5*, 430–441. [CrossRef] [PubMed]

18. Hockema, B.R.; Etxeberria, E. Metabolic contributors to drought-enhanced accumulation of sugars and acids in oranges. *J. Am. Soc. Hortic. Sci.* **2001**, *126*, 599–605.

19. Coleman, H.D.; Yan, J.; Mansfield, S.D. Sucrose synthase affects carbon partitioning to increase cellulose production and altered cell wall ultrastructure. *Proc. Natl. Acad. Sci. USA* **2009**, *106*, 13118–13123. [CrossRef] [PubMed]

20. Baroja-Fernández, E.; Muñoz, F.J.; Li, J.; Bahaji, A.; Almagro, G.; Montero, M.; Etxeberria, E.; Hidalgo, M.; Sesma, M.T.; Pozueta-Romero, J. Sucrose synthase activity in the *sus1/sus2/sus3/sus4* Arabidopsis mutant is sufficient to support normal cellulose and starch production. *Proc. Natl. Acad. Sci. USA* **2012**, *109*, 321–326. [PubMed]

21. Baier, M.C.; Keck, M.; Gödde, V.; Niehaus, K.; Küster, H.; Hohnjec, N. Knockdown of the symbiotic sucrose synthase *MtSucS1* affects arbuscule maturation and maintenance in mycorrhizal roots of *Medicago truncatula*. *Plant Physiol.* **2010**, *152*, 1000–1014. [CrossRef] [PubMed]

22. Xiao, X.H.; Tang, C.R.; Fang, Y.J.; Yang, M.; Zhou, B.H.; Qi, J.; Zhang, Y. Structure and expression profile of the sucrose synthase gene family in the rubber tree: Indicative of roles in stress response and sucrose utilization in the laticifers. *FEBS J.* **2014**, *281*, 291–305. [CrossRef] [PubMed]

23. Barratt, D.H.; Barber, L.; Kruger, N.J.; Smith, A.M.; Wang, T.L.; Martin, C. Multiple, distinct isoforms of sucrose synthase in pea. *Plant Physiol.* **2001**, *127*, 655–664. [CrossRef] [PubMed]

24. Duncan, K.A.; Hardin, S.C.; Huber, S.C. The three maize sucrose synthase isoforms differ in distribution, localization, and phosphorylation. *Plant Cell Physiol.* **2006**, *47*, 959–971. [CrossRef] [PubMed]

25. Baud, S.; Vaultier, M.N.; Rochat, C. Structure and expression profile of the sucrose synthase multigene family in Arabidopsis. *J. Exp. Bot.* **2004**, *55*, 397–409. [CrossRef] [PubMed]

26. Horst, I.; Welham, T.; Kelly, S.; Kaneko, T.; Sato, S.; Tabata, S.; Parniske, M.; Wang, T.L. TILLING mutants of *Lotus japonicus* reveal that nitrogen assimilation and fixation can occur in the absence of nodule-enhanced sucrose synthase. *Plant Physiol.* **2007**, *144*, 806–820. [CrossRef] [PubMed]

27. Hirose, T.; Scofield, G.N.; Terao, T. An expression analysis profile for the entire *sucrose synthase* gene family in rice. *Plant Sci.* **2008**, *174*, 534–543. [CrossRef]

28. Zou, C.; Lu, C.; Shang, H.; Jing, X.; Cheng, H.; Zhang, Y.; Song, G. Genome-wide analysis of the *Sus* gene family in cotton. *J. Integr. Plant Biol.* **2013**, *55*, 643–653. [CrossRef] [PubMed]

29. An, X.; Chen, Z.; Wang, J.; Ye, M.; Ji, L.; Wang, J.; Liao, W.; Ma, H. Identification and characterization of the *Populus* sucrose synthase gene family. *Gene* **2014**, *539*, 58–67. [PubMed]

30. Jiang, S.Y.; Chi, Y.H.; Wang, J.Z.; Zhou, J.X.; Cheng, Y.S.; Zhang, B.L.; Ma, A.; Vanitha, J.; Ramachandran, S. Sucrose metabolism gene families and their biological functions. *Sci. Rep.* **2015**, *5*, 17583. [CrossRef] [PubMed]

31. Bieniawska, Z.; Paul Barratt, D.H.; Garlick, A.P.; Thole, V.; Kruger, N.J.; Martin, C.; Zrenner, R.; Smith, A.M. Analysis of the sucrose synthase gene family in Arabidopsis. *Plant J.* **2007**, *49*, 810–828. [PubMed]

32. Wang, A.Y.; Kao, M.H.; Yang, W.H.; Sayion, Y.; Liu, L.F.; Lee, P.D.; Su, J.C. Differentially and developmentally regulated expression of three rice *sucrose synthase* genes. *Plant Cell Physiol.* **1999**, *40*, 800–807. [CrossRef] [PubMed]

33. Komatsu, A.; Moriguchi, T.; Koyama, K.; Omura, M.; Akihama, T. Analysis of *sucrose synthase* genes in citrus suggests different roles and phylogenetic relationships. *J. Exp. Bot.* **2002**, *53*, 61–71. [CrossRef] [PubMed]

34. Grimplet, J.; Adam-Blondon, A.F.; Bert, P.F.; Bitz, O.; Cantu, D.; Davies, C.; Delrot, S.; Pezzotti, M.; Rombauts, S.; Cramer, G.R. The grapevine gene nomenclature system. *BMC Genom.* **2014**, *15*, 1077. [CrossRef] [PubMed]

35. Shangguan, L.F.; Song, C.N.; Leng, X.P.; Kayesh, E.; Sun, X.; Fang, J.G. Mining and comparison of the genes encoding the key enzymes involved in sugar biosynthesis in apple, grape, and sweet orange. *Sci. Hortic.* **2014**, *165*, 311–318. [CrossRef]

36. Larkin, M.A.; Blackshields, G.; Brown, N.P.; Chenna, R.; McGettigan, P.A.; McWilliam, H.; Valentin, F.; Wallace, I.M.; Wilm, A.; Lopez, R.; et al. Clustal W and Clustal X version 2.0. *Bioinformatics* **2007**, *23*, 2947–2948. [CrossRef] [PubMed]

37. Tamura, K.; Peterson, D.; Peterson, N.; Stecher, G.; Nei, M.; Kumar, S. MEGA5: Molecular Evolutionary Genetics Analysis using Maximum Likelihood, Evolutionary Distance, and Maximum Parsimony Methods. *Mol. Biol. Evol.* **2011**, *28*, 2731–2739. [CrossRef] [PubMed]

38. Yang, Z. PAML 4: Phylogenetic analysis by maximum likelihood. *Mol. Biol. Evo.* **2007**, *24*, 1586–1591.

39. Tajima, F. Unbiased estimation of evolutionary distance between nucleotide sequences. *Mol. Biol. Evol.* **1993**, *10*, 677–688. [PubMed]

40. Vision, T.J.; Brown, D.G.; Tanksley, S.D. The origins of genomic duplications in Arabidopsis. *Science* **2000**, *290*, 2114–2117. [CrossRef] [PubMed]

41. Hughes, A.L. The Evolution of Functionally Novel Proteins after Gene Duplication. *Proc. R. Soc. B-Biol. Sci.* **1994**, *256*, 119–124. [CrossRef] [PubMed]

42. Jaillon, O.; Aury, J.M.; Noel, B.; Policriti, A.; Clepet, C.; Casagrande, A.; Choisne, N.; Aubourg, S.; Vitulo, N.; Jubin, C.; et al. The grapevine genome sequence suggests ancestral hexaploidization in major angiosperm phyla. *Nature* **2007**, *449*, 463–467. [CrossRef] [PubMed]

43. Lunn, J.E. Evolution of sucrose synthesis. *Plant Physiol.* **2002**, *128*, 1490–1500. [PubMed]

44. Yang, Z.; Nielsen, R. Estimating synonymous and nonsynonymous substitution rates under realistic evolutionary models. *Mol. Biol. Evol.* **2000**, *17*, 32–43. [CrossRef] [PubMed]

45. Fasoli, M.; Dal Santo, S.; Zenoni, S.; Tornielli, G.B.; Farina, L.; Zamboni, A.; Porceddu, A.; Venturini, L.; Bicego, M.; Murino, V.; et al. The grapevine expression atlas reveals a deep transcriptome shift driving the entire plant into a maturation program. *Plant Cell* **2012**, *24*, 3489–3505. [CrossRef] [PubMed]

46. Fung, R.W.; Gonzalo, M.; Fekete, C.; Kovacs, L.G.; He, Y.; Marsh, E.; McIntyre, L.M.; Schachtman, D.P.; Qiu, W. Powdery mildew induces defense-oriented reprogramming of the transcriptome in a susceptible but not in a resistant grapevine. *Plant Physiol.* **2008**, *146*, 236–249. [CrossRef] [PubMed]

47. Albertazzi, G.; Milc, J.; Caffagni, A.; Francia, E.; Roncaglia, E.; Ferrari, F.; Tagliafico, E.; Stefania, E.; Pecchionia, N. Gene expression in grapevine cultivars in response to Bois Noir phytoplasma infection. *Plant Sci.* **2009**, *176*, 792–804. [CrossRef]

48. Vega, A.; Gutierrez, R.A.; Pena-Neira, A.; Cramer, G.R.; Arce-Johnson, P. Compatible GLRaV-3 viral infections affect berry ripening decreasing sugar accumulation and anthocyanin biosynthesis in *Vitis vinifera*. *Plant Mol. Biol.* **2011**, *77*, 261–274. [CrossRef] [PubMed]

49. Islam, M.Z.; Hu, X.M.; Jin, L.F.; Liu, Y.Z.; Peng, S.A. Genome-wide identification and expression profile analysis of citrus *sucrose synthase* genes: Investigation of possible roles in the regulation of sugar accumulation. *PLoS ONE* **2014**, *9*, e113623. [CrossRef] [PubMed]

50. Wang, Z.; Wei, P.; Wu, M.; Xu, Y.; Li, F.; Luo, Z.; Zhang, J.; Chen, A.; Xie, X.; Cao, P.; et al. Analysis of the *sucrose synthase* gene family in tobacco: Structure, phylogeny, and expression patterns. *Planta* **2015**, *242*, 153–166. [PubMed]

51. Frugoli, J.A.; McPeek, M.A.; Thomas, T.L.; McClung, C.R. Intron loss and gain during evolution of the catalase gene family in angiosperms. *Genetics* **1998**, *149*, 355–365. [PubMed]

52. Lecharny, A.; Boudet, N.; Gy, I.; Aubourg, S.; Kreis, M. Introns in, introns out in plant gene families: A genomic approach of the dynamics of gene structure. *J. Struct. Funct. Genom.* **2003**, *3*, 111–116. [CrossRef]

53. Tang, H.; Bowers, J.E.; Wang, X.; Ming, R.; Alam, M.; Paterson, A.H. Synteny and collinearity in plant genomes. *Science* **2008**, *320*, 486–488. [CrossRef] [PubMed]

54. Cannon, S.B.; Mitra, A.; Baumgarten, A.; Young, N.D.; May, G. The roles of segmental and tandem gene duplication in the evolution of large gene families in *Arabidopsis thaliana*. *BMC Plant Biol.* **2004**, *4*, 10. [CrossRef] [PubMed]

55. Ghiurcuta, C.G.; Moret, B.M.E. Evaluating synteny for improved comparative studies. *Bioinformatics* **2014**, *30*, 9–18. [CrossRef] [PubMed]

56. Zhang, D.Q.; Xu, B.H.; Yang, X.H.; Zhang, Z.Y.; Li, B.L. The *sucrose synthase* gene family in *Populus*: Structure, expression, and evolution. *Tree Genet Genomes* **2011**, *7*, 443–456. [CrossRef]

57. Flagel, L.E.; Wendel, J.F. Gene duplication and evolutionary novelty in plants. *New Phytol.* **2009**, *183*, 557–564. [CrossRef] [PubMed]

58. Li, M.; Feng, F.; Cheng, L. Expression patterns of genes involved in sugar metabolism and accumulation during apple fruit development. *PLoS ONE* **2012**, *7*, e33055. [CrossRef] [PubMed]

59. Barrero, S.C.; Hernando, A.S.; Gonzalez, M.P.; Carbonero, P. Structure, expression profile and subcellular localisation of four different sucrose synthase genes from barley. *Planta* **2011**, *234*, 391–403. [CrossRef] [PubMed]

60. Kleines, M.; Elster, R.C.; Rodrigo, M.J.; Blervacq, A.S.; Salamini, F.; Bartels, D. Isolation and expression analysis of two stress-responsive *sucrose-synthase* genes from the resurrection plant *Craterostigma plantagineum* (Hochst.). *Planta* **1999**, *209*, 13–24. [CrossRef] [PubMed]

9

Identification of Drought-Responsive MicroRNAs from Roots and Leaves of Alfalfa by High-Throughput Sequencing

Yue Li [†], Liqiang Wan [†], Shuyi Bi, Xiufu Wan, Zhenyi Li, Jing Cao, Zongyong Tong, Hongyu Xu, Feng He and Xianglin Li *

Institute of Animal Sciences, Chinese Academy of Agricultural Sciences, Beijing 100193, China; liyue_s@163.com (Y.L.); wanliqiang@caas.cn (L.W.); bsy9239@163.com (S.B.); xiufuwan@163.com (X.W.); lizhenyily@163.com (Z.L.); pandajing0919@163.com (J.C.); dradon.tong@163.com (Z.T.); xhy_rensheng@163.com (H.X.); hefeng@caas.cn (F.H.)
* Correspondence: lixl@iascaas.net.cn
† These authors contributed equally to this work.

Academic Editor: Bin Yu

Abstract: Alfalfa, an important forage legume, is an ideal crop for sustainable agriculture and a potential crop for bioenergy resources. Drought, one of the most common environmental stresses, substantially affects plant growth, development, and productivity. MicroRNAs (miRNAs) are newly discovered gene expression regulators that have been linked to several plant stress responses. To elucidate the role of miRNAs in drought stress regulation of alfalfa, a high-throughput sequencing approach was used to analyze 12 small RNA libraries comprising of four samples, each with three biological replicates. From the 12 libraries, we identified 348 known miRNAs belonging to 80 miRNA families, and 281 novel miRNAs, using Mireap software. Eighteen known miRNAs in roots and 12 known miRNAs in leaves were screened as drought-responsive miRNAs. With the exception of miR319d and miR157a which were upregulated under drought stress, the expression pattern of drought-responsive miRNAs was different between roots and leaves in alfalfa. This is the first study that has identified miR3512, miR3630, miR5213, miR5294, miR5368 and miR6173 as drought-responsive miRNAs. Target transcripts of drought-responsive miRNAs were computationally predicted. All 447 target genes for the known miRNAs were predicted using an online tool. This study provides a significant insight on understanding drought-responsive mechanisms of alfalfa.

Keywords: alfalfa; drought; microRNA; small RNA; differential expression

1. Introduction

Alfalfa (*Medicago sativa* L.) is an important forage species, with high nutritional quality and high yield [1]. As a legume plant, alfalfa is capable of fixing nitrogen to nitrate in nodules, by establishing a symbiotic relationship with *Rhizobium* in the root system, making it an ideal crop for sustainable agriculture. Thus, planting alfalfa can also improve the condition of the soil and reduce the usage of fertilizer. Alfalfa also has the potential to become a crop for bioenergy resources, with many appropriate attributes including high biomass yield potential [2,3]. However, the yield of alfalfa is often constrained by diverse abiotic stresses, including drought. Drought is a common environmental stress, affecting plants productivity [4,5]. Understanding the molecular mechanism of the alfalfa response to drought stress is a necessary step towards improving the drought tolerance of alfalfa.

Plants respond to stresses by regulating the expression of specific genes to avoid or minimize cellular damage, in order to adapt to stress conditions [6]. Gene regulation occurs at multiple levels, including transcriptional, post-transcriptional, and epigenetic levels. At the transcriptional level, many genes and transcriptional factors related to drought stress responses have been identified, including those involved with abscisic acid (ABA) regulation. Through the action of a number of known transcription factors, ABA-dependent and independent pathways can be induced by drought [7], which results in the activation of Late Embryogenesis-Dependent (LEA) protein synthesis, active oxygen scavenging enzymes, and osmolytes [8,9]. At the epigenetic level, evidence suggests that DNA methylation, histone modification and chromatin remodeling are related to regulating the plant response to drought [10–12].

Recently, the discovery of MicroRNAs (miRNAs) sheds light on post-transcriptional gene regulation. MiRNAs are 21 to 24 nt in length, and are noncoding small RNAs (sRNAs) that negatively regulate gene expression [5,13,14]. In plants, the biogenesis of miRNAs has been reviewed by several articles [5,13,14]. Briefly, the miRNA gene is transcribed to primary miRNAs by polymerase II; then the primary miRNA is 5′capped and 3′polyadenylated to form a classic stem-loop structure that is processed into a pre-miRNA by Dicer-Like Protein 1 (DCL1); subsequently, the pre-miRNA is cleaved into a double strand miRNA:miRNA* duplex by DCL1 in the nucleus, and the duplex is then separated into miRNA and miRNA* by helicase in cytoplasm. MiRNAs repress gene expression through guiding an RNA-induced silencing complex (RISC) to cleave target mRNAs, or inhibit translation of target mRNAs. Plant miRNAs are highly complementary to target mRNAs, and miRNA sequences are usually conserved in related organisms. Unlike miRNAs, small interfering RNAs (siRNAs) are double-stranded RNA, and are processed from double-stranded precursors [15,16]. MiRNAs as well as siRNAs can be incorporated into RISC, thus repressing mRNA expression [17].

It has now been accepted that miRNAs play vital roles on multiple crucial biological and metabolic processes in plants. For example, leaf morphology was severely affected by miR319 overexpression in the *jaw-D* mutant [18]. Changing the expression level of miR172 can affect flower morphology [19]. A group of miRNAs also play pivotal regulatory roles in the plant response to various stresses, such as hypoxia [20], salinity [21], cold [22], UV-B radiation [23], nutrient deficiency [24], heavy metal [25], and drought [4,26].

Studies from sequencing, microarray, quantitative reverse transcription polymerase chain reaction (qRT-PCR) and Northern blot have demonstrated that the expression level of miRNAs was altered by drought in many plant species, including *Medicago truncatula* [27], *Vigna unguiculata* [28], *Manihot esculenta* [29], *Solanum tuberosum* [30], *Gossypium* [31], *Nicotiana tabacum* [32], *populous* [33], *Oryza sativa* [6], *Saccharum officinarum* [34], *Panicum virgatum* [35], *Hordeum vulgare* [36], *Triticum turgidum* L. ssp. *durum* [37,38], and foxtail millet [39]. Importantly, genes associated with the miRNA pathway, such as *DCL1* and *ARGONAUTE* (*AGO*) genes, were up-regulated under drought stress, implying the involvement of miRNA in plant adaptation to drought [40]. Conversely, changing the expression level of miRNAs can affect plants' response to drought stress [41]. All these studies suggest that miRNAs are a potentially powerful tool for modifying drought resistance in alfalfa and other legumes.

In alfalfa, only salinity-regulated and fall dormancy-related miRNAs have been identified so far, both using the publicly available genome of *Medicago truncatula* [1,21]. In 2014, Fan et al. identified some fall dormancy-related miRNAs in two varieties of alfalfa (Maverick and CUF101). In 2015, Long et al. identified many known miRNAs, along with 68 miRNA candidates [21]. However, none of the previous studies have explored drought-responsive miRNAs. Identifying these drought responsive miRNAs is valuable for investigating miRNA-mediated gene regulation in alfalfa. In this context, the goals of our work are: (i) to identify target miRNAs that may regulate stress responses to drought in alfalfa, and explore the underlying mechanisms for miRNA function in the drought stress response of alfalfa; and (ii) to discover novel miRNAs in alfalfa. In this study, 12 sRNAs libraries from the leaves and roots of alfalfa plants in response to control or drought conditions were established and sequenced with the high throughput sequencing Hiseq2500 platform. The data set of 12 sRNAs libraries from alfalfa was analyzed in silico. We identified 348 known miRNAs and predicted 281 novel miRNAs in alfalfa. MiRNA qRT-PCR was also adopted for validation of the expression of selected miRNAs, which was examined by high throughput sequencing. Furthermore, characterization of target genes was performed by using bioinformatic approaches. Through high-throughput sequencing and bioinformatics analysis, known drought stress-responsive miRNAs and miRNA candidates in alfalfa were identified. This study will be very helpful for understanding post-transcriptional regulation under drought stress in alfalfa, and improving the drought tolerance of alfalfa and other legumes.

2. Materials and Methods

2.1. Plant Materials and Experiment Design

Medicago sativa L. cv. Aohan was used in this study. This cultivar was kindly provided by Dr. Liqiang Wan (Institute of Animal Sciences, Chinese Academy of Agricultural Sciences, Beijing, China). Alfalfa plants were grown in pots with a diameter of three inches, containing a mixture of sand:vermiculite (1:1 *v/v*) in a growth chamber at 25–28 °C under a 16 h light/8 h dark photoperiod. Plants were supplied daily with Murashige-Skoog (MS) nutrient solution. Alfalfa plants at the age of eight weeks were then randomly separated into two groups, namely drought treatment and control groups. Drought stress treatment was imposed by withholding water supply for 10 days. The control plants received normal watering throughout the experiment. The volumetric water content of soil (VWCS) was detected before sampling by using a WET Sensor (Delta-T Devices Ltd., Cambridge, UK) which is a soil moisture sensor. The VWCS of control group was approximately 45%, and the VWCS of treatment group at the fifth day was 26.7%, at the tenth day this was 16.8% in the stress treatment.

2.2. Total RNA Isolation

Root and leaf samples from both drought and control plants were collected at the fifth and tenth days during the stress treatment, respectively. For each sample of leaf or root, three biological replicates were prepared, with each biological replicate collected from 10 plants. Samples were fast frozen in liquid nitrogen, stored at −80 °C.

Total RNA samples were extracted and then were prepared for sequencing, reverse transcription PCR and qRT-PCR. Equal quantities of RNA isolated from leaves and/or roots at each stress stage were pooled, using HiPure Plant RNA Mini Kit (Magen, Shanghai, China) according to the manufacturer's instructions.

2.3. Small RNA Library Construction and High-Throughput Sequencing

A total of four groups of RNA samples (WL: leaves with watering, WR: roots with watering, DL: leaves with drought stress, and DR: roots with drought stress), each with three biological replicates, were prepared. For each group, RNA at both five days and 10 days were equally pooled to make one sample. Thus, a total of 12 samples were used to construct sRNA libraries. Since we pool samples at different stress treatment time points (five days and 10 days), only robust and consistent responses could be detected.

Total RNA was isolated by 15% polyacrylamide gel electrophoresis, and RNA molecules that were less than 50 nt in length were enriched and ligated with proprietary adapters. The RNA samples ligated with adapters were reverse-transcribed and amplified by PCR to produce sequencing libraries. The 12 sRNA libraries from alfalfa leaves and roots were sequenced on an Illumina Hiseq 2500 (Santiago, CA, USA) platform at the Guangzhou RiboBio Company, China. The raw data has been deposited in the Sequence Read Archive of NCBI, with a SRA data study accession number of SRP094823.

2.4. Identification of Known and Novel MicroRNAs

The raw sequencing reads were processed to obtain unique sequences and read count/unique reads as per the procedure reported by Hackenberg et al. [36]. First, sRNA reads of 17–45 nt were annotated to Rfam databases (Rfam 11.0, rfam.janelia.org) [42], to identify and eliminate transfer RNA (tRNA), ribosomal RNA (rRNA), small nuclear RNA (snRNA) and small nucleolar RNA (snoRNA) sequences from the sRNA reads. Then we computed the rest of sequences for sequence redundancy, and mapped these sequences to miRBase (release 21, http://www.mirbase.org/) [43] without mismatches to identify known miRNAs. After removal of the known miRNAs, the remaining sequences were used to predict the novel miRNAs. The unique sequences were mapped to the *M. truncatula* genome version 4.0 (http://www.medicagohapmap.org/?genome) using Burrows-Wheeler Alignment (BWA) [44] to get pre-miRNA sequences for prediction of novel miRNAs.

Novel miRNAs were predicted by using Mireap [45]. Novel miRNA candidates were identified according to the criteria reported by [46]. The normalization of reads count and calculation of log Fold change were processed as [21] described.

2.5. MicroRNA Validation by qRT-PCR

The same RNA samples used for Illumina sequencing were employed in qRT-PCR analysis. Total miRNA was reverse transcribed to complementary DNA (cDNA) using the miRcute miRNA First-Strand cDNA Synthesis Kit (Tiangen, Beijing, China). According to the manufacturers' instructions, the miRNAs were polyadenylated and reverse transcribed in one step using miRNA RT Enzyme Mix (*E. coli* Poly(A) Polymerase, RTase and RNasin). The Universal RT Primer was provided in the kit. Then the first-strand cDNA was prepared for qRT-PCR analysis.

qRT-PCR was performed on an Applied Biosystems 7300 Real-Time PCR System (Applied Biosystems, Foster City, CA, USA). The reaction system was constructed according using the miRcute miRNA qRT-PCR Kit (Tiangen, Beijing, China) containing SYBR® Green detection reagents (Applied Biosystems). The cycling parameters were set according to manufacturers' recommendations. Briefly, the cycling parameters were: initial polymerase activation step for 15 min at 95 °C, 40 cycles for 20 s at 94 °C for denaturation, 34 s at 60 °C for annealing and elongation, followed by a disassociation stage. The forward primers were designed according to the miRNA sequences of interest and synthesized by Invitrogen (Carlsbad, CA, USA). The sequences of the forward primers are supplied in Table S8. The melting curves of the PCR products can be found in Figure S1. The Universal qPCR Primer was provided in the kit. The transcript abundance of each miRNA was normalized to U6 snRNA, and the $2^{-\Delta\Delta Ct}$ method was used to calculate relative expression of miRNAs [47]. In order to compare pair-wise differences in expression, a Student's *t*-test was performed by using Statistical Analysis Software (SAS) program.

2.6. MicroRNA Target Prediction and Function Analysis

Target genes of drought-responsive miRNAs in alfalfa were predicted using the psRNATarget online tool (http://plantgrn.noble.org/psRNATarget/). Gene annotation can also be accomplished by using this online tool. psRNATarget is a modified version of miRU. The *M. truncatula* spliced transcript sequences 4.0 V1 was selected as the transcript library for target search. Mature miRNA sequences responsive to drought and identified in alfalfa roots and leaves, were used as custom miRNA sequences. Default parameters for target prediction were used.

3. Results

3.1. Overview of Small RNAs from Alfalfa via High-Throughput Sequencing

A total of 12 sRNA libraries comprising of four samples (WL: leaves with watering, WR: roots with watering, DL: leaves with drought stress, and DR: roots with drought stress) were generated using the Illumina HiSeq 2500 platform, each with three biological replicates. In order to obtain high quality data sets, adaptors and low-quantity reads were removed, and 12 million to 16 million clean reads at 17–45 nt in length were obtained from each of the 12 libraries. The details of raw reads and clean reads for each library are shown in Table 1. We analyzed common/specific sequences between four groups (WL, DL, WR and DR) for the total sRNA sequences. There were 40.6% and 31.1% specific sequences in group DL and DR, respectively (Figure 1), which indicates that there are changes occurring on the molecular level. These changes may be caused by drought stress, or by other phenomena such as plant development.

Table 1. Alfalfa small RNA (sRNA) sequencing datasets. Statistics of sRNA sequences for water and drought stress libraries from *Medicago sativa* leaf and root.

Library	Replicates	Raw Reads	Clean Reads	Reads Mapped to the Genome	Match Known miRNAs
WL	WL1	16,729,829	14,718,375	13,037,667	1,030,262
	WL2	18,529,495	16,118,371	15,226,670	917,066
	WL3	18,683,242	16,026,900	15,129,492	816,007
WR	WR1	17,888,348	15,351,268	11,164,625	1,360,347
	WR2	18,777,136	15,724,912	11,131,597	1,225,474
	WR3	17,354,872	14,048,339	8,964,877	1,472,074
DL	DL1	15,668,523	12,648,491	10,760,915	1,315,694
	DL2	14,266,305	12,378,478	10,236,124	1,269,855
	DL3	14,741,343	12,407,912	10,797,039	1,191,355
DR	DR1	16,484,366	14,594,872	8,586,263	1,814,906
	DR2	14,587,651	12,844,508	8,055,680	1,662,649
	DR3	14,256,856	12,721,466	7,391,569	1,554,103

Notes: WL: leaves with watering, WR: roots with watering, DL: leaves with drought stress, and DR: roots with drought stress.

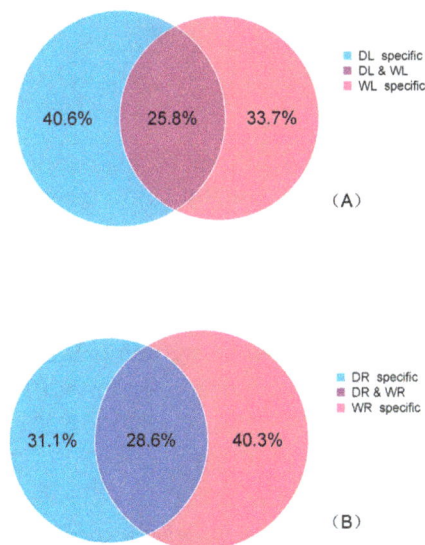

Figure 1. Venn diagram illustrating common and specific smallRNA (sRNA) sequences induced by drought stress. (**A**) Common and specific sRNA sequences in leaves; (**B**) Common and specific sRNA sequences in roots.

In order to classify the sRNA sequenced reads into different categories and identify all the miRNA sequences in the 12 libraries, we mapped the reads to specific databases. Generally, the sequenced sRNA reads mapped to the miRBase database were abundant in drought samples (DL 10.09%; DR 12.53%) compared to control samples (WL 5.93%; WR 9.04%). The percentage of the sRNA reads mapped to the tRNA database was greater in drought samples (leaves 12.71%; roots 10.98%) than in control samples (leaves 8.98%; roots 5.14%). For rRNA, the opposite was true. Both drought and control samples had large numbers of unannotated reads (WR 63.71%, WL 50.74%, DR 60.08%, DL 55.3%) (Figure 2). The number of total sequences that matched *M. truncatula* genome is given in Table 1. Sequences failing to map to the genome ranged from 5.5% to 17.3%, with exception of WR (30.9%) and DR (40.1%). These unmapped reads may have been due to unavailable genome or sequencing errors.

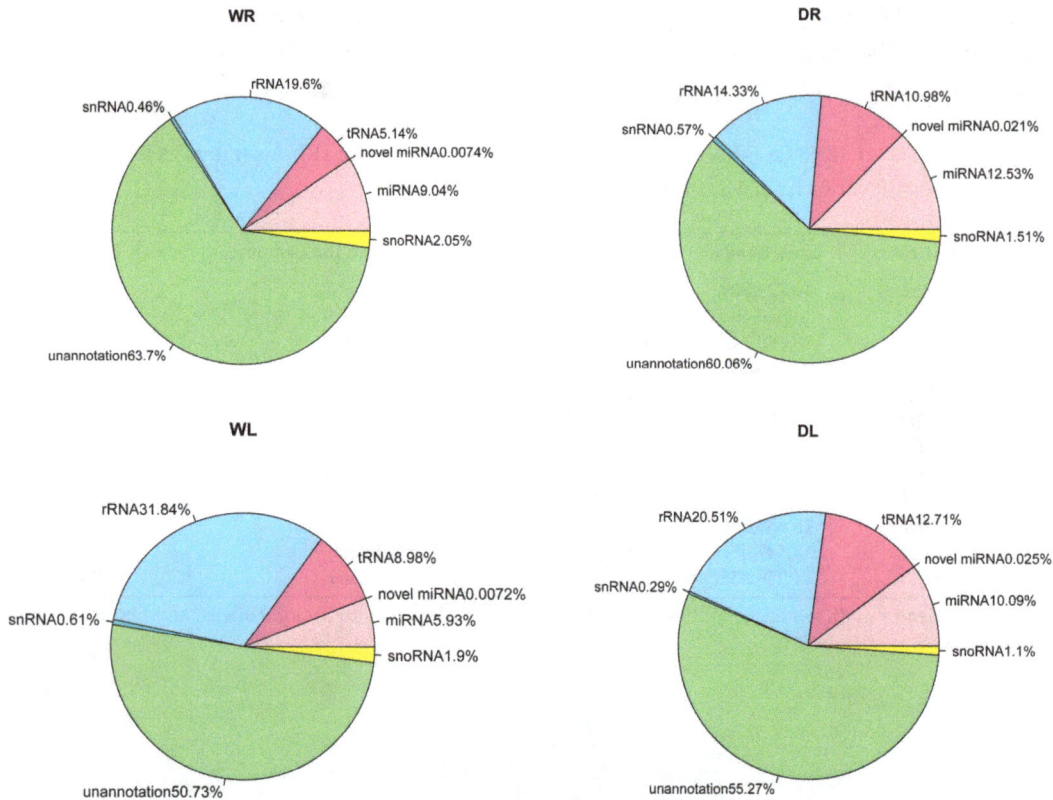

Figure 2. An overview of the frequency of different sRNA species present in the different groups. snoRNA: small nucleolar RNA; miRNA: microRNA; tRNA: transfer RNA; rRNA: ribosomal RNA; snRNA: small nuclear RNA.

3.2. Identification of Known miRNAs

Known miRNAs in alfalfa were identified by mapping sRNA sequences generated from each library to the miRNAs database (miRBase 21, released in June 2014). After a homology search and removal of miRNAs with expression levels less than 10, 287, 314, 204 and 142 miRNAs were identified from the groups WL, DL, WR and DR, respectively. There were 348 known miRNAs belonging to 80 miRNA families that were identified from the 12 libraries. The details of miRNAs of each library are listed in Tables S1 and S2. Among these miRNA families, the miR159 and miR166 families had the most reads, with exception of the replicate WR1 (miR166 and miR398 families). Of these identified miRNAs, miR166 contained the most members, including miR166a-g, miR166i and miR166u. Additionally, the most abundant miRNA was miR5213-5p followed by miR166a-3p in the DR1, DR2 and DR3 libraries. In the other libraries, miR166a-3p was the most abundant, followed by miR5213-5p or miR159 (see Tables S1 and S2).

3.3. Identification of Novel miRNAs

A total of 281 novel miRNAs were identified from the 12 sRNA libraries using Mireap software. For each library, detailed information of predicted novel miRNAs are listed in Table S3, including novel miRNAs sequences, reads length, reads number, GC contents, pre-miRNA sequences, miRNAs loci and pre-miRNA length. The read counts of these novel miRNAs ranged from 10 to 902. A total of 26 out of 281 novel miRNAs were sequenced over 100 times, while only six novel miRNAs were sequenced over 500 times. Eight novel miRNAs were found to exist in at least five libraries (Table 2), indicating that they are miRNA candidates with a higher level of confidence. Their precursor sequences, as well as the stem-loop hairpin secondary structure are shown in Figure 3.

Table 2. Eight novel miRNAs identified from *M. truncatula* genome databases.

Name	Sequences	Length	GC Contents	Loci	Minimum Free Energy (kcal/mol)
msa-N005	GUUGACCGCUCAUACGACCCCUG	23	60.87	NC_016411.2:1711404:1711490:+	−54.00
msa-N017	GUGGUGAUGGAGAUGAGGAG	20	55	NC_016410.2:44927035:44927123:−	−31.90
msa-N033	GCUUUCGGUUGCUGUCCGUCC	21	61.90	NC_016407.2:37807782:37807862:−	−22.80
msa-N043	UCUGACGAGGUUGAGGACCA	20	55	NC_016412.2:20603621:20603715:+	−22.30
msa-N046	UCUAAUCUCUGUUCCCAAUUAC	22	36.36	NC_016411.2:42295745:42295832:−	−33.90
msa-N054	ACAUGAGAUGGGUUAGACCC	20	50.00	NC_016414.2:14441574:14441672:−	−20.70
msa-N060	GGGGAUGUAGCUGAAUUUCGUC	22	50.00	NC_016408.2:4907664:4907746:−	−20.50
msa-N075	AUCCCUAACUAGUAGUUAUC	20	35.00	NC_016414.2:17652459:17652554:+	−20.30

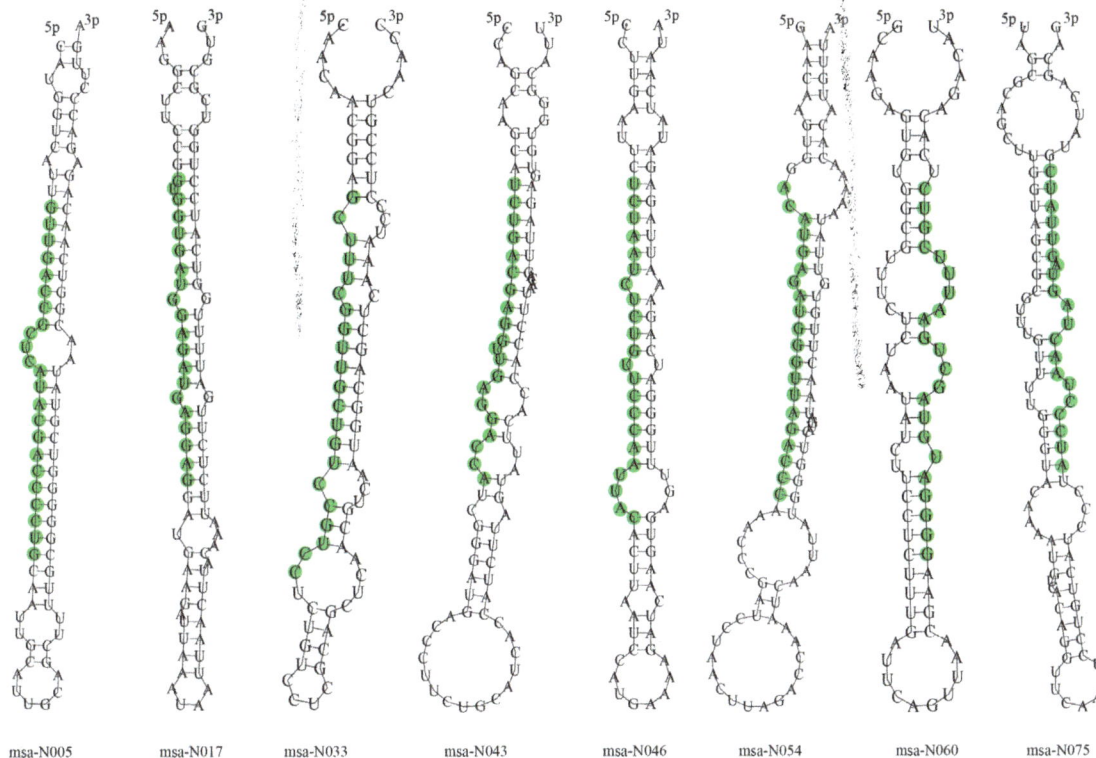

Figure 3. Representatives of precursor hairpin structures for predicted miRNAs from the *M. truncatula* genome database. Mature miRNA sequences are shown in green.

3.4. Drought-Responsive miRNAs Identified in Alfalfa

To identify drought-responsive miRNAs, miRNAs which absent from two of three biological replicates were filtered out, then the normalized expression profiles of known miRNAs in drought-treated samples were compared to the control samples using generalized linear model analysis with the edgeR package [48]. A log Fold Change (logFC) change cut-off of 1 and a *p*-value ≤ 0.05 were used to obtain the differentially expressed miRNAs [49]. These differentially expressed miRNAs between the control and drought treatment group were called as "Drought-responsive miRNAs".

A total of 12 and 18 miRNAs were observed responsive to drought treatment in alfalfa leaves and roots, respectively (Figure 4). Some of drought-responsive miRNAs reported by previous studies were also detected in our analysis. For example, miRNAs miR166 and miR398 were found to be down-regulated in alfalfa roots, while miR319 and miR157 were up-regulated in both roots and leaves. Detail information of the drought-responsive miRNAs such as log fold-change, counts per million, and p-value, are given in Tables S4 and S5. MiR396, miR159/319, miR160, miR482, miR157 and miR1507 in alfalfa roots were also found to be drought-responsive. MiR156, miR3512, miR5368, miR3630, miR6173, miR5213 and miR5294 were drought-responsive in alfalfa leaves.

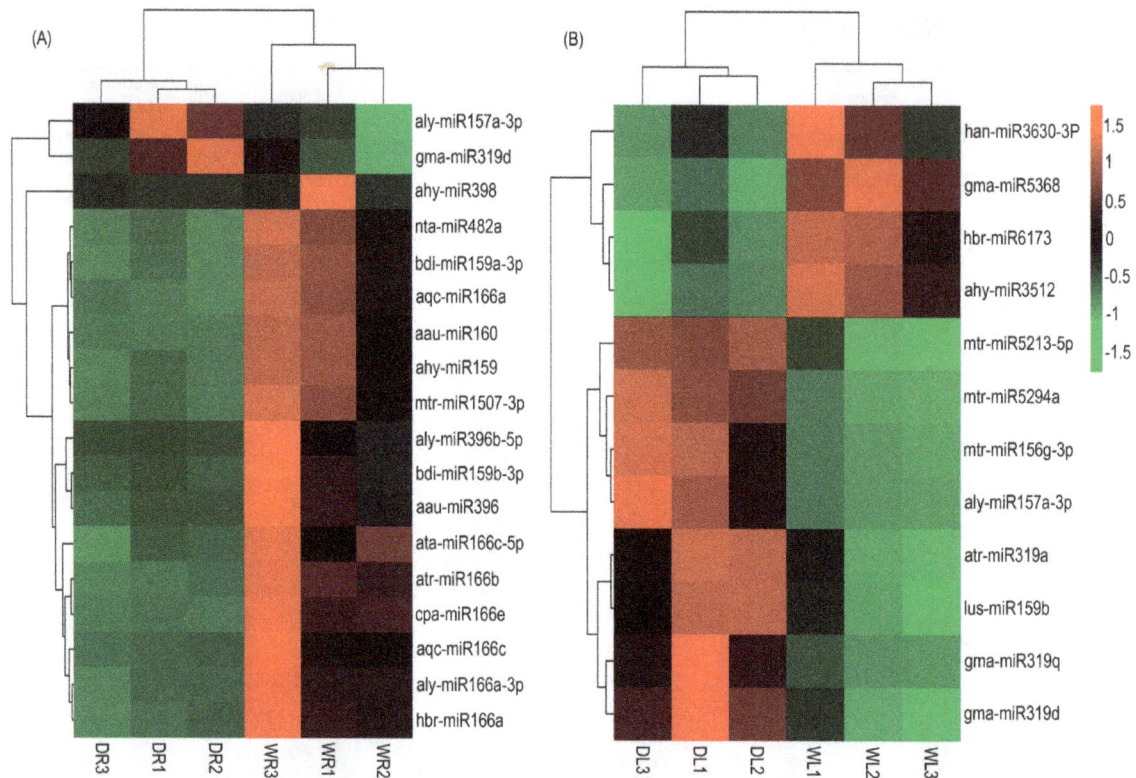

Figure 4. Heatmap representing the expression profile of the drought-responsive miRNAs. Differential expression of drought-responsive miRNAs in leaves (**A**) and roots (**B**). The upregulated miRNAs are showed in red, whereas the downregulated miRNAs are showed in green.

Our results also revealed that miR396, miR159, miR160, miR482, and miR1507 were down-regulated in alfalfa roots; miR3512, miR5368, miR3630 and miR6173 were down-regulated in leaves, whereas miR156, miR157, miR159/319, miR5213 and miR5294 were up-regulated in alfalfa leaves. The expression level of some miRNAs showed inconsistency and very high variation across replications. These miRNAs may have been be falsely identified by software, as the algorithm for discovering differential expressed miRNAs is subject to false positives. Validation of drought-responsive miRNAs by qRT-PCR experiments was therefore necessary. It is worthy to note that the expression of the homologues miRNAs belonging to the same miRNA family was consistently similar. For instance, in alfalfa roots, aqc-miR166a, hbr-miR166a and aly-miR166a-3p were all down-regulated.

We also detected other drought related miRNAs, but the expression levels of these miRNAs did not show significant changes, such as miR168, miR393, miR408 and miR2118. One of the reasons may have been due to the approach of sample pooling used in this study. The expression level of these miRNAs may have changed significantly at 5 days and/or 10 days, but we could not detect such changes in this experiment, because the significant change of expression level at one time-point could be canceled out by the expression level at another time-point after pooling the samples at five days and 10 days together.

3.5. Validation of Drought-Responsive miRNAs by qRT-PCR

To experimentally validate a number of selected miRNAs detected from the Illumina high-throughput sequencing, qRT-PCR technique was employed. Fifteen known miRNAs were tested by qPCR, and the results suggested that ahy-miR398, aau-miR396, mtr-miR1507-3p, aqc-miR166a, aly-miR166a-3p, ahy-miR3512, han-miR3630-3p and mtr-miR156g-3p showed similar expression patterns to those revealed by high-throughput sequencing analysis. However, we found that the expression level of aly-miR396b-5p as detected by qRT-PCR was inconsistent with that of our sequencing results, and the *p*-values of gma-miR319d, bdi-miR159a-3p, nta-miR482a, gma-miR5368 and mtr-miR5294a from qRT-PCR analysis showed discrepancies to the high-throughput sequencing data, these may have been caused by sequencing error or other reasons. The 15 known miRNAs are shown in Table 3. The qRT-PCR results suggest that our sequencing data are credible.

Table 3. Expression levels detected by high-throughput sequencing and quantitative reverse transcription polymerase chain reaction (qRT-PCR) of selected drought-responsive miRNAs.

Name	Normalized Read Count		*p* Value	qRT-PCR		*p* Value
	Control	Stress		Control	Stress	
Root						
gma-miR319d	19.23	43.90	0.00	1.00	1.60 ± 1.16	0.63
ahy-miR398	471.98	11.77	0.03	1.00	0.51 ± 0.11	0.01
aau-miR396	107.62	17.07	0.00	1.00	0.27 ± 0.10	0.00
mtr-miR1507-3p	17.71	2.83	0.00	1.00	0.22 ± 0.07	0.00
aqc-miR166a	11.23	1.94	0.00	1.00	0.21 ± 0.03	0.00
bdi-miR159a-3p	5.26	0.98	0.00	1.00	0.54 ± 0.23	0.12
aly-miR166a-3p	1915.66	673.23	0.02	1.00	0.34 ± 0.12	0.01
nta-miR482a	121.66	37.75	0.01	1.00	0.86 ± 0.11	0.27
aly-miR396b-5p	59.07	12.52	0.03	1.00	1.32 ± 0.54	0.59
Leaf						
ahy-miR3512	7.43	0.96	0.00	1.00	0.67 ± 0.09	0.03
gma-miR5368	30.07	9.89	0.00	1.00	0.63 ± 0.18	0.11
han-miR3630-3p	16.40	5.89	0.00	1.00	0.78 ± 0.07	0.04
mtr-miR5294a	6.39	53.70	0.01	1.00	1.04 ± 0.12	0.76
gma-miR319d	103.05	605.32	0.00	1.00	1.76 ± 0.62	0.28
mtr-miR156g-3p	1.04	17.56	0.01	1.00	2.49 ± 0.21	0.00

3.6. Targets of Drought-Responsive miRNAs and Their Functional Analysis

The mature sequences of the 18 miRNAs in roots and 12 miRNAs in leaves that were modulated by drought, were used to search for their targets in alfalfa (Tables S4 and S5). Target genes of some of the miRNAs identified in this study are already known, such as miR166, miR159/319, miR160, miR396, miR398, miR482, mir156 and miR157. As for the drought responsive miRNAs of which targets are unknown, we used the online tool psRNATarget to predict their targets, by matching the miRNAs to Mt 4.0, and the results are shown in Table 4.

A total of 445 target genes for the drought-responsive miRNAs, and 196 target genes for the eight novel miRNAs were predicted, and detailed information of the target genes including target ID and functional annotation are shown in Tables S6 and S7. We noted that most of these miRNAs had more than one predicted targets. Some miRNAs had no predicted target due to lack of genome information. A number of these predicted targets were involved in metabolism, growth and response to stresses. Some predicted targets play vital roles in abiotic stress responses. For instance, ahy-miR3512 was predicted to target three genes (Table 4), one of which was *spermidine synthase* (Medtr8g063940.1), which is involved in growth and resistance to adverse stresses including heat, salinity and drought.

Table 4. Predicted targets for drought-responsive miRNAs identified from *M. truncatula* genome databases.

miRNA Name	Target Accession	Expectation	UPE	Target Start	Target End	Inhibition	Target Description
han-miR3630-3p	Medtr0168s0060.1	2.5	18.13	35	56	Translation	zein-binding protein
han-miR3630-3p	Medtr0021s0360.1	3	16.35	3893	3913	Cleavage	phospholipid-transporting ATPase-like protein
han-miR3630-3p	Medtr1g015620.1	3	17.66	1832	1852	Cleavage	myosin heavy chain
han-miR3630-3p	Medtr2g039770.1	3	17.55	1184	1203	Cleavage	disease resistance protein (TIR-NBS-LRR class)
hbr-miR6173	Medtr8g009970.1	3	15.51	593	612	Cleavage	splicing factor 3A subunit 2
ahy-miR3512	Medtr8g063940.1	2	21.78	980	999	Cleavage	spermidine synthase
ahy-miR3512	Medtr3g058220.1	2.5	15.17	1732	1751	Cleavage	cytochrome P450 family 71 protein
ahy-miR3512	Medtr7g028160.1	3	13.87	1624	1643	Cleavage	TCP family transcription factor
aly-miR157a-3p	Medtr8g101880.1	3	15.82	23	42	Cleavage	ATP synthase G subunit family protein
mtr-miR5213-5p	Medtr3g025460.1	0	14.58	73	94	Cleavage	neutral/alkaline invertase
mtr-miR5213-5p	Medtr4g014580.1	1.5	13.06	121	142	Cleavage	TIR-NBS-LRR class disease resistance protein
mtr-miR1507-3p	Medtr7g091550.1	2	20.51	883	904	Cleavage	NBS-LRR disease resistance protein
mtr-miR1507-3p	Medtr7g078630.1	2	22.30	655	676	Cleavage	cysteine proteinase superfamily protein
mtr-miR1507-3p	Medtr5g015600.1	3	15.75	1400	1419	Cleavage	condensin-2 complex subunit G2, putative
nta-miR482a	Medtr6g463480.1	1	17.17	329	350	Cleavage	kinesin KIF2A-like protein
nta-miR482a	Medtr6g477950.1	1	13.93	39	60	Cleavage	B3 DNA-binding domain protein
nta-miR482a	Medtr7g088640.1	2	17.84	550	569	Cleavage	NBS-LRR type disease resistance protein
nta-miR482a	Medtr7g026400.1	3	14.25	1263	1283	Cleavage	response regulator receiver domain protein
nta-miR482a	Medtr5g099160.1	3	14.68	67	86	Cleavage	cation/calcium exchanger, putative

Notes: UPE: unpaired energy.

4. Discussion

In this study, we constructed 12 libraries from different tissues of alfalfa treated with drought stress and well watering (control). All libraries were sequenced using the Illumina Hiseq 2500 platform. Small RNA deep sequencing from leaves and roots of alfalfa and comprehensive and systematic analysis were performed. We first identified drought-responsive miRNAs and their targets, and subsequently focused on discovering novel miRNAs.

In general, an RNAseq tag density of 2–5 million reads is sufficient for miRNA expression profiling and discovery applications [49]. In the current study, approximately 14–18 million reads for each library were generated by high-throughput sequencing. Furthermore, we sequenced three biological replicates for WL, DL, WR and DR. Our ur sRNA sequencing depth is sufficient not only for profiling of the miRNAs expression, but also for the discovery of poorly-expressed novel miRNAs.

Approximately 82–94% of the reads sequenced in alfalfa leaves mapped to the *M. truncatula* genome, indicating that most of the miRNAs of *M. truncatula* and *M. sativa* are identical in leaves. However, only 58–73% of the reads sequenced in alfalfa roots matched to the *M. truncatula*. This is similar to several previous studies, on peach [50] and on *M. truncatula* [25]. Known miRNAs comprised 5.1–12.9% of the mapped reads, with DL and DR having higher proportions of known miRNAs (10.1–12.5%) while WL and WR had a lower frequency (5.9–9.0%). This indicates that the drought treatment activated some miRNA-related pathways. Unannotated sequences ranged from 50.7%–63.7%, which strongly implies the existence of large amounts of undiscovered sRNAs in alfalfa.

By In silico analysis, we identified 348 known miRNAs from the 12 libraries. The miR166, miR159, miR482 and miR2118 families were abundantly expressed. MiR166a-3p and miR5213-5p were the most abundant miRNAs. These highly expressed miRNAs may play important regulatory roles in gene expression. For example, miR166 is involved in plant organism morphogenesis, such as shoot apical meristem and floral development [51].

Discovering novel miRNAs is one of the advantages of next-generation sequencing compared to other technologies such as microarrays. In this study, 281 new miRNA candidates were predicted by computational methods. However, only a few miRNAs were commonly expressed in roots and/or leaves. Most of the novel miRNAs were uniquely expressed in each library, thus we cannot profile the differential expression of the novel miRNAs among libraries. For profiling the expression of the novel miRNAs, enhanced sequencing depth may be required. Moreover, the expression level of the predicted miRNA candidates was relatively lower than conserved miRNAs, and this result is in agreement with previous reports [25,50,52,53]. One possible explanation for this result is that the conserved miRNAs regulate target genes which may be involved in many important metabolic processes in Viridiplantae, while the expression of nonconserved miRNAs may be environmentally inducible or tissue-specific, thus the expression level of conserved miRNAs may be higher than non-conserved miRNAs [52,53].

Besides identifying known and novel miRNAs, high-throughput sequencing technology also provides an alternative way for evaluating the expression of miRNA genes. To obtain robust and consistently-expressed drought-responsive miRNAs, we pooled the samples from different time-points of stress together when constructing sRNA libraries according to [25,32,54,55]. To discover drought-responsive miRNAs from the sequencing data, the mature miRNA expression profile of drought-treated leaves and roots was compared with the control group, to identify miRNAs significantly modulated by drought. Finally, 18 known miRNAs in roots and 12 known miRNAs in leaves were screened as drought-responsive miRNAs.

In roots, 16 out of 18 drought-responsive miRNAs belonging to miR396, miR159, miR160, miR482, miR1507, miR166, miR156 and miR398 families were down-regulated. Since miRNAs negatively regulate their target genes, it can be predicted that targets of down-regulated miRNAs during drought stress may play positive roles for drought stress responses. Concurring with expectations, miR166, which down-regulates HD-ZIPIII transcription factors, was down-regulated. This indicates that the target gene of miR166 was up-regulated. HD-ZIPIII transcription factors are important for lateral leaf development, root development and axillary meristem initiation [56,57]. In *Triticum dicoccoides* [58] and in barley [59], miR166 was down-regulated under drought stress. Inversely, miR166 was up-regulated by drought in *M. truncatula* [60], which was different from our results. Similarly, Long et al. [21] previously reported that miRNAs in *M. truncatula* and *M. sativa* presented the opposite expression pattern under salt stress. The difference of expression pattern between *M. truncatula* and *M. sativa* may be caused by their different resistances to stresses, or the fact that the sRNA libraries were pooled from different time-points, which would have complicated the results.

MiR159/319 was also down-regulated in alfalfa root. The MYB family and TCP family are targets families of this miRNA, and both of them were identified by 5′ RACE [18]. Under drought conditions, most active MYB transcription factors are involved in ABA signaling pathway. Previous studies in *Arabidopsis* spp. indicate that some *MYBs*, including *MYB2*, are positive regulators of ABA signaling [61,62]. With the decreased expression of miR159 resulting in the positive regulation of the ABA signaling pathway, down-stream of ABA signaling pathways are stimulated, including root development. Similarly, miR160, which targets auxin response factors (ARFs), also has positive regulatory roles in drought stress responses [63].

Known targets of miR398 are involved in respiration and oxidative stress [64]. We found that drought stress down-regulated miR398 in alfalfa roots, in accordance with the results in maize [65] and *M. truncatula* [27]. However, Trindade et al. [60] found the opposite in *M. truncatula*. The differences in the expression of miR398 showed here may be caused by differences in species responses, duration of drought stress, and the metabolic states of the individual plants in different studies [5].

MiR396, miR482 and miR1507 were down-regulated under drought stress. However, their currently identified roles only include development or disease resistance; therefore, it can be predicted that they may have additional targets that are yet to be identified.

Two of these drought-responsive miRNAs (gam-miR319d and aly-miR157a-3p) were up-regulated in alfalfa roots. In order to conserve water and protect the cell, miRNAs are expected to be up-regulated during drought stress, so that those processes involved in normal growth and metabolism can be shut down. However, we failed to obtain valuable information about stress resistance from their target gene annotation. Interestingly, as a member of miR159 family, gam-miR319d was expected to be down-regulated by drought, to cause the positive regulation of the ABA signaling pathway, but it did not show down-regulation in our study. However, the target gene of gam-miR319d, was the NB-ARC domain protein, a disease resistance protein, instead of the MYB transcription factor, as predicted using the online program psRNATarget. There are two possible reasons to explain this result: (1) gam-miR319d may have other unknown targets that play negative roles in drought adaption in alfalfa; (2) gam-miR319d plays no role in drought adaption, but can be stimulated by drought through an unknown mechanism. With the function of gam-miR319d still being unclear, further studies will be needed to explore the roles of gam-miR319d in drought stress.

In alfalfa leaves, most of the drought-responsive miRNAs were involved in development, substance synthesis and transport. MiR3512, miR5368, miR3630-3p and miR6137, whose targets remain unknown, were down-regulated in this experiment. Target genes of miR3512, miR3630-3p and miR6137 were obtained using the online program psRNATarget. These targets include zein-binding protein, polyol/monosaccharide transporter, spermidine synthase, cytochrome P450 family 71 protein and TCP family transcription factor.

MiR159/319 is down-regulated in roots and up-regulated in leaves, which may indicate that the same miRNA could play different roles in different tissues. In *Arabidopsis* spp., *MYB33* and *MYB101* transcripts are targets of miR159a [66,67]. Under drought conditions, *MYB33* and *MYB101* could modulate stomatal movement by regulating the ABA signal, suggesting that miR159/319 plays a positive role on drought response by decreasing stomatal conductance. Additionally, miR156, miR157, miR5213 and miR5294 were up-regulated significantly ($p < 0.01$), and their targets were involved in development or disease resistance, indicating these miRNAs play some roles under drought stress.

In summary, 348 known miRNAs have been identified from alfalfa leaves and roots, and a number of candidates for drought-responsive miRNAs have been identified, thus this attempt paves the path for better understanding the drought-responsive mechanisms of alfalfa. In addition, the identification of 300 novel miRNAs also provides promising resources for future research in understanding post-transcriptional regulation in alfalfa. Future studies should focus on the verification of novel miRNAs and their predicted targets by experimental approaches, and additionally, the effects of drought-responsive miRNAs on drought tolerance need to be illuminated. To our knowledge, our study is the first systematic and comprehensive identification of drought-responsive miRNAs in an alfalfa species. This study is valuable for improving drought tolerance and systems to mitigate crop losses under drought stress.

Supplementary Materials: The following are available online at www.mdpi.com/2073-4425/8/4/119/s1. Figure S1: Melting curves of qRT-PCR products. Table S1: Members and expression abundance of known miRNAs from leaves of alfalfa treated with drought and irrigation. Table S2: Members and expression abundance of known miRNAs from roots of alfalfa treated with drought and irrigation. Table S3: Mature and precursor sequences of predicted miRNAs and the other information including reads length, reads number, GC contents and minimum free energy. Table S4: Members and expression abundance of drought-responsive miRNAs from roots of alfalfa treated with drought and irrigation. Table S5: Members and expression abundance of drought-responsive miRNAs from leaves of alfalfa treated with drought and irrigation. Table S6: Targets of drought-responsive miRNAs from leaves of alfalfa. Table S7: Targets of drought-responsive miRNAs from roots of alfalfa. Table S8: Sequences of the forward primers.

Acknowledgments: This work was supported by the National Natural Science Foundation of China (No. 31372370) and the China Forage and Grass Research System (CARS-35-12). We thank Lihong Miao and Qixin Zhang for preparing the alfalfa plants used in the experiments.

Author Contributions: Y.L., L.W. and X.L. conceived and designed research. Y.L., S.B. and X.W. conducted experiments. Y.L., Z.L., J.C., H.X., F.H. and Z.T. analyzed data. Y.L. and L.W. wrote the manuscript. All authors read and approved the manuscript.

References

1. Fan, W.; Zhang, S.; Du, H.; Sun, X.; Shi, Y.; Wang, C. Genome-wide identification of different dormant *Medicago sativa* L. microRNAs in response to fall dormancy. *PLoS ONE* **2014**, *9*, e114612. [CrossRef] [PubMed]

2. Barnes, D.K.; Goplen, R.P.; Baylor, J.E. *Alfalfa and Alfalfa Improvement*; American Society of Agronomy, Crop Science Society of America and Soil Science Society of America: Madison, WI, USA, 1988; pp. 1–24.

3. Boller, B.; Posselt, U.K.; Veronesi, F.; Boller, B.; Posselt, U.K.; Veronesi, F. *Fodder Crops and Amenity Grasses*; Springer: New York, NY, USA, 2010; pp. 395–437.

4. Akdogan, G.; Tufekci, E.D.; Uranbey, S.; Unver, T. miRNA-based drought regulation in wheat. *Funct. Integr. Genom.* **2016**, *16*, 221–233. [CrossRef] [PubMed]

5. Ding, Y.; Tao, Y.; Zhu, C. Emerging roles of microRNAs in the mediation of drought stress response in plants. *J. Exp. Bot.* **2013**, *64*, 3077–3086. [CrossRef] [PubMed]

6. Cheah, B.H.; Nadarajah, K.; Divate, M.D.; Wickneswari, R. Identification of four functionally important microRNA families with contrasting differential expression profiles between drought-tolerant and susceptible rice leaf at vegetative stage. *BMC Genom.* **2015**, *16*, 692. [CrossRef] [PubMed]

7. Shinozaki, K.; Yamaguchi-Shinozaki, K. Gene networks involved in drought stress response and tolerance. *J. Exp. Bot.* **2007**, *58*, 221–227. [CrossRef] [PubMed]

8. Ramanjulu, S.; Bartels, D. Drought-and desiccation-induced modulation of gene expression in plants. *Plant Cell Environ.* **2002**, *25*, 141–151. [CrossRef] [PubMed]

9. Battaglia, M.; Olvera-Carrillo, Y.; Garciarrubio, A.; Campos, F.; Covarrubias, A.A. The enigmatic LEA proteins and other hydrophilins. *Plant Physiol.* **2008**, *148*, 6–24. [CrossRef] [PubMed]

10. Liang, D.; Zhang, Z.; Wu, H.; Huang, C.; Shuai, P.; Ye, C.Y.; Tang, S.; Wang, Y.; Yang, L.; Wang, J.; et al. Single-base-resolution methylomes of *Populus trichocarpa* reveal the association between DNA methylation and drought stress. *BMC Genet.* **2014**, *15* (Suppl. 1), S9. [CrossRef] [PubMed]

11. Granot, G.; Sikron-Persi, N.; Gaspan, O.; Florentin, A.; Talwara, S.; Paul, LK.; Morgenstern, Y.; Granot, Y.; Grafi, G. Histone modifications associated with drought tolerance in the desert plant *Zygophyllum dumosum* Boiss. *Planta* **2009**, *231*, 27–34. [CrossRef] [PubMed]

12. Ríos, G.; Gagete, A.P.; Castillo, J.; Berbel, A.; Franco, L.; Rodrigo, M.I. Abscisic acid and desiccation-dependent expression of a novel putative *SNF5*-type chromatin-remodeling gene in *Pisum sativum*. *Plant Physiol. Biochem.* **2007**, *45*, 427–435. [CrossRef] [PubMed]

13. Khraiwesh, B.; Zhu, J.K.; Zhu, J. Role of miRNAs and siRNAs in biotic and abiotic stress responses of plants. *Biochim. Biophys. Acta* **2012**, *1819*, 137–148. [CrossRef] [PubMed]

14. Kumar, R. Role of microRNAs in biotic and abiotic stress responses in crop plants. *Appl. Biochem. Biotechnol.* **2014**, *174*, 93–115. [CrossRef] [PubMed]

15. Chapman, E.J.; Carrington, J.C. Specialization and evolution of endogenous small RNA pathways. *Nat. Rev. Genet.* **2007**, *8*, 884–896. [CrossRef] [PubMed]

16. Mallory, A.C.; Bouché, N. MicroRNA-directed regulation: To cleave or not to cleave. *Trends Plant Sci.* **2008**, *13*, 359–367. [CrossRef] [PubMed]

17. Förstemann, K.; Horwich, M.D.; Wee, L.M.; Tomari, Y.; Zamore, P.D. *Drosophila* microRNAs are sorted into functionally distinct Argonaute protein complexes after their production by Dicer-1. *Cell* **2007**, *130*, 287–297. [CrossRef] [PubMed]

18. Palatnik, J.F.; Allen, E.; Wu, X.; Schommer, C.; Schwab, R.; Carrington, J.C.; Weigel, D. Control of leaf morphogenesis by microRNAs. *Nature* **2003**, *425*, 257–263. [CrossRef] [PubMed]

19. Chen, X. A microRNA as a translational repressor of APETALA2 in *Arabidopsis* flower development. *Science* **2004**, *303*, 2022–2025. [CrossRef] [PubMed]

20. Moldovan, D.; Spriggs, A.; Yang, J.; Pogson, B.J.; Dennis, E.S.; Wilson, I.W. Hypoxia-responsive microRNAs and trans-acting small interfering RNAs in *Arabidopsis*. *J. Exp. Bot.* **2010**, *61*, 165–177. [CrossRef] [PubMed]

21. Long, R.C.; Li, M.N.; Kang, J.M.; Zhang, T.J.; Sun, Y.; Yang, Q.C. Small RNA deep sequencing identifies novel and salt-stress-regulated microRNAs from roots of *Medicago sativa* and *Medicago truncatula*. *Physiol. Plant.* **2015**, *154*, 13–27. [CrossRef] [PubMed]

22. Ci, D.; Song, Y.P.; Tian, M.; Zhang, D.Q. Methylation of miRNA genes in the response to temperature stress in *Populus simonii*. *Front. Plant Sci.* **2015**, *6*, 921. [CrossRef] [PubMed]

23. Zhou, X.; Wang, G.; Zhang, W. UV-B responsive microRNA genes in *Arabidopsis thaliana*. *Mol. Syst. Biol.* **2007**, *3*, 103. [CrossRef] [PubMed]

24. Paul, S.; Datta, S.K.; Datta, K. miRNA regulation of nutrient homeostasis in plants. *Front. Plant Sci.* **2015**, *6*, 232. [CrossRef] [PubMed]

25. Zhou, Z.S.; Zeng, H.Q.; Liu, Z.P.; Yang, Z.M. Genome-wide identification of *Medicago truncatula* microRNAs and their targets reveals their differential regulation by heavy metal. *Plant Cell Environ.* **2012**, *35*, 86–99. [CrossRef] [PubMed]

26. Liu, H.; Able, A.J.; Able, J.A. SMARTER De-Stressed Cereal Breeding. *Trends Plant Sci.* **2016**, *21*, 909–925. [CrossRef] [PubMed]

27. Wang, T.Z.; Chen, L.; Zhao, M.G.; Tian, Q.Y.; Zhang, W.H. Identification of drought-responsive microRNAs and their targets in *Medicago truncatula* by genome-wide high-throughput sequencing and degradome analysis. *BMC Genom.* **2011**, *12*, 367. [CrossRef] [PubMed]

28. Shui, X.R.; Chen, Z.W.; Li, J.X. MicroRNA prediction and its function in regulating drought-related genes in cowpea. *Plant Sci.* **2013**, *210*, 25–35. [CrossRef] [PubMed]

29. Ballen-Taborda, C.; Plata, G.; Ayling, S.; Rodriguez-Zapata, F.; Becerra, L.A.; Duitama, J.; Tohme, J. Identification of cassava MicroRNAs under abiotic stress. *Int. J. Genom.* **2013**, *2013*, 857986. [CrossRef] [PubMed]

30. Zhang, N.; Yang, J.; Wang, Z.; Wen, Y.; Wang, J.; He, W.; Liu, B.; Si, H.; Wang, D. Identification of novel and conserved microRNAs related to drought stress in potato by deep sequencing. *PLoS ONE* **2014**, *9*, e95489. [CrossRef] [PubMed]

31. Wang, M.; Wang, Q.; Zhang, B. Response of miRNAs and their targets to salt and drought stresses in cotton (*Gossypium hirsutum* L.). *Gene* **2013**, *30*, 26–32. [CrossRef] [PubMed]

32. Yin, F.; Gao, J.; Liu, M.; Qin, C.; Zhang, W.; Yang, A.; Xia, M.; Zhang, Z.; Shen, Y.; Lin, H.; et al. Genome-wide analysis of Water-stress-responsive microRNA expression profile in tobacco roots. *Functi. Integr. Genom.* **2014**, *14*, 319–332.

33. Shuai, P.; Liang, D.; Zhang, Z.; Yin, W.; Xia, X. Identification of drought-responsive and novel *Populus trichocarpa* microRNAs by high-throughput sequencing and their targets using degradome analysis. *BMC Genom.* **2013**, *14*, 233. [CrossRef] [PubMed]

34. Ferreira, T.H.; Gentile, A.; Vilela, R.D.; Costa, G.G.; Dias, L.I.; Endres, L.; Menossi, M. MicroRNAs associated with drought response in the bioenergy crop sugarcane (*Saccharum* spp.). *PLoS ONE* **2012**, *7*, e46703. [CrossRef] [PubMed]

35. Xie, F.; Stewart, C.N.; Taki, F.A.; He, Q.; Liu, H.; Zhang, B. High-throughput deep sequencing shows that microRNAs play important roles in switchgrass responses to drought and salinity stress. *Plant Biotechnol. J.* **2014**, *12*, 354–366. [CrossRef] [PubMed]

36. Hackenberg, M.; Gustafson, P.; Langridge, P.; Shi, B.J. Differential expression of microRNAs and other small RNAs in barley between water and drought conditions. *Plant Biotechnol. J.* **2015**, *13*, 2–13. [CrossRef] [PubMed]

37. Liu, H.; Searle, I.R.; Watson-Haigh, N.S.; Baumann, U.; Mather, D.E.; Able, A.J.; Able, J.A. Genome-wide identification of microRNAs in leaves and the developing head of four durum genotypes during water deficit stress. *PLoS ONE* **2015**, *10*, e0142799. [CrossRef] [PubMed]

38. Liu, H.; Able, A.J.; Able, J.A. Water-deficit stress-responsive microRNAs and their targets in four durum wheat genotypes. *Funct. Integr. Genom.* **2017**, *17*, 237–251. [CrossRef] [PubMed]

39. Wang, Y.; Li, L.; Tang, T.; Liu, J.; Zhang, H.; Zhi, H.; Jia, G.; Diao, X. Combined small RNA and degradome sequencing to identify miRNAs and their targets in response to drought in *foxtail millet*. *BMC Genet.* **2016**, *17*, 57. [CrossRef] [PubMed]

40. Capitão, C.; Paiva, J.A.P.; Santos, D.M.; Fevereiro, P. In *Medicago truncatula*, water deficit modulates the transcript accumulation of components of small RNA pathways. *BMC Plant Biol.* **2011**, *11*, 79. [CrossRef] [PubMed]

41. Li, W.X.; Oono, Y.; Zhu, J.; He, X.J.; Wu, J.M.; Iida, K.; Lu, X.Y.; Cui, X.; Jin, H.; Zhu, J.K. The *Arabidopsis* NFYA5 transcription factor is regulated transcriptionally and posttranscriptionally to promote drought resistance. *Plant Cell* **2008**, *20*, 2238–2251. [CrossRef] [PubMed]

42. Burge, S.W.; Daub, J.; Eberhardt, R.; Tate, J.; Barquist, L.; Nawrocki, E.P.; Eddy, S.R.; Gardner, P.P.; Bateman, A. Rfam 11.0: 10 years of RNA families. *Nucl. Acids Res.* **2013**, *41*, D226–D232. [CrossRef] [PubMed]

43. Kozomara, A.; Griffiths-Jones, S. miRBase: Annotating high confidence microRNAs using deep sequencing data. *Nucl. Acids Res.* **2014**, *42*, D68–D73. [CrossRef] [PubMed]

44. Li, H.; Durbin, R. Fast and accurate short read alignment with Burrows-Wheeler transform. *Bioinformatics* **2009**, *25*, 1754–1760. [CrossRef] [PubMed]

45. Li, Y.; Zhang, Z.; Liu, F.; Vongsangnak, W.; Jing, Q.; Shen, B. Performance comparison and evaluation of software tools for microRNA deep-sequencing data analysis. *Nucl. Acids Res.* **2012**, *40*, 4298–4305. [CrossRef] [PubMed]

46. Chen, L.; Wang, T.; Zhao, M.; Tian, Q.; Zhang, W. Identification of aluminum-responsive microRNAs in *Medicago truncatula* by genome-wide high-throughput sequencing. *Planta* **2012**, *235*, 375–386. [CrossRef] [PubMed]

47. Livak, K.J.; Schmittgen, T.D. Analysis of relative gene expression data using real time quantitative PCR and the $2^{-\Delta\Delta CT}$ method. *Methods* **2001**, *25*, 402–408. [CrossRef] [PubMed]

48. Robinson, M.D.; McCarthy, D.J.; Smyth, G.K. edgeR: A Bioconductor package for differential expression analysis of digital gene expression data. *Bioinformatics* **2010**, *26*, 139–140. [CrossRef] [PubMed]

49. Vaz, C.; Wee, C.W.; Lee, G.P. S.; Ingham, P.W.; Tanavde, V.; Mathavan, S. Deep sequencing of small RNA facilitates tissue and sex associated microRNA discovery in zebrafish. *BMC Genom.* **2015**, *16*, 950. [CrossRef] [PubMed]

50. Eldem, V.; Akcay, U.C.; Ozhuner, E.; Bakir, Y.; Uranbey, S.; Unver, T. Genome-wide identification of miRNAs responsive to drought in peach (*Prunus persica*) by high-throughput deep sequencing. *PLoS ONE* **2012**, *7*, e50298. [CrossRef] [PubMed]

51. Jung, J.H.; Park, C.M. MIR166/165 genes exhibit dynamic expression patterns in regulating shoot apical meristem and floral development in *Arabidopsis*. *Planta* **2007**, *225*, 1327–1338. [CrossRef] [PubMed]

52. Rajagopalan, R.; Vaucheret, H.; Trejo, J.; Bartel, D.P. A diverse and evolutionarily fluid set of microRNAs in *Arabidopsis thaliana*. *Genes Dev.* **2006**, *20*, 3407–3425. [CrossRef] [PubMed]

53. Fahlgren, N.; Howell, M.D.; Kasschau, K.D.; Chapman, E.J.; Sullivan, C.M.; Cumbie, J.S.; Givan, S.A.; Law, T.F.; Grant, S.R.; Dangl, J.L.; et al. High throughput sequencing of *Arabidopsis* microRNAs: Evidence for frequent birth and death of MIRNA genes. *PLoS ONE* **2007**, *2*, e219. [CrossRef] [PubMed]

54. Zhang, Y.; Zhu, X.; Chen, X.; Song, C.; Zou, Z.; Wang, Y.; Wang, M.; Fang, W.; Li, X. Identification and characterization of cold-responsive microRNAs in tea plant (*Camellia sinensis*) and their targets using high-throughput sequencing and degradome analysis. *BMC Plant Biol.* **2014**, *14*, 271. [CrossRef] [PubMed]

55. Cao, X.; Wu, Z.; Jiang, F.; Zhou, R.; Yang, Z. Identification of chilling stress-responsive tomato microRNAs and their target genes by high-throughput sequencing and degradome analysis. *BMC Genom.* **2014**, *15*, 1130. [CrossRef] [PubMed]

56. Juarez, M.T.; Kui, J.S.; Thomas, J.; Heller, B.A.; Timmermans, M.C. microRNA-mediated repression of rolled leaf1 specifies maize leaf polarity. *Nature* **2004**, *428*, 84–88. [CrossRef] [PubMed]

57. Boualem, A.; Laporte, P.; Jovanovic, M.; Laffont, C.; Plet, J.; Combier, J.P.; Niebel, A.; Crespi, M.; Frugier, F. MicroRNA166 controls root and nodule development in *Medicago truncatula*. *Plant J.* **2008**, *54*, 876–887. [CrossRef] [PubMed]

58. Kantar, M.; Lucas, S.; Budak, H. MiRNA expression patterns of Triticum dicoccoides in response to shock drought stress. *Planta* **2011**, *233*, 471–484. [CrossRef] [PubMed]

59. Kantar, M.; Unver, T.; Budak, H. Regulation of barley miRNAs upon dehydration stress correlated with target gene expression. *Funct. Integr. Genom.* **2010**, *10*, 493–507. [CrossRef] [PubMed]

60. Trindade, I.; Capitao, C.; Dalmay, T.; Fevereiro, M.P.; Santos, D.M. miR398 and miR408 are up-regulated in response to water deficit in *Medicago truncatula*. *Planta* **2010**, *231*, 705–716. [CrossRef] [PubMed]

61. Abe, H.; Yamaguchi-Shinozaki, K.; Urao, T.; Iwasaki, T.; Hosokawa, D.; Shinozaki, K. Role of *Arabidopsis* MYC and MYB homologs in drought and abscisic acid-regulated gene expression. *Plant Cell* **1997**, *9*, 1859–1868. [CrossRef] [PubMed]

62. Abe, H.; Urao, T.; Ito, T.; Seki, M.; Shinozaki, K.; Yamaguchi-Shinozaki, K. *Arabidopsis* AtMYC2 (bHLH) and AtMYB2 (MYB) function as transcriptional activators in abscisic acid signaling. *Plant Cell* **2003**, *15*, 63–78. [CrossRef] [PubMed]

63. Hu, H.; Dai, M.; Yao, J.; Xiao, B.; Li, X.; Zhang, Q.; Xiong, L. Overexpressing a NAM, ATAF, and CUC (NAC) transcription factor enhances drought resistance and salt tolerance in rice. *Proc. Natl. Acad. Sci. USA* **2006**, *103*, 12987–12992. [CrossRef] [PubMed]

64. Sunkar, R.; Kapoor, A.; Zhu, J.K. Posttranscriptional induction of two Cu/Zn superoxide dismutase genes in *Arabidopsis* is mediated by downregulation of miR398 and important for oxidative stress tolerance. *Plant Cell* **2006**, *18*, 2051–2065. [CrossRef] [PubMed]

65. Wei, L.; Zhang, D.; Xiang, F.; Zhang, Z. Differentially expressed miRNAs potentially involved in the regulation of defense mechanism to drought stress in maize seedlings. *Int. J. Plant Sci.* **2009**, *170*, 979–989. [CrossRef]

66. Allen, R.S.; Li, J.Y.; Alonso-Peral, M.M.; White, R.G.; Gubler, F.; Millar, A.A. MicroR159 regulation of most conserved targets in *Arabidopsis* has negligible phenotypic effects. *Silence* **2010**, *1*, 18. [CrossRef] [PubMed]

67. Reyes, J.; Chua, N. ABA induction of miR159 controls transcript levels of two Myb factors during *Arabidopsis* seed germination. *Plant J.* **2007**, *49*, 592–606. [CrossRef] [PubMed]

10

A Transcriptomic Comparison of Two Bambara Groundnut Landraces under Dehydration Stress

Faraz Khan [1], **Hui Hui Chai** [2], **Ishan Ajmera** [3], **Charlie Hodgman** [3], **Sean Mayes** [1,2,*] **and Chungui Lu** [4]

[1] School of Biosciences, University of Nottingham, Sutton Bonington Campus, Nottingham LE12 5RD, UK; stxfk5@nottingham.ac.uk
[2] Crops for the Future, Jalan Broga, 43500 Semenyih, Selangor Darul Ehsan, Malaysia; huihui.chai@nottingham.edu.my
[3] Centre for Plant Integrative Biology, University of Nottingham, Sutton Bonington Campus, Nottingham LE12 5RD, UK; bhzia@exmail.nottingham.ac.uk (I.A.); charlie.hodgman@nottingham.ac.uk (C.H.)
[4] School of Animal Rural and Environmental Sciences, Nottingham Trent University, Clifton Campus, Nottingham NG11 8NS, UK; chungui.lu@ntu.ac.uk
* Correspondence: sbzsm4@exmail.nottingham.ac.uk

Academic Editor: Qingyi Yu

Abstract: The ability to grow crops under low-water conditions is a significant advantage in relation to global food security. Bambara groundnut is an underutilised crop grown by subsistence farmers in Africa and is known to survive in regions of water deficit. This study focuses on the analysis of the transcriptomic changes in two bambara groundnut landraces in response to dehydration stress. A cross-species hybridisation approach based on the Soybean Affymetrix GeneChip array has been employed. The differential gene expression analysis of a water-limited treatment, however, showed that the two landraces responded with almost completely different sets of genes. Hence, both landraces with very similar genotypes (as assessed by the hybridisation of genomic DNA onto the Soybean Affymetrix GeneChip) showed contrasting transcriptional behaviour in response to dehydration stress. In addition, both genotypes showed a high expression of dehydration-associated genes, even under water-sufficient conditions. Several gene regulators were identified as potentially important. Some are already known, such as *WRKY40*, but others may also be considered, namely *PRR7*, *ATAUX2*-11, CONSTANS-like 1, *MYB60*, *AGL-83*, and a Zinc-finger protein. These data provide a basis for drought trait research in the bambara groundnut, which will facilitate functional genomics studies. An analysis of this dataset has identified that both genotypes appear to be in a dehydration-ready state, even in the absence of dehydration stress, and may have adapted in different ways to achieve drought resistance. This will help in understanding the mechanisms underlying the ability of crops to produce viable yields under drought conditions. In addition, cross-species hybridisation to the soybean microarray has been shown to be informative for investigating the bambara groundnut transcriptome.

Keywords: Bambara groundnut; landraces; dehydration stress; cross-species microarray analysis

1. Introduction

Dehydration is one of the major stresses that inhibits plant growth and can reduce crop productivity. Hence, drought resistance is a key target in helping to ensure global food supply. Plants respond to dehydration stress in three broad approaches: (1) Dehydration escape; (2) Dehydration avoidance; and (3) Dehydration tolerance. Such mechanisms are seen in a range of leguminous species, including the mung bean [1] and pigeon pea [2]. Dehydration escape is the ability of plants to complete their growth cycle and reach maturity with successful reproduction before the shortage of water reaches damaging levels [3]. Mechanisms of avoidance include improved root traits for a greater extraction of soil moisture, stomatal closure, a decreased radiation absorption through leaf rolling, a decreased leaf area for reduced water loss, and the accumulation of osmoprotectants such as proline, trehalose, and dehydrins [4]. Dehydration tolerance allows plants to survive through improved water-use efficiency, i.e., performing all of the biological, molecular, and cellular functions with minimal water. Numerous studies on the effects of dehydration stress on staple crops have been reported [1,2,4–10].

Reduced water availability causes the production of abscisic acid (ABA), the phyto-hormone which initiates stomatal closure and influences other aspects of plant growth and physiology. It is responsible for regulating a broad range of genes during the dehydration response. The SNF1-related protein kinase, AREB (ABA-responsive element)/ABF are the key regulators of ABA signalling [11]. Improving the dehydration tolerance has also been linked to a reduction in shoot growth, while root growth is maintained, leading to an altered partition between the root and shoot. This process is achieved by cell-wall synthesis and remodelling. The formation of reactive oxygen species (ROS) and lignin peroxidases are the key steps involved in cell wall thickening.

Stomatal closure limits the CO_2 uptake by leaves, which leads to a reduction in photosynthesis as the leaf's internal CO_2 is depleted. Severe dehydration stress also limits photosynthesis by down-regulating the expression of ribulose-1, 5-bisphosphate carboxylase/oxygenase (Rubisco), fructose-1,6-bisphosphatase (FBPase), phosphoenolpyruvate carboxylase (PEPCase), pyruvate orthophosphate dikinase (PPDK), and NADP-malic enzyme (NADP-ME) [12]. Plant responses to dehydration affect vegetative growth by reducing the leaf-area expansion and total dry matter, which in turn decreases light interception [13]. Under dehydration stress, wheat (*Triticum dicoccoides*) shows a reduction in the number of grains, grain yield, shoot dry weight, and harvest index [8]. In soybean specimens (*Glycine max*), the loss of seed yield was reported to be greatest when dehydration appeared during anthesis and the early reproductive stages [6–9].

A range of dehydration stress-related genes have been identified in *Arabidopsis thaliana*, rice (*Oryza sativa*), and other model plants [14]. These can be classified into two main groups: (i) Effector proteins, whose role is to alleviate the effect of the stress (such as water channel proteins, detoxification enzymes, LEA proteins, chaperones, and osmoprotectants); and (ii) Regulatory proteins, which alter the expression or activity of effector genes and modify plant growth, such as the transcription factors DREB2 and AREB, and also protein kinases and phosphatases [15].

In recent years, plant breeders have turned to landraces (i.e., locally adapted genetically mixed populations) for trait improvement in various crops, including barley [16], sorghum [17], sesame [18], and soybean [19]. An early attempt to investigate the use of landraces in addressing the problem of dehydration tolerance has been carried out in wheat [20], although this did not delve into the specific genetics conferring the desirable traits. An alternative approach to identifying the genes conferring dehydration avoidance and tolerance is to study species that are already resilient under arid conditions. In this regard, bambara groundnut (*Vigna subterranea* (L) Verdc.) is a potential candidate. It is an underutilised, drought-resistant African legume, which is mainly grown in sub-Saharan Africa [5–21] and is sometimes used as an intercrop with major cereals, such as maize, because of its nitrogen fixing potential [22]. Bambara groundnut is considered as a drought resistant crop with a reasonable protein content (18% to 22%), a high carbohydrate content (65%), and some level of lipids (6.5%) [23], with a similar overall composition to chickpea. A number of bambara groundnut landraces have well-developed tap roots which grow up to a height of 30–35 cm [24].

From the results of Mabhaudhi et al. [25], bambara groundnut has been shown to adopt dehydration-escape mechanisms, including a shortened vegetative growth period, early flowering, a reduced duration of the reproductive stage, and early maturity under dehydration stress. Such responses are likely to be employed where the initial plant growth is based on stored soil water, but further rain is unlikely. It has been reported that bambara groundnut responds to dehydration stress by partitioning more assimilate into the root, relative to the shoots, so that a greater soil volume can be exploited [26,27]. Nyamudeza [27] also observed that bambara groundnut allocated a greater fraction of its total dry weight to the roots than the groundnut, irrespective of the available soil moisture. This would suggest that bambara groundnut commits a greater supply of assimilates to root growth, irrespective of the soil moisture status. This strategy may have clear advantages when water subsequently becomes limited, but there could be a trade-off with the yield under benign environments. A greater root dry-weight was also reported when the bambara landrace, Burkina, was subjected to dehydration stress [28]. Dehydration-avoidance traits have also been observed, especially the accumulation of proline [21] and a reduced leaf area [29].

This study aims to investigate the effects of dehydration on gene expression in this reportedly drought-resistant species. The transcriptomes of two genotypes (DipC and Tiga Nicuru (TN)) were sampled, to identify what is common and how they differ in their response to a prolonged, but slowly intensifying, dehydration treatment. The climatic conditions in their native regions (Botswana and Mali, respectively) suggest that they are likely to have evolved in regions which would select for drought resistance, while potentially exhibiting some variation in the mechanisms employed to deal with dehydration, as they are morphologically and phenologically distinct [30]. Chai et al. [30,31] reported that transgressive segregation was observed in the segregating F_5 population derived from the TNxDipC cross. The contrast between the two parental lines for a number of traits such as the days-to-maturity, stomatal conductance, 100-seed weight, leaf area, internode length, peduncle length, pod number per plant, and leaf carbon (delta C^{13}) isotope analysis, suggest that some of these mechanisms for adaptation to dehydration could be non-identical in the two genotypes. For example, delta C^{13} was associated with a higher yield as observed in DipC, compared to TN [30]. In addition, the results showed that there were lines in the segregating population that performed better in terms of the ability to produce higher yields under drought conditions than the parental genotypes. Hence, evaluating the transcriptome of the two parental lines under dehydration stress could be a good indicator to investigate the molecular mechanism occurring in the two genotypes and its relationship to phenology and phenotype.

As a complete genome sequence is not available and microarray tools are still to be developed in this species, cross-species hybridisation with the Affymetrix Glycine-max microarray was investigated to test if this approach is acceptable for bambara groundnut transcriptomics, as it has been successful for other species [32–34].

2. Materials and Methods

2.1. Plant Materials

In this study, the experiment was conducted in the FutureCrop controlled tropical glasshouses at the School of Biosciences, Sutton Bonington Campus, University of Nottingham, UK. Two genotypes of bambara groundnut, DipC and TN, were planted in both 'Water-limited' and 'Water-sufficient' control plots.

2.2. Site Descriptions and Experimental Design

Plants were grown over a period of five months. A 12-hour photoperiod was created using an automated blackout system (Cambridge Glasshouses, Newport, UK), with day and night temperatures set at 28 °C and 23 °C respectively. Trickle tape irrigation with PVC micro-porous tubing was placed beside each plant row. The plants were irrigated at 06:00 h and 18:00 h for 20 min, with a measured flow rate of 1 L/h per tube, and each tube was 5 m in length. Two independent soil pits (5 m × 5 m × 1 m) containing sandy loam soil were used in the glasshouses. These were isolated from the surrounding soil by a Butyl liner and concrete pit structure with gravel drainage for separate water-limited and water-sufficient plots. The PR2 water profile probe (Delta-T devices, Cambridge, UK) was used to measure the soil moisture content. A randomised block design (RBD) with three blocks for each soil pit was implemented for this experiment. Three replicate plants for the water-sufficient plot (continuously irrigated) and four replicates for the water-limited treatment plot were used. Three seeds were sown per replicate at a depth of 3–4 cm with a spacing of 25 cm × 25 cm between each final plant position and multiple plants were later thinned to one plant per replicate at 20 days after sowing (DAS). Figure S1 shows the treatment regime. The irrigation system for the water limited treatment plot was turned off at 50 DAS and resumed at 92 DAS for plant recovery (in total, six weeks of treatment after 100% flowering). Normal irrigation continued for the water-sufficient plot throughout. The water-limited treatment was continued until an average of a 50% reduction in *stomatal conductance* was observed. Leaves from water-sufficient and water-limited plants were collected at 92 DAS before recommencing irrigation, while those from 'recovered' plants were collected at 107 DAS after watering was resumed at 92 DAS. Labelled aluminium foil was used to wrap the harvested leaves, which was then transferred into liquid nitrogen for long term storage. All samples were stored in a −80 °C freezer before RNA extraction. DNA extraction from the two parental genotypes was completed using the DNA extraction Qiagen kit handbook.

2.3. RNA Extraction

RNA was extracted using the RNeasy Qiagen kit (Qiagen, Manchester, UK), according to the manufacturer's instructions. DNA was eliminated using DNase. A total of 80 µL of DNase I incubation mix, containing 10 µL DNase I stock solution and 70 µL buffer RDD, was added and incubated at room temperature for 15 min. Nanodrop readings and gel electrophoresis were performed to check the quality and quantity of RNA, as RNA samples required 100 ng/µL for 10 µL for microarray analysis. To make sure that the samples were free from active RNAse, 0.63 µL of 40 U/µL RNasin (Promega, Southampton, UK) was added for every 25 µL of the RNA sample. All samples were tested on an Nanodrop and Agilent bioanalyser for integrity (looking at the quality (ratio of 2.0) and integrity (a ratio of 2 for 28S/18S) for respective quantitation) before preparation for the microarray.

2.4. cRNA and Genomic DNA Affymetrix Labelling and Hybridisation

The above RNA extracts were reverse transcribed to synthesize double stranded complementary DNA (cDNA). After purification of the double-stranded cDNA products, the sample was transcribed in vitro to generate Biotinylated complementary RNAs (cRNAs), followed by purification and fragmentation. The purified and fragmented cRNAs were then hybridised to the Affymetrix Soybean Gene Chip array (ThermoFisher Scientific, Lutterworth, UK). The scanned arrays produced CEL raw data files that were loaded onto Genespring GX version 13.1 (Agilent Genomics, Santa Clara, CA, USA) for further analysis. The extraction of genomic DNA (gDNA) from the two genotypes was performed using the DNA extraction Qiagen kit according to manufacturer's instructions. Extracted DNA was labelled and hybridised to the Affymetrix Soybean TEST3 array and resulted in the generation of gDNA cell-intensity files (CEL files), after scanning. To identify probe pairs that efficiently hybridise to the gDNA, a series of user defined threshold values were evaluated for the signal intensity. The perfect match (PM) probes were selected for interpreting the GeneChip arrays challenged with RNA from the species of interest [35].

2.5. Probe Selection and Identification of Differentially Expressed Genes

The soybean array contained 37,500 probe sets, each containing 11 probe pairs per probe-set. For each genotype, custom CDF files were obtained, with reference to their gDNA hybridisation signal strength [36] for a subsequent estimation of the transcript levels. RNA CEL files were normalised in GeneSpring [37] using the Robust Multi-array Average. Differentially expressed genes (DEGs) were calculated using a *t*-test test (corrected by Benjamin Hochberg false discovery rate (FDR) multiple testing). Probe-sets with a FDR corrected *p*-value ≤ 0.05 and fold change of >2 were considered to be differentially expressed (either up or down regulated). Principal Component Analysis (PCA) was also carried out in GeneSpring and Bioconductor package "prcomp". BINGO was used for discovering (from input gene lists) over-represented terms from the Gene Ontology [38].

2.6. Construction of the Co-Expression Network

Co-expression network analysis was carried out using the DeGNserver [39] and cytoscape 3.4 [40]. Separate networks were generated for each genotype. The input probe-sets were restricted to those that were differentially expressed between each pair of treatments (water-limited, water-sufficient and recovery) and RMA (Robust Multi-Array Average)-normalised values were used across all samples. Links were assigned between pairs of nodes (i.e., probe-sets) when their Spearman's Rank correlation was 0.9. The co-expression network was imported into cytoscape for visual representation and network analysis. For each genotype, another input file was made which, for each probe-set, defined the parent (DipC or TN), the direction of differential expression caused by dehydration (up or down), and the role identified through homology in relation to drought resistance. This aided the interpretation of the combined network derived from both genotypes.

2.7. Expression Validation of Differentially Expressed Genes Using Real-Time qPCR

Four genes which were potential candidate dehydration-associated genes (based on their functional annotations) with a differential expression level of >2-fold change and FDR corrected p-value ≤ 0.05 from the differential expression analysis, were chosen for quantitative PCR (qPCR) validation. The actin-11 from the available bambara groundnut transcriptome sequence was used as a housekeeping gene. The actin-11 gene is known to be one of the most stable reference genes for gene expression normalisation and has been used in soybean and rice specimens [41,42]. PCR forward and reverse primers were designed using Primer-BLAST [43] for the chosen genes. The primers were designed in three steps. Firstly, the target gene sequence to which the primers needed to be designed was downloaded from the soybean database. Secondly, the soybean-specific target gene sequence was blasted against a bambara groundnut transcriptome generated from RNA-sequencing data for a low-temperature stress experiment [44], by creating a BLAST database. Thirdly, the target gene sequence obtained from the bambara groundnut BLAST database was used to search through the BLAST database at NCBI to add weight to the selection of this sequence. Once the gene sequence was identified in the BLAST database, it was utilised to design primers with an appropriate primer size, GC content, and melting temperature (Tm) using Primer-BLAST. PCR was performed to check the quality of all the primers designed for the four dehydration-associated genes. PCR analysis was performed using the 7000 Sequence Detection System (Applied Biosystems, Cheshire, UK). The annealing temperature was set to 60 °C for the primer designed for the genes for *PAL1* (Phenylalanine ammonia-lyase 1) and *COMT* (3-Caffeic acid o methyltransferase), and 58 °C for the Beta-fructofuranosidase and *UBC-2* (ubiquitin conjugating enzyme-2) genes. The cycling parameters were set as: 95 °C for 10 min, 40 cycles of denaturing at 95 °C for 30 s, annealing at 60 °C/58 °C for 30 s, and extension at 72 °C for 30 s. First strand cDNA synthesis for all the RNA samples was carried out using a SuperScript III First-Strand Synthesis kit (ThermoFisher Scientific, Lutterworth, UK). The first-strand cDNA was prepared for analysis by qPCR using PerfeCta SYBR Green SuperMix (Quantabio, Beverly, MA, USA) containing 2X reaction buffer (with optimized concentrations of MgCl$_2$), dNTPs (dATP, dCTP, dGTP, dTTP), AccuStart Tag DNA Polymerase (Quantabio, Beverly, MA, USA) SYBR Green 1 dye, and stabilizers. The synthesized cDNA was cleaned from the remaining RNA using the enzyme mix included in the kit (*Escherichia coli* RNase H). The qPCR components were prepared for 10 µL reactions and Melt-curve analysis was performed. The sample cycle threshold (Ct) was standardized for each template based on the actin-11 gene control amplicon behaviour. The $2^{-\Delta\Delta Ct}$ method was used to analyse the relative changes in gene expression from the qRT-PCR experiment [45]. To validate whether the right PCR product was generated for the expression studies, the desired fragment of intact cDNA for all genes was sent for sequencing after the gel extraction using a QIAquick Gel Extraction Kit (Qiagen, Manchester, UK).

3. Results

3.1. Probe Selection Based on gDNA

The genomic DNA of both genotypes was individually hybridised to the Affymetrix Soybean GeneChip array to study the global genome hybridisation for probe selection. The numbers of retained probe-pairs and probe-sets are shown in Table 1. With increasing threshold values, the number of probe-pairs retained in the probe mask file started decreasing rapidly (Figure 1), while the number of probe-sets (representing genes) decreased at a slower rate. This suggests that, even at higher gDNA hybridisation thresholds, at least some of the gene-designed oligonucleotides are cross-hybridising for many of the probe-sets and that the cross-species array approach is a reasonable approach for bambara groundnut transcriptomics.

Table 1. Retained probe-sets and probe-pairs at different threshold values.

Threshold Value	Number of Probe Sets (Soybean Chip Hyb. to DipC gDNA)	Number of Probe Sets (Soybean Chip Hyb. to TN gDNA)	Number of Probe Pairs (Soybean Chip Hyb. to DipC gDNA)	Number of Probe Pairs (Soybean Chip Hyb. to TN gDNA)	Number of DEGs in DipC	Number of DEGs in TN
20	61,072	61,072	670,388	670,388	6165	6165
60	60,877	60,895	479,538	482,352	6927	6814
100	**59,782**	**59,835**	**302,834**	**304,708**	**7183**	**7159**
150	56,266	56,511	190,570	193,522	7036	7159
200	51,071	51,319	129,806	132,521	6638	6731
300	37,813	38,000	66,907	68,106	5275	5345
500	17,469	18,176	23,464	24,693	2784	2911
600	12,258	12,930	15,701	16,650	2089	2170
700	8896	9566	11,193	12,061	1574	1673
800	6687	7208	8415	9070	1195	1291
900	5140	5657	6559	7140	958	1057
1000	4085	4482	5304	5733	802	877

gDNA: genomic DNA; DEG: differentially expressed gene; TN: Tiga Nicuru; Hyb: Hybridisation.

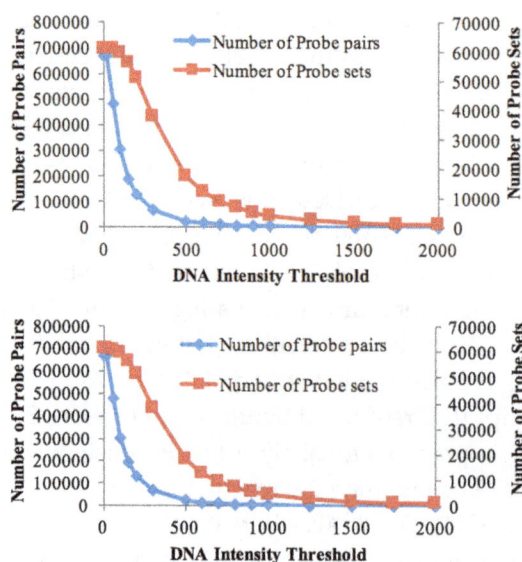

Figure 1. Effect of intensity thresholds. Number of probe pairs (blue line) and probe sets (magenta line) retained for DipC (**top**) and Tiga Nicuru (TN) (**bottom**) respectively at different genomic DNA (gDNA) intensity thresholds.

The number of retained probe-sets and probe-pairs on the Soybean chip for both the DipC and TN gDNA hybridisations were determined, corresponding to each threshold value (Table 1). A custom CDF file with a threshold of 100 was chosen for differential expression analysis in both genotypes, as it allowed for a good sensitivity to detect the maximum number of differentially-expressed transcripts (Table 1). Furthermore, both genotypes were found to be highly similar in terms of the probe-sets detected at this threshold. A total of 59,533 probe-sets were common to both genotypes at the threshold of 100, while 249 and 302 probe-sets were specific to DipC and TN, respectively. These results therefore suggest a high sequence similarity (>99%) at this level of sequence sampling.

3.2. Principal Component Analysis

The PCA plot (Figure 2) shows that, under water-sufficient treatment, the two genotypes appear to have similar transcriptomes. The first two Principal Components account for 25.45% and 17.11% of the variance, respectively, suggesting that it is due to a range of hybridisation/expression differences between the chips. Recovery after dehydration stress, however, caused the most variation and suggests that the recovery transcriptome does not return to the water-sufficient state (control). The DipC water-limited treatment sample 'D.DipC.Rep2' could be a potential outlier and this needs to be borne in mind in further analysis. The 3D PCA plots of genotype-specific data showed a good separation of the three treatments (conditions) and better PCA scores (see Figure S2).

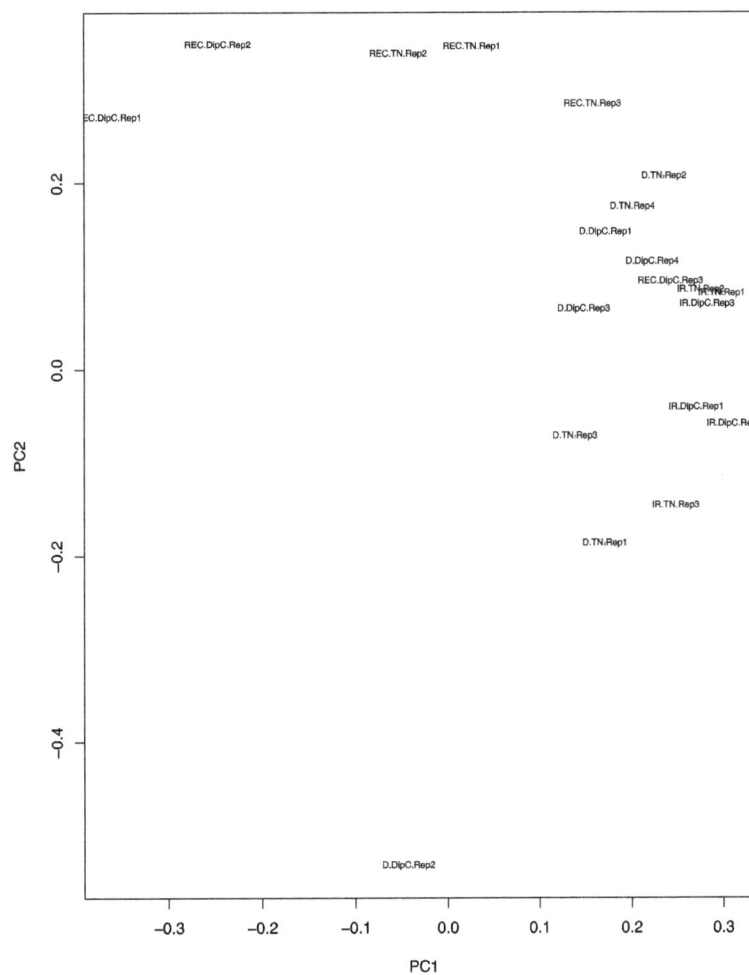

Figure 2. Principal component analysis (PCA) plot of the expression data from the microarrays. The principal components PC1 and PC2 values for each chip have been placed on a scatter plot. Each chip result is defined by a three-part character string consisting of the treatment, genotype, and replicate number. IR, D, and REC refer to water-sufficient, water-limited, and recovery treatment, respectively; the genotypes are DipC and TN; and Rep1–4 refers to the specific biological replicate. Note, the water-sufficient and recovery treatments have only three replicates, while dehydration has four.

3.3. Gene Expression Under Water-Sufficient Conditions

It is pertinent to consider the state of the genotype transcriptomes before any dehydration treatment has taken place. However, owing to the high background noise in microarray studies, it is unclear what intensity level defines a gene as being transcribed. Figure S3 shows that the ranked intensity values follow a roughly sigmoidal curve. The point of inflection (at which the declining gradient is at its shallowest) covers the top two-thirds of the probe-sets, and corresponds to an RMA value of 0.97. This may be a stringent cut-off, given that an RMA value of one corresponds to the average across all probe-sets on the array, but it ensures that there were few, if any, false positives. This left 39,855 probe-sets for DipC and 39,890 for TN. There are 26,496 probe-sets in common between the two genotypes, suggesting differences in the general transcriptional behaviour of the two genotypes.

Each genotype had a little over 90 probe sets with functional annotations related to ABA signalling and dehydration responses (see Tables S1 and S2), of which 60 were common to both. These include homologues of much of the ABA synthesis and response network, the DREB1 transcription factor, Early-Response to Dehydration proteins 3, 4, 8, 14–16, and 18, four osmoprotectant genes, two dehydration-response genes influencing photosynthesis, and 21 other probe sets corresponding to dehydration-associated proteins of an unknown function (see Table S3). Table S4 lists the genes differentially expressed between the two water-sufficient treated genotypes, but at this stage, nothing stands out as remarkable.

3.4. Identification of Differentially Expressed Genes

For DipC and TN, the numbers of genes differentially expressed as a result of the dehydration and recovery treatments, and detected by the cross-species microarray approach, are shown in Table 2, with the full lists of probe-sets and functional annotations presented in Tables S5–S8. The top upregulated and downregulated genes in DipC and TN are shown in Tables 3 and 4, respectively. The numbers for DipC were consistently higher than for TN, and the water-limited treatment caused more down- than upregulation, while recovery had the reverse effect.

Table 2. Differentially expressed gene numbers.

	Water-Limited versus Water-Sufficient				Water-Limited versus Recovery	
	Up-Regulated under Dehydration	Down-Regulated under Recovery	Down-Regulated under Dehydration	Up-Regulated under Recovery	Up-Regulated	Down-Regulated
DipC	80	68	109	94	340	146
Tiga Nicuru	28	22	53	42	294	97

Recovery led to many more differentially expressed genes (486 and 391) than dehydration stress (189 and 81). There were six possible system effects that can be gleaned from these data (Figure S4). The upregulated genes under the water-limited treatment that returned to a water-sufficient state on recovery and the downregulated genes that returned to a normal expression at recovery are the strictly dehydration-responsive genes (~75% in both genotypes), while those that significantly changed and did not return to the pretreatment levels (~25%) correspond to a dehydration-induced state change. The latter may be due to epigenetic effects, such as a change in the methylation state of gene-regulatory regions. The larger numbers of differentially expressed genes from water-limited conditions to recovery may be accounted for by aging and other highly variable factors (see Figure 2), such as the soil conditions in each pit.

The fold changes of the upregulated genes under dehydration stress in both genotypes are relatively small (mostly < 4-fold). Furthermore, there were only nine differentially expressed genes which were common to both genotypes (see Table 5). The only common upregulated gene was beta-fructofuranosidase, which hydrolyses sucrose to provide more glucose, hence playing a potential role in osmoprotection and energy production. In contrast, half of the common downregulated genes were related to transcription and also play roles in stomatal regulation. Excluding the potential outlier, 'D.DipC.Rep2' had little effect upon the common gene analysis (Table S9), so it has been included in subsequent analyses.

Mostly, the upregulated genes under dehydration stress in DipC relate to the secondary metabolism of cell-wall components, while the TN genes include transcription-related factors, most notably a CONSTANS-like gene. Furthermore, GO term overrepresentation analysis for both DipC and TN showed an emphasis on various metabolic processes related to cellular amino acids and their derivatives, secondary metabolites and carbohydrates (Table S10). Hence, despite the genomic hybridisation mask demonstrating that the pure hybridisation was very similar between the two genotypes, there is a very different transcriptional response to dehydration stress by each genotype. Microarray data has a limited dynamic range, even when within species, so it is important to validate a small set of microarray observations. Hence, validation through qRT-PCR was performed.

Table 3. Top upregulated genes in DipC and TN.

Gene Name	FDR	Fold Change	Gene Description	References
UP-Regulated Genes in DipC				
PAL1 (Phenylalanine ammonia-lyase 1)	0.018	3.901	Key enzyme involved in the biosynthesis of isoprenoid antioxidative and polyphenol compounds such as lignin and is involved in defense mechanism.	[53]
ATEP3/AtchitIV	0.001	3.845	Encodes an EP3 chitinase that is stimulated under abiotic stress.	[54]
TXR1(Thaxtomin A resistant 1)/ ATPAM16	6.87×10^{-5}	3.718	TXR1 is a component of a dispensable transport mechanism. Involved in negative regulation of defense responses by reducing reactive oxygen species (ROS).	[55]
Acetyl-CoA C-acyltransferase, putative/ 3-ketoacyl-CoA thiolase	0.001	3.554	Functions in Jasmonic acid synthesis which plays a role in plant response to mechanical and abiotic stress.	[56]
UBC-2 (ubiquitin-conjugating enzyme 2)	0.004	3.407	Ubiquitination plays a part in increasing rate of the protein breakdown. Arabidopsis plants overexpressing UBC-2 were more tolerant to dehydration stress compared to the control plants.	[57]
Rho GDP dissociation inhibitor 2	0.001	3.348	Involves in the regulation of Rho protein and small GTPase mediated signal transduction.	[58]
Histidine amino acid transporter (LHT1)	0.001	3.256	Amino acid transmembrane transporter involved in apoplastic transport of amino acids in leaves.	[59]
COMT (3-Caffeic acid o methyltransferase)	0.006	3.234	Involved in lignin biosynthesis. High activation of lignifying enzymes was found in dehydration-stressed white clover (*Trifolium repens* L.), which lead to reduced forage growth.	[60]
Glycine decarboxylase complex H	0.005	3.113	Functions in photo respiratory carbon recovery. Carbon dioxide is found to be low in plants subjected to dehydration stress due to the closing of stomata in order to prevent water loss.	[61]
Up-Regulated Genes in TN				
Clp amino terminal domain-containing protein, putative	0.035	3.778	Protein and ATP binding.	
CONSTANS-LIKE 1	0.025	3.294	Transcription factor regulating flower development and response to light stimulus.	[62]
DRB3 (DSRNA-BINDING PROTEIN 3)	0.020	2.984	Assists in miRNA-targeted RNA degradation.	[63]
SIGE (SIGMA FACTOR E)	0.032	2.808	Responds to effects of abiotic stresses. Phosphorylation of major sigma factor SIG1 in *Arabidopsis thaliana* inhibits the transcription of the *psaA* gene, which encodes photosystem-I (PS-I). This disturbs photosynthetic activity.	[64,65]
Reticulon family protein	0.029	2.772	Playing a role in promoting membrane curvature	
Cytochrome c oxidase family protein	0.025	2.727	Essential for the assembly of functional cytochrome oxidase protein.	
DNA-binding S1FA family protein	0.049	2.717	Binds to the negative promoter element S1F.	
DNA photolyase	0.032	2.667	DNA repair enzyme.	
Zinc knuckle (CCHC-type) family protein	0.040	2.567	Zinc ion binding.	
Monosaccharide transporter	0.025	2.547	Plays a role in long-distance sugar partitioning or sub-cellular sugar distribution.	
Nodulin MtN3 family protein	0.025	2.376	Key role in the establishment of symbiosis.	
Serine acetyltransferase, N-terminal	0.040	2.302	Catalyzes the formation of a cysteine precursor.	

Table 4. Top downregulated genes in DipC and TN.

Gene name	FDR	Fold Change	Gene Description	References
Down-Regulated Genes in DipC				
Dihydroxyacetone kinase	0.003	6.489	Glycerone kinase activity.	
Phosphoglucomutase, putative/glucose phosphomutase, putative	0.007	6.471	Involved in controlling photosynthetic carbon flow and plays essential role starch synthesis. Down regulation of photosynthesis-related gene will lead to significant reduction in plant growth.	[66]
Auxin-induced protein 22D AUXX-IAA	0.003	4.627	Involved in stress defense response. Many AUXX-IAA genes were found to be down-regulated in Sorghum bicolor under drought conditions.	[67]
CP12-1, putative	0.014	4.390	Involved in calvin cycle, therefore linked to photosynthesis. Most drastic down-regulated genes which were photosynthesis-related was observed in barley (*Hordeum vulgare* L.).	[68]
PHS2 (ALPHA-GLUCAN PHOSPHORYLASE 2).	0.014	4.375	Encodes a cytosolic alpha-glucan phosphorylase.	
APRR5 (PSEUDO-RESPONSE REGULATOR 5), Pseudo ARR-B family	0.001	4.145	Linked to cytokinin-mediated regulation.	
Thiamine biosynthesis family protein	0.002	4.132	Catalyses the activation of small proteins, such as ubiquitin or ubiquitin-like proteins.	
Zinc finger (C3HC4-type RING finger)	0.007	3.611	Mediate ubiquitin-conjugating enzyme (UBC-2) dependent ubiquitation.	[69]
WRKY40	0.033	3.104	Regulator of ABA signalling. It inhibits the expression of ABA-responsive genes ABF4, AB14, AB15, DREB1A, MYB2 and RAB18.	[70]
Down-Regulated Genes in TN				
AGL83 (AGAMOUS-LIKE 83)	0.025	4.374	DNA-binding transcription factor.	
CRR23 (chlororespiratory reduction 23)	0.025	3.625	A subunit of the chloroplast NAD(P)H dehydrogenase complex, involved in PS-I cyclic electron transport. Located on the thylakoid membrane. Mutant has impaired NAD(P)H dehydrogenase activity. Part of dehydration repressing photosynthesis.	[71]
MYB30 (MYB DOMAIN PROTEIN 30)	0.032	3.250	Acts as a positive regulator of hypersensitive cell death and salicylic acid synthesis. Involved in the regulation of abscisic acid (ABA) signalling.	[72]
Photosystem II family protein, putative	0.029	3.158	Linked to photosynthesis. Down-regulation of photosynthesis-related genes during dehydration stress was observed in maize (*Zea mays*), which in turn leads to significant reduction in plant growth.	[73]
Phosphoesterase	0.047	3.136	Hydrolase activity, acting on ester bonds.	
Zing-finger (C3HC4-type)	0.045	2.947	Mediate ubiquitin-conjugating enzyme (UBC-2) dependent ubiquitation.	[69]
NHX2 (Sodium proton exchanger 2)	0.040	2.742	Involved in antiporter activity. Also involved in potassium ion homoeostasis and regulation of stomatal closure. Involved in the accumulation of K^+ that drives the rapid stomatal opening. Down-regulation of genes related to stomatal regulation has been observed in soybean, which appears to be a part of dehydration response, leading to a reduction in the amount of stomata in leaves.	[74]
Inositol 1,3,4-trisphosphate 5/6-kinase	0.035	2.090	Part of IP3 signal transduction pathway.	[75]

Table 5. Overlapping up- and downregulated genes.

Gene Name	FDR	Fold Change	Gene Description	References
Up-Regulated Genes				
Beta-fructofuranosidase	8.90×10^{-4}	3.193	Catalyses the hydrolysis of sucrose. A rise in monosaccharide content caused by the Beta-fructofuranosidase can compensate for the decline in photosynthetic carbon assimilation indicated by the decrease in net photosynthesis.	[46,47]
Down-Regulated Genes				
MEE59 (maternal effect embryo arrest 59)	8.94×10^{-4}	8.580	Embryo development ending in seed dormancy.	
Calcineurin-like phosphoesterase family protein (CPPED1)	6.72×10^{-4}	5.857	Plays inhibitory role in glucose uptake. Down-regulation of CPPED1 improves glucose metabolism.	[48]
Putative lysine-specific demethylase JMJD5 Jumonji/Zinc-finger-class domain containing protein	0.003	4.971	Plays role in a histone demethylation mechanism that is conserved from yeast to human. Down-regulation may lead to an increase in methylated histones and hence general down-regulation of transcription.	[49]
MYB-like transcription factor	0.024	4.103	Arabidopsis homolog is known to regulate stomatal opening, flower development, and plays role in circadian rhythm.	[50]

Table 5. *Cont.*

Gene Name	FDR	Fold Change	Gene Description	References
Down-Regulated Genes				
F-box family protein (FBL14)	0.001	3.744	Functions in signal transduction and regulation of cell cycle.	
BRH1 (BRASSINOSTEROID-RESPONSIVE RING-H2)	0.007	2.899	BRH1 is known to influence stomatal density.	[51]
Bundle-sheath defective protein 2 family/bsd2 family	0.003	2.441	Protein required for post-translational regulation of Rubisco large subunit (rbcL).	[52]
Mitochondrial substrate carrier family protein.	0.030	2.435	Involved in energy transfer.	

FDR: false discovery rate.

3.5. Confirmation of Candidate Dehydration-Associated Genes by Real-Time qRT-PCR

Four differentially expressed genes (*PAL1*, Beta-fructofuranosidase, *COMT* and *UBC-2*) were chosen for further analysis, as they showed high levels of expression under water-limited treatment [46–53,57–60] (Tables 4 and 5) and are dehydration-associated genes based on their functional annotations. Figure 3 shows the results of qPCR analysis. The transcript levels of Beta-fructofuranosidase, *COMT*, and *UBC-2* confirmed the expression trends seen in the microarray data. *PAL1* showed the expected increase in DipC, and an increase in TN was observed, which was not observed in the microarray results. The reason for this is unclear.

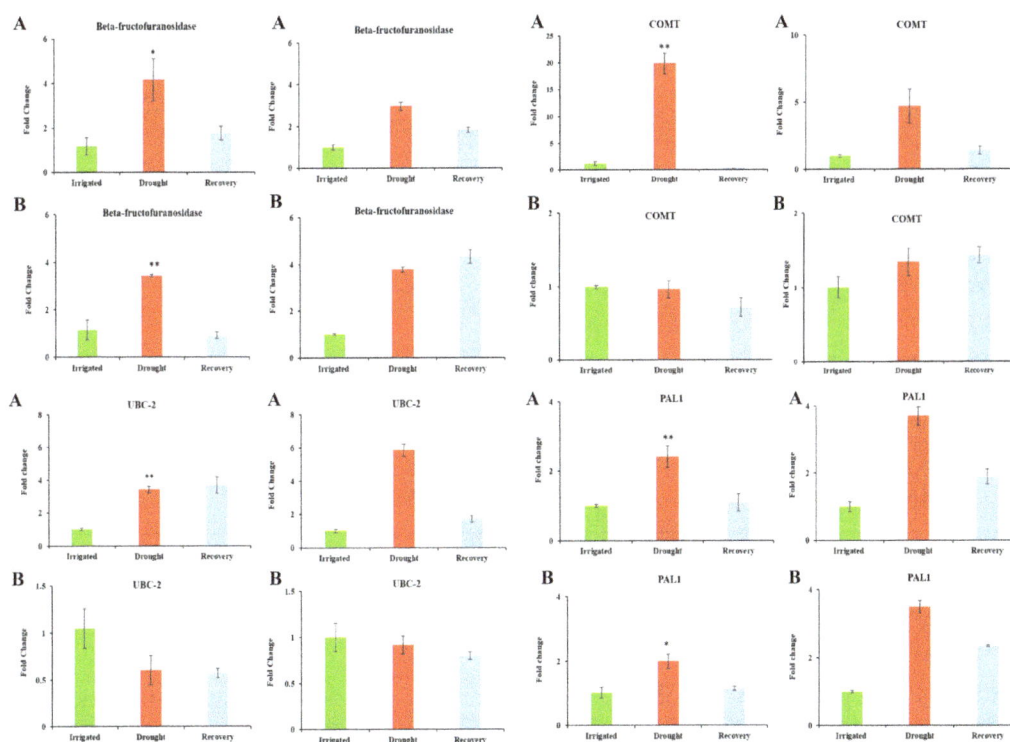

Figure 3. Comparison of qPCR and microarray intensity values: Rows (**A**) and (**B**) respectively refer to results for DipC and TN. The left- and right-hand pairs of columns correspond to the qPCR and microarray values for DipC and TN, respectively. The gene under study is named at the top of each panel. In order, the investigated genes are Beta-fructofuranosidase, *COMT*, *UBC-2*, and *PAL1*. qPCR and Microarray values are shown as fold changes with respect to the water-sufficient treatment (Irrigated). Error bars denote the standard error. Single and double asterisks indicate that *p*-value is less than 0.05 and 0.01, respectively, which was assessed by the paired *t*-test between groups. Irrigated and Drought refer to water-sufficient and water-limited treatments, respectively.

3.6. Transcription Factors Associated with Dehydration Stress

The DEGs genes from both genotypes identified various transcription-related factors (TFs). Common to both genotypes are the downregulation of *BRH1*, an *MYB*, *MEE59*, and *JMJD5*. The latter is a histone demethylase, so could suggest changes at the epigenetic level of gene expression. Its downregulation could result in the indirect repression of multiple genes. On top of the common genes, DipC shows the upregulation of two TFs (*WRKY51* and a *bHLH* TF) and the downregulation of four others (*ATAUX2-11*, *WRKY40*, a C2H2 Zn-finger, and three probe-sets for GIGANTEA). TN, on the other hand, shows the upregulation of genes for CONSTANS1-like, S1FA DNA-binding, and a double-strand RNA-binding protein (which can aid microRNA-mediated RNA degradation). The downregulated TFs in TN are *MYB60* and a second *MEE59*.

Co-expression networks were individually built for DipC and TN (see Tables S11 and S12), and the dehydration-specific network of each were merged. This resulted in more TFs being included, which are features of recovery treatment. By looking at the number of links that each node has in the genotype-specific and merged networks, it is possible to rank the potential importance of the different TFs (see Table 6). The DipC TFs had a higher number of links than TN. In the case of DipC, *WRKY40* stands out as being the TF with the most co-expressed genes, with *ATAUX2-11*, *PRR7*, and a Zinc-finger protein (GmaAffx.33796.3.S1_at) also looking relevant. For TN, however, the TFs are not ranked so highly, with CONSTANS-like 1 and *MYB60* showing the greatest involvement. For this genotype, the differentially expressed TFs in common with DipC seem almost as important.

Table 6. Vertex degrees of differentially expressed transcription factors.

DipC				TN			
Probe-Set	Name	V°Whole	V°Drought	Probe-Set	Name	V°Whole	V°Drought
Gma.16733.1.S1_at	WRKY40	68	17	GmaAffx.45249.1.S1_at	CONSTANS-like 1	16	3
Gma.6670.1.S1_at	PRR7	49	7	GmaAffx.84566.2.S1_at	MYB60	8	3
GmaAffx.33796.3.S1_at	Zinc-finger like C2H2	45	7	GmaAffx.86517.1.S1_at	AGL83	6	1
GmaAffx.92679.1.S1_s_at	ATAUX2-11	41	9	Gma.1576.1.S1_at	Zinc-finger C3HC4	5	1
GmaAffx.35309.1.S1_s_at	GRF2	35	6				
GmaAffx.65059.1.S1_at	bHLH	32	7				
GmaAffx.90399.1.S1_at	C3HC4 Zinc-finger	31	9				
Gma.15774.1.S1_at	Zinc-finger C3HC4	26	3				
GmaAffx.53180.1.S1_at	PRR7	25	9				
GmaAffx.80492.1.S1_at	PRR5	9	2				
GmaAffx.73009.2.S1_at	WRKY51	7	5				
Common TFs							
GmaAffx.60283.1.S1_at	BRH1	42	6				
GmaAffx.9286.1.S1_s_at	MYB	27	4				
Gma.17248.1.A1_at	JMJD5	26	3				
GmaAffx.10162.1.S1_at	MEE59	13	3				

V° refers to the number of links of each transcription factor (TF) node, in either the whole genotype-specific network, or merged dehydration-specific network.

4. Discussion

Landraces are a potentially valuable resource for finding genes conferring useful agricultural and processing traits. Bambara groundnut is an underutilised African legume whose landraces are adapted, in many cases, to arid conditions. We have developed single genotypes derived from landraces for analysis. There have been several dehydration studies carried out on bambara groundnut, but the molecular mechanisms of how the crop responds and adapts to dehydration stress are still under investigation. This study has carried out transcriptomic comparisons in two genotypes of bambara groundnut, DipC and TN, in an attempt to identify potential genes conferring advantageous traits for crop growth and yields in marginal environments.

Cross-species hybridisation to the soybean microarray has been shown to be informative for investigating the bambara groundnut transcriptome, as good gene (probe-set) retention was observed at high gDNA hybridisation thresholds. In support of the results, Bonthala et al. [44], reported a high correlation between cross-species microarray data and RNA-sequencing approaches for detecting differentially expressed genes under a cold temperature stress experiment in bambara groundnut. Probe-sets retained by the mask after genomic hybridisation are almost identical (>99%), suggesting that, at this level of resolution, the two genotypes are highly similar at the sequence level. Four known dehydration-associated genes, seen to be differentially expressed in these data, were subjected to qPCR, and supported the notion that the observed trends in the microarray data are valid.

The 26,496 probe sets common between the two genotypes, under irrigated conditions, (with a RMA cut-off of 0.97), include some sixty dehydration- and ABA-related genes. The latter include genes for producing osmoprotectants. They might provide two components of the dehydration avoidance capability of these genotypes, by retaining normal cell functioning when water access becomes limited. Clearly, if the plant has already activated part of the dehydration response, it could have multiple effects. The presence of osmoprotectants might draw in even more water than otherwise might be the case, and there will be a greater proportion of biomass devoted to root growth, resulting in even deeper roots that are better able to survive dehydration later on. Bambara groundnut is known to allocate a greater fraction of its dry weight to the roots than to the shoots, irrespective of the soil moisture status [27]. This strategy may have clear advantages when water subsequently becomes limited, suggesting an adaptation to harsh environments and a decision to prioritise survival. In addition, as bambara groundnut is grown in harsh environments and has not undergone intensive breeding for the yield and above ground biomass, this suggests that it still allocates more effort to developing root architecture to handle dehydration when it happens. Moreover, Nayamudeza [27] also stated that the fraction of total dry weight allocated to the roots in bambara groundnut is greater than that allocated to the groundnut. In addition, a relatively higher expression of dehydration-associated genes in both genotypes under water-sufficient treatment including *ABI1* (ABA Insensitive 1), *ABF1* (ABRE binding factor 1), *ERD4* (Early responsive to dehydration 4), and *RD19* (Response to dehydration 19), compared to other species such as Soybean [76] (see Figure S5), suggest that bambara groundnut could at least be in a partially ready state for dehydration, even in the absence of dehydration stress. However, further research is needed to validate this hypothesis.

Given that 59,782 and 59,835 probe-sets were used to evaluate the transcriptome changes after probe-masking in DipC and TN, respectively, there were only very small numbers of genes significantly differentially expressed (189 in DipC and 81 in TN) under water-limited treatment. It could be speculated that the slow and progressive dehydration stress might not cause significant shock to the plants.

The upregulated genes in both genotypes were subdivided into ~75% dehydration responsive (with expression levels returning to normal after recovery) and ~25% dehydration perturbed (where the expression levels remained altered). In the case of downregulated genes, 80–85% of the expression levels returned to being comparable with the non-stressed state. The dehydration-perturbed expression levels might be caused by changes at the chromatin level, through DNA methylation or histone modification, and it is therefore interesting to note that a protein-lysine demethylase is repressed by dehydration.

The above observations show that the two genotypes appear to be very similar in terms of their genotype (validating the comparability of the transcriptome data compared using the microarray), while exhibiting differences in their general transcriptional behaviour in water-sufficient conditions and in response to dehydration stress. However, when the sets of differentially expressed genes are compared, there is almost no overlap. Out of 189 and 91 genes differentially expressed in DipC and TN, respectively, only nine were common between the two genotypes, suggesting that some of the mechanisms for adaption to dehydration are substantially different in the two genotypes. Of these, Beta-fructofuranosidase contributes to osmoprotection [46,47], an MYB gene is associated with the stomatal opening in Arabidopsis thaliana [50], *BRH1* affects the stomatal density [51], and bsd2 affects photosynthesis in maize [52], while *JMJD5* plays an epigenetic role [49], as mentioned above. Figure 4 illustrates how two genotypes with very similar genomes may have adapted to achieve dehydration response traits (transcriptional and hormone signalling to affect cell-wall modification, lignin synthesis, photosynthesis, transporters, hormone signalling, osmoprotection, oxidative stress) through largely different sets of effector genes.

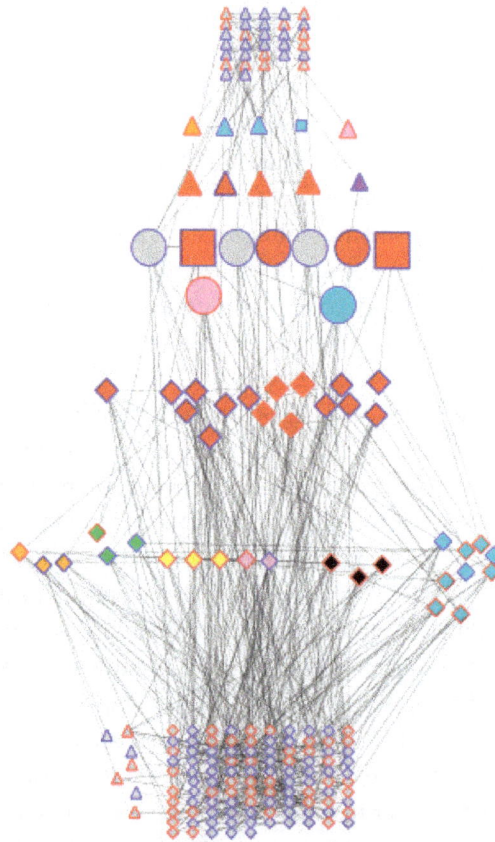

Figure 4. Comparison of genotype co-expression networks. Cytoscape has been used to layout the merged dehydration-responsive network of co-expressed probe-sets. Node shapes are triangles, diamond circles, and squares, respectively, for the differential expression of the probe-sets of TN, DipC, both (i.e., common), and both but affecting stomata. They have been coloured according to their activity in relation to the dehydration response: red (transcription), orange (cell wall), yellow (lignin synthesis), green (photosynthesis), blue (transporters), indigo (hormone signalling), pink (osmoprotection), black (oxidative stress), and grey (others). Node borders have been coloured red and blue to denote up- and downregulation under stress. Nodes have been arranged in seven horizontal bands with probe-sets in common in the middle flanked by TFs and hormone-signalling genes, other genes that play various roles in response to dehydration, and others. Nodes have been linked by the criteria of the co-expression analysis.

Several transcription factors that seem likely to play a role in the bambara groundnut dehydration response and which are common to both genotypes are *BRH1* and an MYB transcription factor, which are known to affect the stomata in *Arabidopsis thaliana* [50], and *JMJD5*. DipC shows a more significant response, with changes to *WRKY40*, and is of particular interest. It is a well-known member of plant dehydration-response networks [67] and is the most highly linked TF node in the co-expression networks. For DipC, the network also reveals the importance of *PRR7*, a core circadian clock component known to play a complex role in abiotic stresses [77]. It is somewhat surprising that TN does not show a >2-fold change in the expression of *WRKY40*, but it may have roles for CONSTANS-like 1 (another clock-related gene associated with flowering in rice that may be associated with abiotic stress in bambara groundnut [78]) and *MYB60*, which affect stomatal closure in *A. thaliana* [79], and *AGL-83*, a MADS-Box protein with an uncertain role.

5. Conclusions

Understanding the mechanisms underlying the ability of crops to produce viable yields under drought conditions is a priority for global food security. This study has examined the transcriptomic reponse to dehydration and recovery in two genotypes derived from landraces of bambara groundnut, in an attempt to investigate the molecular mechanisms occurring in the two landraces. In addition, this study also tested whether the cross-species hybridisation to the soybean microarray is suitable for investigating the bambara groundnut transcriptome. It was shown that many potential dehydration-responsive genes are expressed, even under water-sufficient conditions, in both landraces, suggesting that bambara groundnut could at least be in a partially ready state for dehydration, even in the absence of dehydration stress. In terms of differential expression, there were only a very small number of genes differentially expressed under water-limited treatment in both landraces, suggesting that the slow and progressive dehydration stress might not cause a significant shock to the plants. Although the transcription factors and dehydration-response genes were largely different between the two landraces, they may achieve the same effect in terms of survival under drought conditions. The DipC genotype displayed the differential expression of some well-known dehydration-associated transcriptions factors (especially *WRKY40*), while TN showed the differential expression of CONSTANS-LIKE 1 and *MYB60*. Cross-species hybridisation to the soybean microarray has been shown to be informative for investigating the bambara groundnut transcriptome, as good gene retention was observed at high gDNA hybridisation thresholds.

Supplementary Materials: The following are available online at www.mdpi.com/2073-4425/8/4/121/s1. Figure S1: Treatment Regimes, Figure S2: Genotype specific 3D PCA scatter plots, Figure S3: Ranked mean intensities of water-sufficient genotype samples, Figure S4: Categories of system response in gene expression, Figure S5: Comparison between bambara groundnut and soybean for expression levels for dehydration-associated genes under water-sufficient conditions. Table S1: Dehydration-associated genes expressed under water-sufficient condition in DipC, Table S2: Dehydration-associated genes expressed under water-sufficient condition in TN, Table S3: Dehydration-associated genes expressed under water-sufficient condition in both DipC and TN, Table S4: Comparison of water-sufficient conditions transcriptomes of DipC and TN, Table S5: Upregulated DipC genes as a result of water-limited treatment (p-value <= 0.05, Abs. F.C > 2), Table S6: Downregulated DipC genes as a result of water-limited treatment (p-value <= 0.05, Abs. F.C > 2), Table S7: Upregulated TN genes as a result of water-limited treatment (p-value <= 0.05, Abs. F.C > 2), Table S8: Downregulated TN genes as a result of water-limited treatment (p-value <= 0.05, Abs. F.C > 2), Table S9: List of differentially expressed genes (water-limited versus water-sufficient) common to DipC and TN when sample D.DipC.rep2 is excluded from the analysis, Table S10: GO-term overrepresentation of all the gene-sets in Soybean GeneChip array to compare DipC, TN, and Soybean datasets, Table S11: DipC Co-expression network, Table S12: TN Co-expression network.

Acknowledgments: Faraz Khan thank Kevin Ryan and Jiang Lu for their guidance and comments on the manuscript. Faraz Khan thank Crops for the Future (CFF) and the University of Nottingham for providing funding for the research. Faraz Khan thank the NASC arabidopsis centre for Microarray hybridisation and Presidor Kendabie for the help in the dehydration experiment. Charlie Hodgman thanks the support from the European Research Council Advanced Grant funding (FUTUREROOTS 294729).

Author Contributions: F.K. analysed the microarray data, created custom CDF files, performed qPCR experiments, constructed and analysed the co-expression network, interpreted the results, and drafted the manuscript. C.H. created the merged dehydration-responsive network, co-analysed the co-expression network, interpreted the results, and critically reviewed the manuscript. H.H.C. performed the dehydration experiment and RNA extraction. S.M. and C.L. conceived the project and critically reviewed the manuscript. I.A. helped with the network and microarray analysis.

References

1. Ocampo, E.; Robles, R. Drought tolerance in mungbean. I. Osmotic adjustment in drought stressed mungbean. *Philipp. J. Crop Sci.* **2000**, *25*, 1–5.

2. Subbarao, G.V.; Chauhan, Y.S.; Johansen, C. Patterns of osmotic adjustment in pigeonpea—Its importance as a mechanism of drought resistance. *Eur. J. Agron.* **2000**, *12*, 239–249. [CrossRef]

3. Kooyers, N.J. The evolution of drought escape and avoidance in natural herbaceous populations. *Plant Sci.* **2015**, *234*, 155–162. [CrossRef] [PubMed]

4. Harb, A.; Krishnan, A.; Ambavaram, M.M.R.; Pereira, A. Molecular and physiological analysis of drought stress in Arabidopsis reveals early responses leading to acclimation in plant growth. *Plant Physiol.* **2010**, *154*, 1254–1271. [CrossRef] [PubMed]

5. Al Shareef, I.; Sparkes, D.; Azam-Ali, S. Temperature and drought stress effects on growth and development of bambara groundnut (*Vigna subterranea* L.). *Exp. Agric.* **2013**, *50*, 72–89. [CrossRef]

6. Eslami, S.V.; Gill, G.S.; McDonald, G. Effect of water stress during seed development on morphometric characteristics and dormancy of wild radish (*Raphanus raphanistrum* L.) seeds. *Int. J. Plant Prod.* **2012**, *4*, 159–168.

7. Gaur, M.P.; Krishnamurthy, L.; Kashiwagi, J. Improving drought-avoidance root traits in chickpea (*Cicer arietinum*); current status of research at ICRISAT. *Plant Prod. Sci.* **2008**, *11*, 3–11. [CrossRef]

8. Gupta, N.K.; Gupta, S.; Kumar, A. Effect of water stress on physiological attributes and their relationship with growth and yield of wheat cultivars at different stages. *J. Agron. Crop Sci.* **2001**, *186*, 55–62. [CrossRef]

9. Liu, F.; Jensen, C.R.; Andersen, M.N. Hydraulic and chemical signals in the control of leaf expansion and stomatal conductance in soybean exposed to drought stress. *Funct. Plant Biol.* **2003**, *30*, 65–73. [CrossRef]

10. Ludlow, M. Strategies of response to water stress. In *Structural and Functional Responses to Environmental Stresses: Water Shortage*; Kreeeb, K.H., Richter, H., Hinckley, T.M., Eds.; SPB Academic publishing: The Hague, The Netherlands, 1989; pp. 269–282.

11. Nakashima, K.; Yamaguchi-Shinozaki, K.; Shinozaki, K. The transcriptional regulatory network in the drought response and its crosstalk in abiotic stress responses including drought, cold, and heat. *Front Plant Sci.* **2014**, *5*, 170. [CrossRef] [PubMed]

12. Farooq, M.; Wahid, A.; Kobayashi, N.; Fujita, D.; Basra, S.M.A. Plant drought stress: Effects, mechanisms and management. *Agron. Sustain. Dev.* **2009**, *29*, 185–212. [CrossRef]

13. Mwale, S.S.; Azam-Ali, S.N.; Massawe, F.J. Growth and development of bambara groundnut (*Vigna subterranea*) in response to soil moisture. *Eur. J. Agron.* **2007**, *26*, 345–353. [CrossRef]

14. Umezawa, T.; Fujita, M.; Fujita, Y.; Yamaguchi-Shinozaki, K.; Shinozaki, K. Engineering drought tolerance in plants: Discovering and tailoring genes to unlock the future. *Curr. Opin. Biotechnol.* **2006**, *17*, 113–122. [CrossRef] [PubMed]

15. Todaka, D.; Shinozaki, K.; Yamaguchi-Shinozaki, K. Recent advances in the dissection of drought-stress regulatory networks and strategies for development of drought-tolerant transgenic rice plants. *Front Plant Sci.* **2015**, *6*, 84. [CrossRef] [PubMed]

16. Poets, A.M.; Fang, Z.; Clegg, M.T.; Morrell, P.L. Barley landraces are characterized by geographically heterogeneous genomic origins. *Genome Biol.* **2015**, *16*, 173. [CrossRef] [PubMed]

17. Lasky, J.R.; Upadhyaya, H.D.; Ramu, P.; Deshpande, S.; Hash, C.T.; Bonnette, J.; Juenger, T.E.; Hyma, K.; Acharya, C.; Mitchell, S.E.; et al. Genome-environment associations in sorghum landraces predict adaptive traits. *Sci. Adv.* **2015**, *1*, e1400218. [CrossRef] [PubMed]

18. Wei, X.; Zhu, X.; Yu, J.; Wang, L.; Zhang, Y.; Li, D.; Zhou, R.; Zhang, X. Identification of sesame genomic variations from genome comparison of landrace and variety. *Front. Plant Sci.* **2016**, *7*, 1169. [CrossRef] [PubMed]

19. Valliyodan, B.; Nguyen, H.T. Understanding regulatory networks and engineering for enhanced drought tolerance in plants. *Curr. Opin. Plant Biol.* **2006**, *9*, 189–195. [CrossRef] [PubMed]

20. Dodig, D.; Zorić, M.; Kandić, V.; Perović, D.; Šurlan-Momirović, G. Comparison of responses to drought stress of 100 wheat accessions and landraces to identify opportunities for improving wheat drought resistance. *Plant Breed.* **2012**, *131*, 369–379. [CrossRef]

21. Collinson, S.T.; Clawson, E.J.; Azam-Ali, S.N.; Black, C.R. Effects of soil moisture deficits on the water relations of bambara groundnut (*Vigna subterranea* L. Verdc.). *J. Exp. Bot.* **1997**, *48*, 877–884. [CrossRef]

22. Linnemann, A.; Azam-ALI, S. *Bambara groundnut (Vigna subterraneanea)*; Williams, J.T., Ed.; Chapman and Hall: London, UK, 1993.

23. Mazahib, A.M.; Nuha, M.O.; Salawa, I.S.; Babiker, E.E. Some nutritional attributes of bambara groundnut as influenced by domestic processing. *Int. Food Res. J.* **2013**, *20*, 1165–1171.

24. Heller, J. *Bambara Groundnut: Vigna subterranea (l.) Verdc. Promoting the Conservation and Use of Under-Utilized and Neglected Crops*; IPGRI: Harare, Zimbabwe, 1997.

25. Mabhaudhi, T.; Modi, A.T.; Beletse, Y.G. Growth, phenological and yield responses of a bambara groundnut (*Vigna subterranea* L. Verdc) landrace to imposed water stress: II. Rain shelter conditions. *Afr. Crop Sci. J.* **2013**, *39*, 191–198. [CrossRef]

26. Collinson, S.T.; Azam-Ali, S.N.; Chavula, K.M.; Hodson, D.A. Growth, development and yield of bambara groundnut (*Vigna subterranea*) in response to soil moisture. *J. Agric. Sci.* **1996**, *126*, 307. [CrossRef]

27. Nayamudeza, P. Crop water use and the root systems of bambara groundnut (*Vigna subterranea* (L.) verdc.) and groundnut (*Arachis hypogaea* (L.)) in response to irrigation and drought. Master's Thesis, The University of Nottingham, Nottingham, UK, 1989.

28. Berchie, J.N. Evaluation of five bambara groundnut (*Vigna subterranea* (L.) verdc.) landraces to heat and drought stress at Tono-Navrongo, Upper East Region of Ghana. *Afr. J. Agric. Res.* **2012**, *7*, 250–256. [CrossRef]

29. Vurayai, R.; Emongor, V.; Moseki, B.; Emongor, V.; Moseki, B. Physiological responses of bambara groundnut (*Vigna subterranea* L. Verdc) to short periods of water stress during different developmental stages. *Asian J. Agric. Sci.* **2011**, *3*, 37–43.

30. Chai, H.H. Developing new approaches for transcriptomics and genomics—Using major resources developed in model species for research in crop species. Ph.D. Thesis, The University of Nottingham, Nottingham, UK, 2014.

31. Chai, H.H.; Massawe, F.; Mayes, S. Effects of mild drought stress on the morpho-physiological characteristics of a bambara groundnut segregating population. *Euphytica* **2015**, *208*, 225–236. [CrossRef]

32. Buckley, B.A. Comparative environmental genomics in non-model species: Using heterologous hybridization to DNA-based microarrays. *J. Exp. Biol.* **2007**, *210*, 1602–1606. [CrossRef] [PubMed]

33. Davey, M.W.; Graham, N.S.; Vanholme, B.; Swennen, R.; May, S.T.; Keulemans, J. Heterologous oligonucleotide microarrays for transcriptomics in a non-model species; a proof-of-concept study of drought stress in musa. *BMC Genomics* **2009**, *10*, 436. [CrossRef] [PubMed]

34. Pariset, L.; Chillemi, G.; Bongiorni, S.; Spica, V.R.; Valentini, A. Microarrays and high-throughput transcriptomic analysis in species with incomplete availability of genomic sequences. *New Biotechnol.* **2009**, *25*, 272–279. [CrossRef] [PubMed]

35. Hammond, J.P.; Broadley, M.R.; Craigon, D.J.; Higgins, J.; Emmerson, Z.F.; Townsend, H.J.; White, P.J.; May, S.T. Using genomic DNA-based probe-selection to improve the sensitivity of high-density oligonucleotide arrays when applied to heterologous species. *Plant Methods* **2005**, *1*, 10. [CrossRef] [PubMed]

36. Graham, N.S.; May, S.T.; Daniel, Z.C.; Emmerson, Z.F.; Brameld, J.M.; Parr, T. Use of the affymetrix human genechip array and genomic DNA hybridisation probe selection to study ovine transcriptomes. *Animal* **2011**, *5*, 861–866. [CrossRef] [PubMed]

37. Chu, L.; Scharf, E.; Kondo, T. Genespring tm: Tools for analyzing data microarray expression. *Genome Inform.* **2001**, *12*, 227–229.

38. Maere, S.; Heymans, K.; Kuiper, M. Bingo: A cytoscape plugin to assess overrepresentation of gene ontology categories in biological networks. *Bioinformatics* **2005**, *21*, 3448–3449. [CrossRef] [PubMed]

39. Li, J.; Wei, H.; Zhao, P.X. DeGNServer: Deciphering genome-scale gene networks through high performance reverse engineering analysis. *Biomed. Res. Int.* **2013**, *2013*, 856325. [CrossRef] [PubMed]

40. Smoot, M.E.; Ono, K.; Ruscheinski, J.; Wang, P.-L.; Ideker, T. Cytoscape 2.8: New features for data integration and network visualization. *Bioinformatics* **2011**, *27*, 431–432. [CrossRef] [PubMed]

41. Hu, R.; Fan, C.; Li, H.; Zhang, Q.; Fu, Y.-F. Evaluation of putative reference genes for gene expression normalization in soybean by quantitative real-time RT-PCR. *BMC Mol. Biol.* **2009**, *10*, 93. [CrossRef] [PubMed]

42. Jain, M.; Nijhawan, A.; Tyagi, A.K.; Khurana, J.P. Validation of housekeeping genes as internal control for studying gene expression in rice by quantitative real-time pcr. *Biochem. Biophys. Res. Commun.* **2006**, *345*, 646–651. [CrossRef] [PubMed]

43. Ye, J.; Coulouris, G.; Zaretskaya, I.; Cutcutache, I.; Rozen, S.; Madden, T.L. Primer-blast: A tool to design target-specific primers for polymerase chain reaction. *BMC Bioinform.* **2012**, *13*, 134. [CrossRef] [PubMed]

44. Bonthala, V.S.; Mayes, K.; Moreton, J.; Blythe, M.; Wright, V.; May, S.T.; Massawe, F.; Mayes, S.; Twycross, J. Identification of gene modules associated with low temperatures response in bambara groundnut by network-based analysis. *PLoS ONE* **2016**, *11*, e0148771. [CrossRef] [PubMed]

45. Livak, K.J.; Schmittgen, T.D. Analysis of relative gene expression data using real-time quantitative PCR and the $2-\Delta\Delta ct$ method. *Methods* **2001**, *25*, 402–408. [CrossRef] [PubMed]

46. Li, Y.C.; Meng, F.R.; Zhang, C.Y.; Zhang, N.; Sun, M.S.; Ren, J.P.; Niu, H.B.; Wang, X.; Yin, J. Comparative analysis of water stress-responsive transcriptomes in drought-susceptible and -tolerant wheat (*Triticum aestivum* L.). *J. Plant Biol.* **2012**, *55*, 349–360. [CrossRef]

47. Majlath, I.; Darko, E.; Palla, B.; Nagy, Z.; Janda, T.; Szalai, G. Reduced light and moderate water deficiency sustain nitrogen assimilation and sucrose degradation at low temperature in durum wheat. *J. Plant Physiol.* **2016**, *191*, 149–158. [CrossRef] [PubMed]

48. Vaittinen, M.; Kaminska, D.; Kakela, P.; Eskelinen, M.; Kolehmainen, M.; Pihlajamaki, J.; Uusitupa, M.; Pulkkinen, L. Downregulation of cpped1 expression improves glucose metabolism in vitro in adipocytes. *Diabetes* **2013**, *62*, 3747–3750. [CrossRef] [PubMed]

49. Tsukada, Y.-i.; Fang, J.; Erdjument-Bromage, H.; Warren, M.E.; Borchers, C.H.; Tempst, P.; Zhang, Y. Histone demethylation by a family of JMJC domain-containing proteins. *Nature* **2006**, *439*, 811–816. [CrossRef] [PubMed]

50. Ding, Y.; Liu, N.; Virlouvet, L.; Riethoven, J.-J.; Fromm, M.; Avramova, Z. Four distinct types of dehydration stress memory genes in *Arabidopsis thaliana*. *BMC Plant Biol.* **2013**, *13*, 229. [CrossRef] [PubMed]

51. Kim, T.W.; Michniewicz, M.; Bergmann, D.C.; Wang, Z.Y. Brassinosteroid regulates stomatal development by GSK3-mediated inhibition of a MAPK pathway. *Nature* **2012**, *482*, 419–422. [CrossRef] [PubMed]

52. Brutnell, T.P.; Sawers, R.J.H.; Mant, A.; Langdale, J.A. Bundle sheath defective2, a novel protein required for post-translational regulation of the RBCL gene of maize. *Plant Cell* **1999**, *11*, 849–864. [CrossRef] [PubMed]

53. Gholizadeh, A. Effects of drought on the activity of phenylalanine ammonia lyase in the leaves and roots of maize inbreds. *Aust. J. Basic Appl. Sci.* **2011**, *5*, 952–956.

54. Gerhardt, L.B.d.A.; Magioli, C.; Perez, A.B.U.C.M.; Margis, R.; Sachetto-Martins, G.; Margis-Pinheiro, M. AtCHITIV gene expression is stimulated under abiotic stresses and is spatially and temporally regulated during embryo development. *Genetics Mol. Biol.* **2004**, *27*, 118–123. [CrossRef]

55. Huang, Y.; Chen, X.; Liu, Y.; Roth, C.; Copeland, C.; McFarlane, H.E.; Huang, S.; Lipka, V.; Wiermer, M.; Li, X. Mitochondrial atPAM16 is required for plant survival and the negative regulation of plant immunity. *Nat. Commun.* **2013**, *4*, 2558. [CrossRef] [PubMed]

56. Wasternack, C. Jasmonates: An update on biosynthesis, signal transduction and action in plant stress response, growth and development. *Ann. Bot.* **2007**, *100*, 681–697. [CrossRef] [PubMed]

57. Zhou, L.; Liu, Y.; Liu, Z.; Kong, D.; Duan, M.; Luo, L. Genome-wide identification and analysis of drought-responsive micrornas in *Oryza sativa*. *J. Exp. Bot.* **2010**, *61*, 4157–4168. [CrossRef] [PubMed]

58. Scherle, P.; Behrens, T.; Staudt, L.M. Ly-gdi, a gdp-dissociation inhibitor of the rhoa GTP-binding protein, is expressed preferentially in lymphocytes. *Proc. Natl. Acad. Sci. USA* **1993**, *90*, 7568–7572. [CrossRef] [PubMed]

59. Chen, L.; Bush, D.R. Lht1, a lysine- and histidine-specific amino acid transporter in arabidopsis. *Plant Physiol.* **1997**, *115*, 1127–1134. [CrossRef] [PubMed]

60. Hu, Y.; Li, W.C.; Xu, Y.Q.; Li, G.J.; Liao, Y.; Fu, F.L. Differential expression of candidate genes for lignin biosynthesis under drought stress in maize leaves. *J. Appl. Genetics* **2009**, *50*, 213–223. [CrossRef] [PubMed]

61. Srinivasan, R.; Oliver, D.J. H-protein of the glycine decarboxylase multienzyme complex: Complementary DNA encoding the protein from *Arabidopsis thaliana*. *Plant Physiol.* **1992**, *98*, 1518–1519. [CrossRef] [PubMed]

62. Ledger, S.; Strayer, C.; Ashton, F.; Kay, S.A.; Putterill, J. Analysis of the function of two circadian-regulated constans-like genes. *Plant J.* **2001**, *26*, 15–22. [CrossRef] [PubMed]

63. Eamens, A.L.; Wook Kim, K.; Waterhouse, P.M. Drb2, drb3 and drb5 function in a non-canonical microRNA pathway in *Arabidopsis thaliana*. *Plant Signal. Behav.* **2012**, *7*, 1224–1229. [CrossRef] [PubMed]

64. Ades, S.E.; Grigorova, I.L.; Gross, C.A. Regulation of the alternative sigma factor σ(e) during initiation, adaptation, and shutoff of the extracytoplasmic heat shock response in *Escherichia coli*. *J. Bacteriol.* **2003**, *185*, 2512–2519. [CrossRef] [PubMed]

65. Shimizu, M.; Kato, H.; Ogawa, T.; Kurachi, A.; Nakagawa, Y.; Kobayashi, H. Sigma factor phosphorylation in the photosynthetic control of photosystem stoichiometry. *Proc. Natl. Acad. Sci. USA* **2010**, *107*, 10760–10764. [CrossRef] [PubMed]

66. Periappuram, C.; Steinhauer, L.; Barton, D.L.; Taylor, D.C.; Chatson, B.; Zou, J. The plastidic phosphoglucomutase from arabidopsis. A reversible enzyme reaction with an important role in metabolic control. *Plant Physiol.* **2000**, *122*, 1193–1200. [CrossRef] [PubMed]

67. Singh, D.; Laxmi, A. Transcriptional regulation of drought response: A tortuous network of transcriptional factors. *Front. Plant Sci.* **2015**, *6*, 895. [CrossRef] [PubMed]

68. Wedel, N.; Soll, J.; Paap, B.K. Cp12 provides a new mode of light regulation of Calvin cycle activity in higher plants. *Proc. Natl. Acad. Sci. USA* **1997**, *94*, 10479–10484. [CrossRef] [PubMed]

69. Lorick, K.L.; Jensen, J.P.; Fang, S.; Ong, A.M.; Hatakeyama, S.; Weissman, A.M. Ring fingers mediate ubiquitin-conjugating enzyme (E2)-dependent ubiquitination. *Proc. Natl. Acad. Sci. USA* **1999**, *96*, 11364–11369. [CrossRef] [PubMed]

70. Lindemose, S.; Shea, C.; Jensen, K.M.; Skriver, K. Structure, function and networks of transcription factors involved in abiotic stress responses. *Int. J. Mol. Sci.* **2013**, *14*, 5842–5878. [CrossRef] [PubMed]

71. Peng, L.; Yamamoto, H.; Shikanai, T. Structure and biogenesis of the chloroplast NAD(P)H dehydrogenase complex. *BBA Bioenerg.* **2011**, *1807*, 945–953. [CrossRef] [PubMed]

72. Liu, L.; Zhang, J.; Adrian, J.; Gissot, L.; Coupland, G.; Yu, D.; Turck, F. Elevated levels of MYB30 in the phloem accelerate flowering in arabidopsis through the regulation of flowering locus t. *PLoS ONE* **2014**, *9*, e89799. [CrossRef] [PubMed]

73. Marshall, A.; Aalen, R.B.; Audenaert, D.; Beeckman, T.; Broadley, M.R.; Butenko, M.A.; Cano-Delgado, A.I.; de Vries, S.; Dresselhaus, T.; Felix, G.; et al. Tackling drought stress: Receptor-like kinases present new approaches. *Plant Cell* **2012**, *24*, 2262–2278. [CrossRef] [PubMed]

74. Tripathi, P.; Rabara, R.C.; Reese, R.N.; Miller, M.A.; Rohila, J.S.; Subramanian, S.; Shen, Q.J.; Morandi, D.; Bucking, H.; Shulaev, V.; et al. A toolbox of genes, proteins, metabolites and promoters for improving drought tolerance in soybean includes the metabolite coumestrol and stomatal development genes. *BMC Genomics* **2016**, *17*, 102. [CrossRef] [PubMed]

75. Xia, H.J.; Guang, Y. Inositol 1,4,5-trisphosphate 3-kinases: Functions and regulations. *Cell Res* **2005**, *15*, 83–91. [CrossRef] [PubMed]

76. Le, D.T.; Nishiyama, R.; Watanabe, Y.; Tanaka, M.; Seki, M.; Ham le, H.; Yamaguchi-Shinozaki, K.; Shinozaki, K.; Tran, L.S. Differential gene expression in soybean leaf tissues at late developmental stages under drought stress revealed by genome-wide transcriptome analysis. *PLoS ONE* **2012**, *7*, e49522. [CrossRef] [PubMed]

77. Grundy, J.; Stoker, C.; Carré, I.A. Circadian regulation of abiotic stress tolerance in plants. *Front. Plant Sci.* **2015**, *6*, 648. [CrossRef] [PubMed]

78. Liu, J.; Shen, J.; Xu, Y.; Li, X.; Xiao, J.; Xiong, L. Ghd2, a constans-like gene, confers drought sensitivity through regulation of senescence in rice. *J. Exp. Bot.* **2016**, *67*, 5785–5798. [CrossRef] [PubMed]

79. Oh, J.E.; Kwon, Y.; Kim, J.H.; Noh, H.; Hong, S.-W.; Lee, H. A dual role for myb60 in stomatal regulation and root growth of *Arabidopsis thaliana* under drought stress. *Plant Mol. Biol.* **2011**, *77*, 91–103. [CrossRef] [PubMed]

11

The Ageing Brain: Effects on DNA Repair and DNA Methylation in Mice

Sabine A. S. Langie [1,*,†], Kerry M. Cameron [2], Gabriella Ficz [3], David Oxley [4],
Bartłomiej Tomaszewski [1], Joanna P. Gorniak [1], Lou M. Maas [5], Roger W. L. Godschalk [5],
Frederik J. van Schooten [5], Wolf Reik [6,7], Thomas von Zglinicki [2] and John C. Mathers [1]

[1] Centre for Ageing and Vitality, Human Nutrition Research Centre, Institute of Cellular Medicine, Newcastle University, Campus for Ageing and Vitality, Newcastle upon Tyne NE4 5PL, UK; tomaszewski.bartlomiej@gmail.com (B.T.); joanna.p.gorniak@googlemail.com (J.P.G.); john.mathers@ncl.ac.uk (J.C.M.)

[2] The Ageing Biology Centre and Institute for Cell and Molecular Biology, Newcastle University, Campus for Ageing and Vitality, Newcastle upon Tyne NE4 5PL, UK; kerry.m.cameron@effem.com (K.M.C.); t.vonzglinicki@newcastle.ac.uk (T.v.Z.)

[3] Barts Cancer Institute, Queen Mary University, London EC1M 6BQ, UK; g.ficz@qmul.ac.uk

[4] Mass Spectrometry Laboratory, Babraham Institute, Cambridge CB22 3AT, UK; david.oxley@babraham.ac.uk

[5] Department of Pharmacology & Toxicology, School for Nutrition and Translational Research in Metabolism (NUTRIM), Maastricht University, 6200 MD Maastricht, The Netherlands; l.maas@maastrichtuniversity.nl (L.M.M.); r.godschalk@maastrichtuniversity.nl (R.W.L.G.); f.vanschooten@maastrichtuniversity.nl (F.J.v.S.)

[6] Epigenetics Programme, Babraham Institute, Cambridge CB22 3AT, UK; wolf.reik@babraham.ac.uk

[7] Wellcome Trust Sanger Institute, Hinxton CB10 1SA, UK

* Correspondence: sabine.langie@vito.be

† Current address: Environmental Risk and Health unit, Flemish Institute of Technological Research (VITO), Boeretang 200, 2400 Mol, Belgium.

Academic Editor: Dennis R. Grayson

Abstract: Base excision repair (BER) may become less effective with ageing resulting in accumulation of DNA lesions, genome instability and altered gene expression that contribute to age-related degenerative diseases. The brain is particularly vulnerable to the accumulation of DNA lesions; hence, proper functioning of DNA repair mechanisms is important for neuronal survival. Although the mechanism of age-related decline in DNA repair capacity is unknown, growing evidence suggests that epigenetic events (e.g., DNA methylation) contribute to the ageing process and may be functionally important through the regulation of the expression of DNA repair genes. We hypothesize that epigenetic mechanisms are involved in mediating the age-related decline in BER in the brain. Brains from male mice were isolated at 3–32 months of age. Pyrosequencing analyses revealed significantly increased *Ogg1* methylation with ageing, which correlated inversely with *Ogg1* expression. The reduced *Ogg1* expression correlated with enhanced expression of methyl-CpG binding protein 2 and ten-eleven translocation enzyme 2. A significant inverse correlation between *Neil1* methylation at CpG-site2 and expression was also observed. BER activity was significantly reduced and associated with increased 8-oxo-7,8-dihydro-2′-deoxyguanosine levels. These data indicate that *Ogg1* and *Neil1* expression can be epigenetically regulated, which may mediate the effects of ageing on DNA repair in the brain.

Keywords: DNA methylation; epigenetics; base excision repair; ageing; brain; gene regulation

1. Introduction

Ageing is associated with the accumulation of oxidative DNA damage resulting from increased exposure to reactive oxygen species (ROS) from exogenous and endogenous sources [1]. DNA is subject to constant attack by DNA-damaging agents, and the accumulation of unrepaired DNA damage has profound effects on cell function. This may cause the characteristic features of ageing, including changes in gene expression, genome instability, changes in cell replication, cell senescence and cell death [2]. The DNA base guanine is particularly sensitive to oxidation by ROS to form 8-oxo-7,8-dihydro-2′-deoxyguanosine (8-oxodG) due to its low redox potential (reviewed by [3]). The DNA glycosylase oxoguanosine 1 (OGG1) is the major base excision repair (BER) enzyme, which recognizes and removes 8-oxodG from DNA. Other DNA glycosylases of interest are: (i) Nei endonuclease VIII-like 1 (NEIL1), which recognizes and incises 8-oxodG lesions located near the 3′-end of single strand breaks, DNA bubble structures and single-stranded structures where OGG1 has limited activity [4]; and (ii) muty homolog (MUTYH), which has a specificity for adenine opposite 8-oxoguanine [5]. The DNA ligase and X-ray repair cross-complementing protein 1 (XRCC1) complex complete the repair process by sealing the nick. Interestingly, OGG1 interacts with XRCC1, enhancing its incision activity [6]. Whilst most oxidative damage is removed by BER and other DNA repair mechanisms, it is thought that these mechanisms become less effective with ageing, resulting in the accumulation of DNA lesions, loss of genome stability and altered gene expression that contribute to age-related degenerative diseases. The role of DNA glycosylases dysfunction in ageing may be important, as shown by age-related accumulation of oxidative damage in liver from *Ogg1* KO mice [7]. Furthermore, the activity of human OGG1 declines with age in lymphocytes [8] and is lower in Alzheimer's disease brains [9,10]. *Neil1* is highly expressed in the brain [11]. Binding between the DNA damage sensor protein poly(ADP-ribose) polymerase 1 (PARP-1) and NEIL1 was diminished in older mice compared with younger mice, supporting the idea of impaired DNA repair during aging [12].

The mechanisms responsible for age-related decline in DNA repair capacity are uncertain, but growing evidence suggests that epigenetic events, including aberrant DNA methylation, contribute to the ageing process and may be functionally important through dysregulation of gene expression of, e.g., DNA repair genes [13–17]. Epigenetics defines processes and genomic markers, including DNA methylation, covalent histone modifications and non-coding RNAs, that result in changes in gene expression and phenotype without a corresponding alteration in DNA sequence, thus providing a process for genome regulation. DNA methylation is the most widely-studied epigenetic mechanism and is achieved by the addition of a methyl group to a cytosine (5mC) in CpG dinucleotides by DNA methyltransferases [18]. CpGs are often densely packed in or close to promoter regions, forming so-called "CpG islands", which are normally unmethylated in expressed genes [19]. Whilst some epigenetic markers are established during embryonic and foetal development and remain relatively stable during adulthood, the methylation status of some genomic loci is labile and changes over time and in response to environmental exposures [20,21]. Aberrant DNA methylation of CpG sites can inhibit the opportunity for transcription factors (TF) to bind, which can lead to gene silencing [18]. In addition, ROS can cause oxidation of 5mC to 5-hydroxymethylcytosine (5hmC) [22], and although the specific biological role of 5hmC is unclear, it may counteract transcriptional repression, making 5hmC important for gene regulation (reviewed by [23]). Alternatively, ten-eleven translocation (TET) enzymes can convert 5mC to 5hmC and play a role in active DNA demethylation [22–24].

Evidence has been demonstrated for age-related changes in DNA methylation in studies of young and older monozygotic twins [25]. Despite the identical genotypes, in older twins, there was greater inter-twin variability in the epigenomes compared with younger twins, and this was accompanied by greater inter-twin diversity in gene expression portraits. Interestingly, global DNA demethylation has been accompanied by hypermethylation of specific gene promoters. Few studies have reported investigations of promoter-specific methylation of genes involved in DNA repair [13–15,17,26–28]; these have been mainly in relation to cancer, and none have studied epigenetic regulation of BER-related genes in the ageing brain.

In the present study, the brain was selected as the target tissue since it is particularly vulnerable to the deleterious effects of ROS due to its high oxygen utilization and relatively low antioxidants levels [29,30]. DNA damage may be especially harmful in post-mitotic neuronal brain cells, which have limited capacity to regenerate. Thus, oxidative DNA damage may play a key role in age-associated loss of brain neurons; hence, cumulative unrepaired DNA damage may be responsible for the underlying cellular dysfunction [31]. For these reasons, the proper functioning of DNA repair mechanisms is important for neuronal survival.

We hypothesize that epigenetic mechanisms are involved in mediating the age-related decline in DNA repair in the brain (Figure 1). Thus, through altered gene expression, changes in the methylation status of promoters of genes encoding components of DNA repair systems may impact on neuronal DNA repair. This may lead to the accumulation of oxidative DNA damage and mutations across the whole genome, causing genome instability and increasing the risk of age-related degenerative neurological diseases. To test this hypothesis, we studied: (i) the formation of oxidative DNA damage and global DNA methylation levels; (ii) BER gene expression and the methylation status of CpGs in TF binding sites that influence the transcription of BER genes; and (iii) correlations with the resulting phenotypic BER-related incision activity in male mice across most of the adult lifespan (3–32 months old).

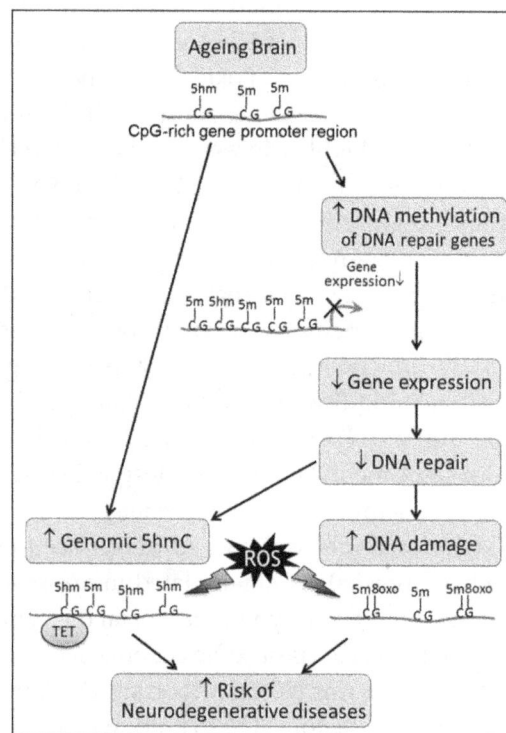

Figure 1. Schematic overview of the study hypothesis. Abbreviations used: 5mC, 5-methylcytosine; 5hmC, 5-hydroxymethycytosine; 8-oxodG, 8-oxo-7,8-dihydro-2′-deoxyguanosine; ROS: reactive oxygen species; TET: ten-eleven translocation enzymes.

2. Materials and Methods

2.1. Animals and Design of the Study

Mice were obtained from a long-established colony of the C57/BL (ICRFa) strain, which had been selected for use in studies of intrinsic ageing because it is free from specific age-associated pathologies and, thus, provides a good general model of ageing [32]. Mice were housed in standard cages of different sizes depending on housing density (between 1 and 6 mice per cage). Mice were housed at $20 \pm 2\,^{\circ}C$ under a 12-h light/12-h dark photoperiod with lights on at 7 a.m. Mice were provided with sawdust and paper bedding and had ad libitum access to water and food (CRM (P) Special Diet Services; BP Nutrition Ltd., Essex, U.K.). All work complied with the U.K. Home Office Animals (Scientific procedures) Act of 1986 (project licence PPL60/3864).

To study the effects of ageing, whole brains were collected from ad libitum fed male mice at ages 3, 6, 12, 24, 28, 31 and 32 months ($n = 4$ per age group, except 31 and 32 months, $n = 3$ per group), immediately snap frozen in liquid nitrogen and stored at $-80\,^{\circ}C$. When required, frozen brain tissues were ground and aliquoted and stored at $-80\,^{\circ}C$ until further analysis.

2.2. Determination of 8-oxodG

Frozen ground brain tissues (~30–80 mg, $n = 3$–4 per group) were thawed, and genomic DNA was isolated using standard phenol extraction [33]. The DNA extraction procedure was optimized to minimize artificial induction of 8-oxodG, by using radical-free phenol, minimizing exposure to oxygen and by the addition of 1 mM deferoxamine mesylate and 20 mM TEMPO (2,2,6,6-tetramethylpiperidine-N-oxyl; Aldrich, Steinheim, Germany), according to the recommendations made by the European Standards Committee on Oxidative DNA Damage (ESCODD [34]). To detect the base oxidation product 8-oxodG, HPLC with electrochemical detection (ECD) was performed as described earlier [35].

2.3. Assessment of Genomic 5mC and 5hmC

Nucleosides were derived from DNA samples ($n = 3$–4 per group) by digestion with DNA Degradase Plus (Zymo Research, Cambridge Bioscience, Cambridge, U.K.) according to the manufacturer's instructions and were analysed by LC-MS/MS on an LTQ Orbitrap Velos mass spectrometer (Thermo Scientific, Cramlington, UK) fitted with a nanoelectrospray ion-source (Proxeon/Thermo Scientific; Amsterdam, The Netherlands). Mass spectral data for 5hmC, 5mC and C were acquired in selected reaction monitoring (SRM) mode, monitoring the transitions $258 \rightarrow 142.0611$ (5hmC), $242 \rightarrow 126.0662$ (5mC) and $228 \rightarrow 112.0505$ (C). Parent ions were selected for SRM with a 4 mass unit isolation window and fragmented by Higher-energy Collisional Dissociation (HCD) with a relative collision energy of 20%, with $R > 14,000$ for the fragment ions. Peak areas from extracted ion chromatograms of the relevant fragment ions were quantified by external calibration relative to authentic standards.

2.4. Gene-Specific Methylation Studies, Using Pyrosequencing of Bisulphite Converted DNA

Selection of transcription factor (TF) binding sites and Primer design: Genomatix's "Gene2Promoter" tool (Genomatix; Munich, Germany) was used to retrieve the target genes' (i.e., *Ogg1*, *Neil1*, *Mutyh* and *Xrcc1*) promoter sequence. Using the free, downloadable CpG Island Explorer 2.0 software ([36]; http://www.soft82.com/download/windows/cpg-island-explorer/), CpG-rich regions were identified in the gene promoters. Next, the CpG-island was screened for TF binding sites by means of the Genomatix "MatInspector" tool. The following selection criteria were applied: core similarity >0.75 and matrix similarity >0.70 (with 1 being a perfect match), and optimized matrix threshold >0.7 (to minimize the number of false positive matches). As advised by Genomatix, both (+)- and (−)-strand matches have been considered equally, since most TF binding sites can occur in both orientations in promoters or enhancers. Moreover, methylation patterns on the (+)- and (−)-strand are believed to be identical since hemi-methylated DNA is restored to the fully-methylated state during DNA replication [37]. All TF binding sites were subsequently filtered based on the association of their TF family with specific tissues (based on MatInspector output); selecting those who are ubiquitously expressed or specifically expressed in the brain/central nervous system/neurons. As a second screening, only those TF that have at least one CpG di-nucleotide in their binding sequence and preferably in their core sequence (i.e., the highest conserved, consecutive positions of the TF) were selected. Based on these screening steps, CpG-sites located in TF binding sites with potential to influence promoter function were selected (Supplementary Materials Figures S1–S4). Using the PSQ Software program (Qiagen, Manchester, U.K.), primers were designed for several amplicons to include all selected CpG sites (Table 1 and Supplementary Materials Figures S1–S4).

Bisulphite conversion: DNA was extracted and purified (including RNase treatment) from ~20 mg of ground tissue using standard chloroform:isoamyl alcohol extraction. Bisulphite conversion of DNA was performed using the EZ DNA Methylation Gold™ kit (Zymo Research, Cambridge Bioscience, Cambridge, U.K.) according to the manufacturer's protocol.

Pyrosequencing: Bisulphite pyrosequencing was used to quantify methylation at individual CpG sites within the specific TF binding sites. About 50 ng of bisulphite-treated DNA were added as a template in PCR reaction containing 12.5 μL Hot Start Taq Master Mix (Qiagen, Manchester, U.K.), 400 nM forward primer and 400 nM biotin-labelled reverse primer in a total volume of 25 μL. The primer sequences and PCR conditions are summarized in Table 1. Amplification was carried out in a Bio-Rad thermocycler (Bio-Rad, Hertfordshire, U.K.) using the following protocol; 95 °C 15 min, then 50 cycles of 95 °C 15 s, annealing temperature for 30 s (Table 1), 72 °C for 30 s, followed by 72 °C for 5 min. Next, the biotin-labelled PCR products were captured with Streptavidin Sepharose beads (GE Healthcare, Amersham U.K.) and made single stranded using a Pyrosequencing Vacuum Prep Tool (Qiagen). Sequencing primer (Table 1) was annealed to the single-stranded PCR product by heating to 80 °C, followed by slow cooling. Pyrosequencing was then carried out on a Pyromark MD system (Qiagen). Each sample was run in duplicate, and cytosine methylation was quantified by the Pyro Q CpG 1.0.6 software (Qiagen, Manchester, U.K.). If poor quality data were obtained for both duplicates or the assay failed (flagged in red by the software), that sample was omitted from further data analysis, which was the case for 3 samples when running the *Ogg1* pyrosequencing analyses and 4 samples in the case of *Neil1*.

2.5. Gene Expression Analyses

Total RNA was extracted from brain samples (~20 mg) using TRIzol reagent (Ambion, Life Technologies, Paisley, U.K.) according to the manufacturer's protocol. Next, 500 ng of total DNase-treated RNA were used for reverse transcription with the RevertAid™ H Minus First Strand cDNA Synthesis Kit (Fermentas, Thermo Scientific, Cramlington, U.K.) at 45 °C for 1 h.

Real-time quantitative reverse transcription (RT)-PCR of cDNAs derived from specific transcripts was performed in a Light Cycler 480 (Roche Diagnostics, Mannheim, Germany) using the respective pairs of oligonucleotide primers (Table 2). cDNA (~25 ng) was mixed with 12.5 μL Maxima™ SYBR Green qPCR Master Mix (Fermentas), 0.75 pmol of the forward and 0.75 pmol of the reverse primer of our genes of interest, and RNase-free water was added to achieve an end volume of 25 μL. Amplification was carried out using the following protocol; 95 °C for 10 min, then 45 cycles of 95 °C for 15 s and annealing temperature 60 °C for 1 min, with signal acquisition at extension steps. To confirm amplification specificity, the PCR products from each primer pair were subjected to melt curve analysis and agarose gel electrophoresis. Each sample was analysed in triplicate, and LightCycler 480 software release 1.5.0 (Roche Diagnostics, Mannheim, Germany) was used for data analysis. Expression of *Ogg1*, *Neil1*, *Mutyh* and *Xrcc1* was normalized relative to that of control transcripts *HPRT* and *β-microglobulin*, while the expression of *Tet1–3* and *Mecp2* was normalized to *Atp5b* and *Gapdh*. Levels of expressions, also called Relative quantification (RQ) values, were obtained by the $2^{-\Delta\Delta Ct}$ method. These RQ values were subsequently log2 transformed to give the symmetric fold change.

Table 1. Overview of primers and sequences to run pyrosequencing to study gene-specific methylation.

Gene	Amplicon	PCR Primers	Sequencing Primers	Product (bp)	Annealing Temperature (°C)	Sequence to Run on Pyrosequencer	Length (bp)
Oggf1	1	Fw: 5′-GGTTTATTTTTTTGAGATAGA-3′ / Rev: 5′-BIO-ACTAAAACCACATCATTA-3′	5′-TTTAGTTAAGTTTTTAAA-3′	134	43	C/TGTGTTTTC/TGTTTTTGTTTATC/TGAGTTTTGGGAC/TGATC/TGGTGTGTATTATTAC/TGTTTC/TG	60
	2	Fw: 5′-GTAGGTTTTGAGATTGTAT-3′ / Rev: 5′-BIO-ATTTAACCCTAAAAATAAC-3′	5′-GAAAGTTTTGAAAATGGTAGA-3′	184	43	GTG/TGGGTTTTGGTAGTAATG/TGTTAAGTAGC/TGAGGTTAGTAGGTTAATC/TGTTTTTATTTTATAGGTTC/TGTTATTTC/TG	79
Neil1	1	Fw: 5′-TGAGGTAGTAGTTAGTAAGG-3′ / Rev: 5′-BIO-ACTCTACTCACAATTCTTT-3′	5′-GTAGTTAGTAAGGGGTTAAT-3′ / 5′-GAATGGAGTTTTTATTTATGA-3′	220	52	TTTAGTAGTTGTC/TGAATTTAGAGTAC/TGTTGGG GAATTC/TGGGTGTGGGTAACTTTGGACTAGTC/TGC/TGTAATTC/TGGAGGTGAC/TGAA	34 / 55
	2	Fw: 5′-AGAATTGTGAGTAGAGTTTTGT-3′ / Rev: 5′-BIO-ATCTTAAATCCCCAAAAATTA-3′	5′-GTTTTAGTTATTTTAGATTATA-3′	186	52	C/TGTTAGTAGTC/TGGAAAC/TGGC/TGTTGTGTAGGAGTTATAAG TAGTTGTATGC/TGAGG	53
Mutyh	1	Fw: 5′-GGATGGTTATAGAAGTTTAAG-3′ / Rev: 5′-BIO-TCACTACTCCACTCTACAA-3′	5′-GTTTTAGTTATTTTAGATTATA-3′	164	46.6	ATTTTTAGTGTGTAGC/TGC/TGTGTAATTGTAAAAATTC/TG	36
Xrcc1	1	Fw: 5′-AGGTTTTAGGAAATTTTTAGTT-3′ / Rev: 5′-BIO-CCCTTAACAACAAACATTC-3′	5′-TTTAATGATTAGGGTAAA-3′	228	50	TTATAC/TGTAGGATTAATTATTGAGGTC/TGTTTTTTGTTGT TAGGTTTT AGGAGTC/TGAGTTTTTAG/TG	67
	2	Fw: 5′-TGTTTGTTGTTAAGGGAATT-3′ / Rev: 5′-BIO-CTCAAAAAACCCCTATCT-3′	5′-GGAGAGGTTTAATYGAGTAT-3′ / 5′-GGGGTTTTTTYGGAGTTGTAA-3′	328	50	GC/TGTAGTGTTGAC/TGTGTGC/TGTGTC/TGGC/TGTGC/TGTGTC/TGC/TGGTT TGAAAGGTTC/TGAGTTTTGC/TGC/TGTTTGC/TGT / TTTTTTTTTTTTATTTTTTGGAC/TGGTC/TGGGC/TGTTTAC/TGGGC/TG TGGATATGTC/TGGAGATTAGTTTTC/TGTTAC/TGTC/TGT	65 / 76

BIO = indicates the biotin label on the reverse primers; Y = indicates the presence of an internal C/T wobble in the primer with a non-defined ratio.

Table 2. Overview qPCR primers to study gene expression.

Primer set 1		Primer set 2	
Gene	q-PCR primers	Gene	q-PCR primers
Ogg1	Fw: 5′-TGGCTTCCCAAACCTCCAT-3′	Mecp2	Fw: 5′-GAGGAGGCGAGGAGGAGAGA-3′
	Rev: 5′-GGCCCAACTTCCTCAGGTG-3′		Rev: 5′-AACTTCAGTGGCTTGTCTCTGAGG-3′
Neil1	Fw: 5′-GACCCTGAGCCAGAAGATCAG-3′	Tet1	Fw: 5′-CCATTCTCACAAGGACATTCACA-3′
	Rev: 5′-AGCTGTGTCTCCTGTGACTT-3′		Rev: 5′-GCAGGACGTGGAGTTGTTCA-3′
Mutyh	Fw: 5′-CTGTCTCCCCATATCATCTCTT-3′	Tet2	Fw: 5′-GCCATTCTCAGGAGTCACTGC-3′
	Rev: 5′-TCACGCTTCTCTTGGTCATAC-3′		Rev: 5′-ACTTCTCGATTGTCTTCTCTATTGAGG-3′
Xrrc1	Fw: 5′-CTTCTCAAGGCGGACACTTA-3′	Tet3	Fw: 5′-GGTCACAGCCTGCATGGACT-3′
	Rev: 5′-ATCTGCTCCTCCTTCTCCAA-3′		Rev: 5′-AGCGATTGTCTTCCTTGGTCAG-3′:
B2m	Fw: 5′-ATGCTGAAGAACGGGAAAAAAA-3′	Atp5b	Fw: 5′-GGCCAAGATGTCCTGCTGTT-3′
	Rev: 5′-CAGTGTGAGCCAGGATATAGAA-3′		Rev: 5′-AACTTTGGCATTGTGGAAGG-3′
Hprt	Fw: 5′-AGGAGAGAAAGATGTGATTGATATT-3′	Gapdh	Fw: 5′-AACTTTGGCATTGTGGAAGG-3′
	Rev: 5′-TCCACTGAGCAAAACCTCTT-3′		Rev: 5′-ATGCAGGGATGATGTTCTGG-3′

2.6. Measurement of BER-Related DNA Incision Activity

BER activity in brain tissues was assessed using a modified comet-based repair assay that was recently optimized for the use of tissue extracts [38]. This assay measures the ability of BER-related enzymes that are present in tissue extracts to recognize and incise substrate DNA containing 8-oxodG lesions that were induced by the photosensitizer Ro 19-8022 plus light.

The protocol has been described in full detail before [38]. Briefly, to prepare tissue extracts, ~30-mg aliquots of ground tissue were incubated with 75 µL Buffer A (45 mM HEPES, 0.4 M KCl, 1 mM EDTA, 0.1 mM dithiothreitol, 10% glycerol, adjusted to pH 7.8 (all purchased from Sigma, Dorset, U.K.)), vortexed vigorously, snap frozen in liquid nitrogen and immediately defrosted. Next, 30 µL of 1% Triton X-100 in Buffer A (Sigma, Dorset, U.K.) was added per 100-µL aliquot and incubated on ice for 10 min. After, centrifugation at $14,000 \times g$ for 5 min at 4 °C to remove cell debris, the supernatant was collected, and protein concentrations were determined by the Bio-Rad DC Protein Assay Kit using bovine serum albumin as a standard and controlling for the presence of Triton X-100. Final protein extracts were diluted with 0.23% Triton X-100 in Buffer A to a concentration of 1 mg/mL before further use in the repair incubation. Comets were visualized using an Olympus BX51 fluorescence microscope, and 50 comets/slide selected at random were analysed using the Comet assay IV software program (Perceptive Instruments, Haverhill, U.K.). %DNA in the tail (also known as tail intensity (TI)) was used for further calculations. After subtracting background levels from all data, the final repair capacity was calculated according to Langie et al. [38].

2.7. Statistical Analysis

Results are presented as the mean values ± standard error. Grubbs' test, also called the ESD method (extreme Studentized deviate), was used to determine significant outliers within the 8-oxodG dataset; data from 5 samples were omitted. Differences in levels of genomic 5mC and 5hmC, gene-specific DNA methylation, gene expression, BER-related incision activity and oxidative DNA damage were analysed by ANOVA, using Dunnett's t-test when comparing treatments. Relationships between variables were assessed by regression analyses, conducting multiple linear stepwise regression analysis when studying the effect of the various individual CpG-sites in TF binding sites on gene expression. Statistical analysis was performed using SPSS v.19.0 (IBM), and $p < 0.05$ was considered statistically significant.

3. Results

Our findings are structured according to the steps indicated in the schematic overview of our study hypothesis (Figure 1).

3.1. The Effect of Ageing on Genome Stability, DNA Damage and DNA Methylation

Ageing was associated with decreased global DNA methylation (5mC) (Figure 2A) and increased 5hmC (Figure 2B), though these trends were not statistically significant. However, there was a significant effect of age on the 5hmC/5mC ratio (Figure 2C; P_{ANOVA} = 0.027), which increased significantly in the older mice (R^2 = 0.382, p = 0.008). In parallel, levels of 8-oxodG increased with age (Figure 3A; R^2 = 0.854, p = 0.025). Notably, higher 5hmC/5mC ratios were significantly associated with higher levels of 8-oxodG (R^2 = 0.785, p = 0.046).

Methylation of BER-related gene promotors tended to increase with age, especially in the oldest mice (28 months) (Figure 3B). Methylation of the *Ogg1* promotor (averaged across 27 CpG-sites; Figure S1) increased significantly with age (P_{ANOVA} = 0.026; R^2 = 0.416, p = 0.005), with the highest effect observed in 28-month-old mice (p = 0.015 vs. three-month-old mice). The average methylation level in the *Xrcc1* promotor was also significantly affected by ageing (P_{ANOVA} = 0.023; R^2 = 0.233, p = 0.031): methylation levels in 28-month-old mice were increased compared with three-month-old mice (P_{ANOVA} = 0.041). Methylation of four specific CpG-sites in the *Xrcc1* promotor was affected differentially by age (Figure S2). Averaged across 12 CpG-sites, *Neil1* promotor methylation was not significantly affected by age, but methylation of three individual CpG-sites was modulated during ageing (Figure S3). Similarly, *Mutyh* average promotor methylation was not affected by age, but methylation at CpG-site 3 was significantly decreased (Figure S4; P_{ANOVA} = 0.014). See Supplementary Materials Figures S1–S4 for details of the individual CpG sites analysed and related TF binding sites for these genes.

3.2. Effect of Ageing on Gene Expression in Mouse Brain

For the ease of comparison, gene expression was expressed as the fold change (calculated as log2 of RQ values) compared with expression in three-month-old mice. In line with our hypothesis, there was an overall decrease in BER-related gene expression with age (Figure 3C). This trend was not statistically significant, except for *Ogg1*, where expression decreased significantly with age (P_{ANOVA} < 0.001; $P_{(24 \text{ vs. 3 months})}$ < 0.001; $P_{(28 \text{ vs. 3 months})}$ < 0.001; and R^2 = 0.692, p < 0.001). This expression change correlated inversely with the average *Ogg1* promoter methylation (Figure S5A; R^2 = 0.320, p = 0.018). Although *Neil1* expression was not significantly affected by age (P_{ANOVA} = 0.080), there was a significant inverse correlation between *Neil1* expression and methylation levels at CpG-site 2 (Figure S5B; R^2 = 0.545, p = 0.001). No correlations were observed between gene expression and methylation for *Xrcc1* and *Mutyh*.

3.3. Phenotypic Effects in the Ageing Brain

BER-related incision activity in the brain decreased significantly with age (Figure 3D; P_{ANOVA} = 0.021). Although not statistically significant, a trend of lower BER activity with decreasing levels of *Ogg1* expression was observed (Figure S6A; R^2 = 0.173, p = 0.068), which seemed to result in higher levels of 8-oxodG lesions (Figure S6B; R^2 = 0.149, p = 0.156). In addition, we observed weak associations between these lower levels of BER-related incision activity and 5hmC levels (Figure S6C; R^2 = 0.229, p = 0.052), as well as 5hmC/5mC ratios (Figure S6D; R^2 = 0.210, p = 0.064).

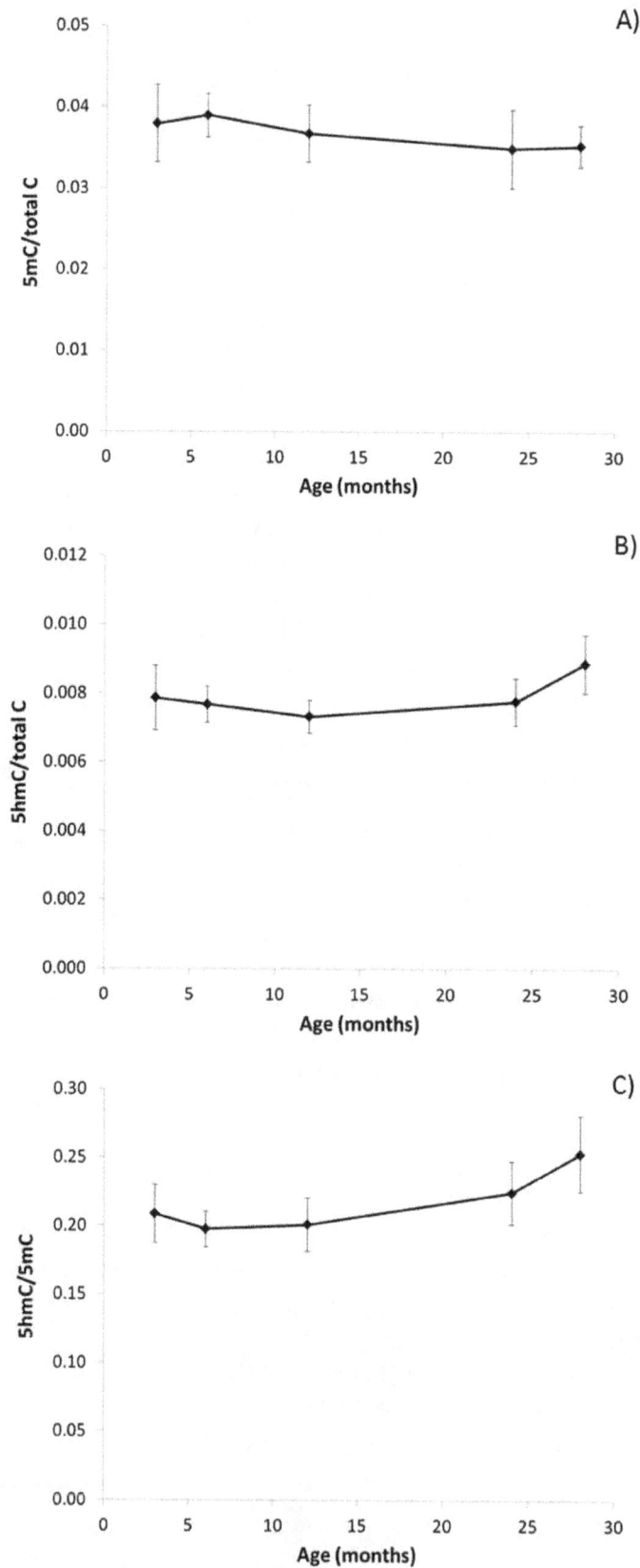

Figure 2. Effect of ageing on genomic 5mC and 5hmC levels. Levels of: (**A**) genomic 5mC over total cytosine; (**B**) levels of 5hmC over total cytosine; and (**C**) ratio of 5hmC/mC ($P_{ANOVA} = 0.027$; $R^2 = 0.382$, $p = 0.008$) in the brain of male mice. Percentages shown are calculated from the mean values ($n = 4$), and error bars represent standard errors of the mean (SEM).

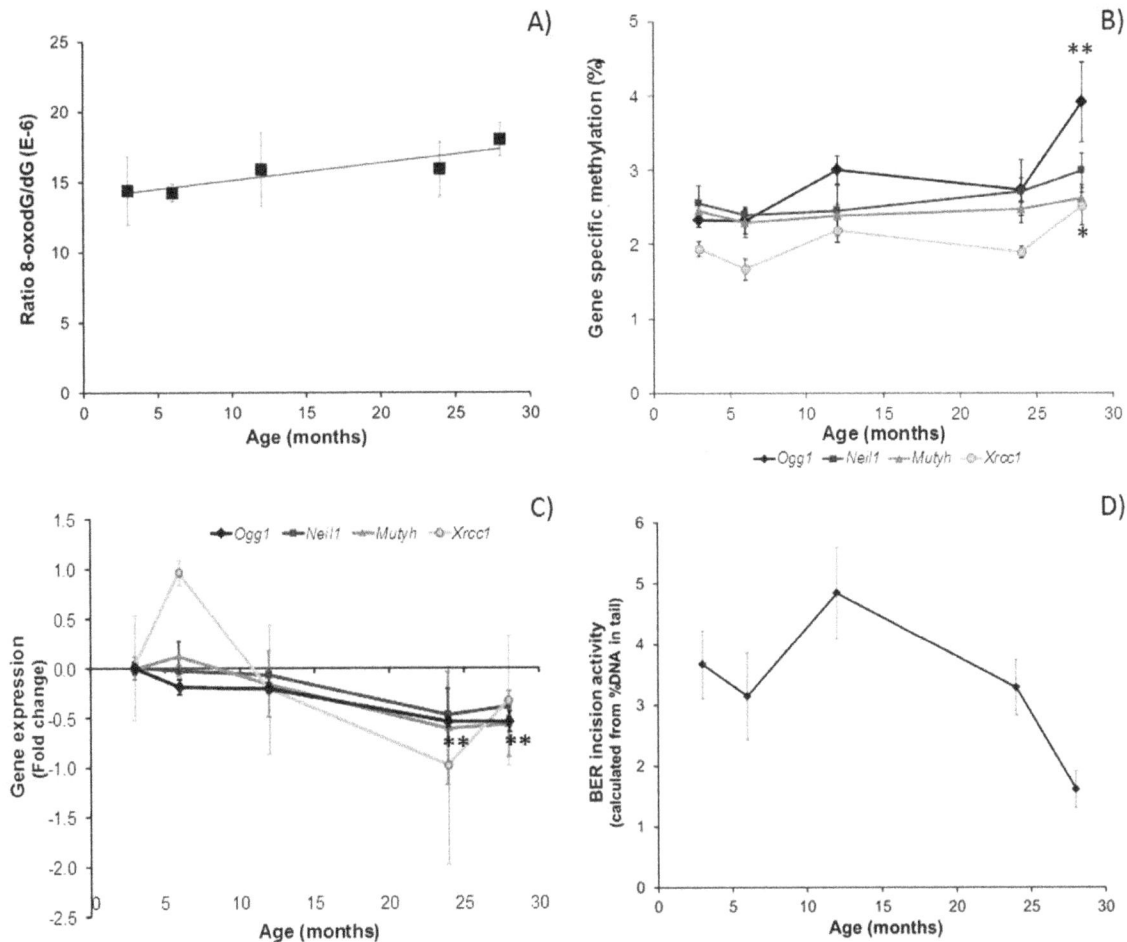

Figure 3. Effect of ageing on: (**A**) levels of 8-oxodG ($R^2 = 0.854$, $p = 0.025$); (**B**) BER-related gene promotor methylation ($Ogg1$: $P_{ANOVA} = 0.026$, ** $P_{(28 \text{ vs. } 3 \text{ months})} = 0.015$; $Xrcc1$: $P_{ANOVA} = 0.023$, * $P_{(28 \text{ vs. } 3 \text{ months})} = 0.041$); (**C**) BER-related gene expression ($Ogg1$: $P_{ANOVA} < 0.001$, ** $P_{(24 \text{ vs. } 3 \text{ months})} < 0.001$, ** $P_{(28 \text{ vs. } 3 \text{ months})} < 0.001$); and (**D**) BER-related incision activity ($P_{ANOVA} = 0.021$) in mouse brain. Data are presented as the mean values ($n = 3$–4), and bars indicate SEM.

3.4. Involvement of Tet Enzymes and Methyl-CpG Binding Proteins

To investigate the involvement of TET enzymes and methyl-CpG binding protein 2 (MECP2) in the effect of ageing on genomic DNA methylation, gene-specific methylation and gene expression, we measured the expression of $Tet1$–3 and $Mecp2$ genes (Figure 4). $Tet2$ ($R^2 = 0.332$, $p = 0.002$), and $Mecp2$ ($P_{ANOVA} = 0.003$, $P_{(28 \text{ vs. } 3 \text{ months})} = 0.050$, $P_{(31 \text{ vs. } 3 \text{ months})} = 0.039$; $R^2 = 0.241$, $p = 0.011$) expression increased significantly with age, while $Tet1$ ($R^2 = 0.154$, $p = 0.047$) and $Tet3$ ($P_{ANOVA} = 0.033$; $P_{(24 \text{ vs. } 3 \text{ months})} = 0.032$, $P_{(28 \text{ vs. } 3 \text{ months})} = 0.050$, $P_{(32 \text{ vs. } 3 \text{ months})} = 0.026$; $R^2 = 0.228$, $p = 0.014$) expression decreased with age.

There was an inverse correlation between 5mC levels and $Tet2$ expression (Figure S7B; $R^2 = 0.327$, $P_{ANOVA} = 0.016$) and lower levels of 5hmC with increasing $Tet1$ expression (Figure S7A; $R^2 = 0.444$, $P_{ANOVA} = 0.003$). Increased levels of $Ogg1$ promotor methylation correlated significantly with increased $Mecp2$ expression (Figure S7C; $R^2 = 0.245$, $p = 0.043$). Interestingly, $Mecp2$ expression also correlated inversely with $Ogg1$ expression (Figure S7D; $R^2 = 0.546$, $p < 0.001$).

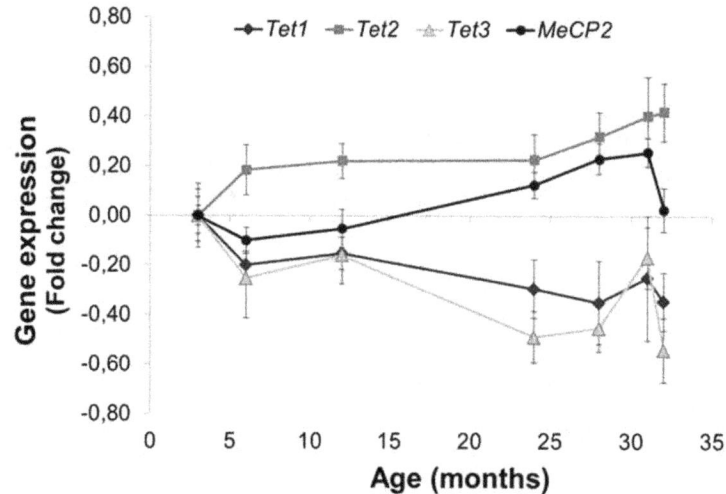

Figure 4. Expression of *Tet1–3* and *Mecp2* with ageing in mouse brain (* $p = 0.050$, ** $p < 0.040$, *** $p = 0.026$). Data are presented as the mean values ($n = 3$–4), and bars indicate standard errors of the mean.

4. Discussion

Several lines of evidence link ageing and genome maintenance pathways. Accelerated ageing is observed in mice defective in DNA repair pathways [39,40], and DNA repair deficiency in mature neural tissue has been linked with ageing and common neurodegenerative diseases. For example, certain forms of ataxia are caused by mutations in the BER-associated gene *APTX*, and mutations in NER-related genes (*XPA-XPG*) result in xeroderma pigmentosum, both syndromes that are accompanied by higher rates of neurodegeneration [30]. Such syndromes, which exhibit signs of premature ageing, have been very important in identifying molecular mechanisms that contribute to physiological ageing.

To date, there have been some descriptive studies of epigenetic changes in the ageing brain of Alzheimer patients, but little is known about the role of epigenetic processes in causing the cellular dysfunction, which is characteristic of neurological disorders [41,42]. The study of young versus older monozygotic twins has shown that there are greater inter-twin differences in DNA methylation in older twins, which were associated with greater inter-twin diversity in gene expression profiles [25]. Few studies have investigated promoter-specific methylation of DNA repair-related genes [13–15,17,26–28], but, for example, Agrelo et al. [28] found that the gene encoding the WRN protein, a RecQ helicase, which is involved in DNA repair, is repressed by promoter hypermethylation in human cancer.

4.1. Age-Related Increase in 5hmC, a Result of TET2 or Decreased BER?

In the current study, we did not observe statistically-significant changes in global 5mC levels, though there was a trend of a decline with age in mouse brain. This trend can be confirmed by previous reports of decreased genomic methylation with time observed in cultured cells and with increasing age in tissues from fish, rats, mice and humans (reviewed by [43]). Interestingly the ratio of 5hmC/5mC increased with ageing, which could be a result of increased oxidative stress during ageing (Figure 1). Indeed, levels of 5hmC have previously been reported to increase with age in the brain in animals [23,44] and humans [45], but the specific molecular role of 5hmC is still unclear. Since 5hmC is uniquely enriched in the brain, it is believed to function independently from 5mC and may play a role in mental health and disease [46].

The conversion of 5mC to 5hmC, as an intermediate in the demethylation process, is believed to occur via TET enzyme activity [46]. Interestingly, absolute *Tet1* and *Tet2* expression levels were

4–8-times higher than *Tet3* expression levels (data not shown), which has been reported before in adult cells [47]: *Tet1* and *Tet2* are believed to be important in maintaining pluripotency and adult neurogenesis, while *Tet3* is associated with cell differentiation and mainly involved in pre-natal development [48,49]. In addition, TET3 was reported to mediate increased gene expression that was associated with rapid behavioural adaptation, while TET1 was observed to alter 5hmC patterns during adaptation to longer term stressful environmental exposures (reviewed by Madrid et al., 2016 [46]). Furthermore, TET1, but not TET2 or TET3, may be involved in 5hmC production, which appears to be essential for Purkinje cell viability and the prevention of ataxia-telangiectasia-like symptoms in mice (reviewed by Madrid et al., 2016 [46]).

In the present study, lower 5mC levels were observed with increasing *Tet2* expression, while higher *Tet1* expression seemed to correlate with lower levels of 5hmC. In general, TET enzymes are the major enzymes catalysing conversion of 5mC to 5hmC, and thus, higher levels of 5hmC would be expected to correlate with increased *Tet1* expression, as was the case for *Tet2*. However, TET enzymes also convert 5hmC to 5-formylcytosine (5fC) and 5-carboxylcytosine (5caC) [22,50,51], which could explain the reduced levels of 5hmC observed in our study with increasing *Tet1* expression. In our current study, *Tet1* expression decreased with ageing and can therefore not explain the accumulation of 5hmC with ageing. However, *Tet2* has recently been shown to catalyse the stepwise oxidation of 5mC oxidation and is able to generate 5fC and 5caC from a single encounter with 5mC [51]. However, stalling at 5hmC was observed rather than progression to 5fC and 5caC. In addition, TET2 induction in vitro resulted in increased levels of 5hmC [52]. Although TET2 has been studied most extensively in various types of leukaemia [49], in the current study, *Tet2* expression increased with ageing and was inversely correlated with 5mC levels, which suggests that *Tet2* might be responsible for the age-related accumulation of 5hmC.

Alternatively, the conversion of 5hmC to cytosine may occur via deamination to 5-hydroxymethyluracil (5hmU), resulting in 5hmU-G mismatches that can be excised by BER DNA glycosylases, such as thymine DNA glycosylase (TDG) and NEIL1 [49,53]. Although, *Neil1* expression decreased with age while 5hmC levels increased, we did not observe direct significant associations between DNA glycosylases and 5hmC. In addition, the increased 5hmC levels and 5hmC/5mC ratios with age were significantly associated with lower BER activity, which implies that the higher levels of 5hmC in the ageing brain can be explained, at least partly, by the decreased BER-related incision activity observed in the oldest mice. Recently, in TDG-/- cells, TET2-induced 5-hmC accumulation was observed to result in GC > AT transitions [52], suggesting a mutagenic potential of 5-hmC metabolites if not removed/repaired and which may increase the risk of developing neurological disorders.

4.2. Epigenetic Regulation of Ogg1 Plays a Role in Age-Related Decline in DNA Repair

Interestingly, age-related global DNA demethylation has been reported to occur concomitantly with hypermethylation of specific CpG sites in the genome [44,54]. Of particular relevance for the development of age-related disease, site-specific hypermethylation of promoter regions, and transcriptional silencing, of tumour-suppressor genes can occur during aging [44]. Indeed, when using pyrosequencing to quantify methylation at CpG sites in TF binding sites that can influence promoter function, we observed significantly increased methylation of the BER-related *Ogg1* gene promoter with ageing. In addition, and possibly as a consequence, *Ogg1* expression decreased significantly in the oldest mice (Figure 3), and a significant inverse correlation between *Ogg1* expression and promotor methylation was observed (Figure S5). The lack of significant associations between increased promotor methylation and reduced expression for the other BER-related genes could be explained by the fact that bisulphite sequencing does not discriminate between 5mC and 5hmC (reviewed by [55]). Thus, increased methylation levels might be explained by increased 5hmC levels, which may inhibit the binding of methyl-CpG binding proteins and thereby counteract transcriptional repression of 5mC [23,47,56].

Multiple linear regression analysis revealed an inverse correlation between *Neil1* expression and methylation levels of CpG-site 2 located in the binding sites of STAF (or zinc finger protein 143 (ZNF143)) and ZIC2 (Figure S2; CpG-site 2). Enhanced expression of ZNF143, a human homolog of *Xenopus* transcriptional activator Staf, occurs in response to treatment with DNA-damaging agents [57] and induces the expression of DNA repair genes, including the BER-related gene *Fen-1*. ZIC2 activates the transcription of several genes and plays a crucial role in brain development [58].

Our observation of an inverse correlation between *Mecp2* and *Ogg1* expression is further evidence for the epigenetic regulation of this gene (Figure S6). MeCP2 is a methyl-CpG binding protein that can suppress transcription [59]. In the adult brain, MeCP2 binds to methylated DNA and plays a crucial role in normal brain functioning.

In line with our hypothesis, the hypermethylation and decreased gene expression of BER-related genes, especially *Ogg1*, with age was associated with reduced BER-related incision activity. We recently confirmed this association in an independent ageing mice group, where we observed a 43% decrease in *Ogg1* expression and 20% decrease in BER activity in association with increased *Ogg1* promotor methylation in the brain [60]. Several studies have reported lower DNA repair activity in older animals [11,61,62], although others have reported conflicting results [63–65]. The lack of direct statistically-significant associations between the expression of the BER-related genes and BER-related incision activity can be explained by the fact that studying gene expression does not necessarily give any indication of enzyme activity. Nonetheless, the age-related decline in BER activity that we observed was paralleled by a significant increase in 8-oxodG levels. This confirms earlier reports of the accumulation of oxidative lesions with ageing [66], which may contribute to the development of a senescent phenotype [67] and increase the risk of neurodegenerative diseases.

4.3. Concluding Remarks

We acknowledge the limitation of the size of this study, i.e., the relatively small numbers of mice at each time point (3–4 mice), which, in some cases, may have led to high variability and which may have reduced our ability to detect statistically-significant effects of ageing. Therefore, we are cautious to extrapolate the weak associations observed between BER and 5hmC or 8-oxodG levels. In particular, note that we analysed brain tissue from the oldest mice (31 and 32 months) for *Tet* and *Mecp2* expression only. Furthermore, the brain has a heterogenic mix of cell types with different ratios of neurons to glia cells in the various brain regions. There is evidence that different brain regions have distinct methylation profiles and may respond differently to ageing, which may influence the development of neurodegenerative diseases. Recent studies have indicated a potential role for 5hmC in various neurodegenerative diseases [46], so it will be important to investigate age-related changes in DNA methylation patterns in the various brain regions. Indeed, in an earlier study of brain from six-month-old mice, we reported region-specific patterns of 5mC and 5hmC in the cortex and cerebellum [68]. In addition, we observed differential responses in 5mC and the 5hmC/5mC ratio in these two brain regions following dietary intervention [68]. Although overall 5hmC levels have been reported to increase with age in the brain, depletion of 5hmC was found in the hippocampus, cerebellum and entorhinal cortex of patients suffering from Alzheimer's disease (AD) [46], while enrichment of 5hmC in the frontal and mid-temporal gyrus was positively correlated with the hallmarks of AD. A recent investigation of the genome-wide distribution of 5hmC in a mouse model of Huntington's disease (HD) found reduced levels of 5hmC in the mouse striatum and cortex tissues.

In terms of gene-specific methylation profiles, age-related DNA methylation changes are most often observed in CpG islands (as studied in the gene promotors of BER-related genes in the current study), while tissue-specific differences are observed more frequently outside those sites [54]. Future studies of the mechanisms underlying these marked effects of ageing on brain function could also include other genomic domains, e.g., CpG island shores, which are susceptible to altered methylation in response to environmental exposures and which may be important in regulating expression of the corresponding genes. A number of reports show that DNA methylation at intragenic regions (reviewed in Kulis et al., 2013 [69]), CpG island shores [70], partially-methylated domains [71] or long hypomethylated domains [72,73] can influence gene expression. In addition, it may be informative to investigate gene-specific 5hmC and 5mC levels, since *Tet1*-assisted bisulphite sequencing has become an established method for 5hmC detection [46]. Moreover, it will be important to broaden the enquiry to include other epigenetic marks and post-translational modifications, since these work together in a coordinated manner to regulate gene expression.

In addition to the investigation of different brain regions, it would be interesting to investigate the effects of ageing on specific sub-cellular fractions. Increased levels of 8-oxodG lesions during ageing have been reported for nuclear DNA (nDNA) and to a higher extent for mitochondrial DNA (mtDNA). Indeed, the degree of mtDNA oxidative damage in neuronal tissue appears to be inversely related to the maximum life span potential in mammals (reviewed by Gredilla et al., 2010 [74]). Since *Ogg1* is one of the most extensively-investigated mitochondrial DNA glycosylases and recent reports also show the presence of *Neil1* in brain mitochondria (reviewed by Gredilla et al., 2010 [74]), changes in BER-related gene methylation and expression may have a bigger impact on mitochondrial than on nuclear DNA damage and function and, subsequently, on brain ageing. Interestingly, altered mtDNA methylation profiles in human brain have recently been linked to AD and Parkinson's disease [75]. Although, quantification of mtDNA methylation is still challenging [76], immunoprecipitation methods (e.g., methylated DNA immunoprecipitation (MeDIP)) in combination with microarray hybridization, as well as mass spectrometry-based analysis have been proven very useful in the detection of mtDNA methylation (reviewed by Castegna et al., 2015 [77]).

Although, an increasing number of studies focus on elucidating the molecular events that lead from accumulation of DNA damage, to loss of cellular function and, ultimately, neurodegeneration, further studies are needed to understand the molecular mechanisms underlying age-related changes in DNA repair capacity. The current study helps to solve part of the puzzle and provides evidence that epigenetic mechanisms, i.e., increased *Ogg1* promoter methylation and the involvement of *Tet* enzymes and *Mecp2*, may affect gene expression in the ageing mammalian brain, which could impact the capacity for neuronal DNA repair. Overall, our data suggest that the accelerated accumulation of oxidative DNA damage may be mediated by epigenetic dysregulation of BER activity, causing genome instability and increasing the risk of age-related degenerative neurological diseases.

Supplementary Materials: The following are available online at www.mdpi.com/2073-4425/8/2/75/s1: Figure S1: Overview of the *Ogg1* sequence. Figure S2: Overview of the *Neil1* sequence. Figure S3: Overview of the *Mutyh* sequence. Figure S4: Overview of the *Xrcc1* sequence. Figure S5: Correlation plots between gene expression and DNA methylation profiles. Figure S6: Correlation plots with BER-related incision activity. Figure S7: Linear regression plots.

Acknowledgments: The Centre for Ageing & Vitality is funded by the MRC and BBSRC (Grant Reference MR/L016354/1). This work was further supported by the Centre for Integrated Systems Biology of Ageing and Nutrition funded by the BBSRC and EPSRC (G0700718). Part of the work was supported by BBSRC Grant BB/K010867/1. We thank Hoffmann-La Roche (Basel) for supplying Ro 19-8022 and Sofia Lisanti for providing us the primer sequences for the qRT-PCR reference genes. We also would like to thank Adele Kitching, Satomi Miwa, Liz Nicolson and Julie Wallace for the care of the animals and assistance with dissections.

Author Contributions: Conceived of and designed the experiments: S.A.S.L., K.M.C., T.v.Z., J.C.M. Performed the experiments: S.A.S.L., G.F., D.O., B.T., J.P.G., L.M. Analysed the data: S.A.S.L., G.F., B.T., J.P.G., R.W.L.G. Contributed reagents/materials/analysis tools: F.J.v.S., W.R., T.v.Z., J.M.M. Wrote the paper: S.A.S.L., G.F., R.W.L.G., J.C.M. Evaluated and interpreted the results: S.A.S.L., G.F., R.W.L.G., J.C.M. Evaluated the manuscript text: S.A.S.L., K.M.C., G.F., D.O., B.T., J.P.G., L.M.M., R.W.L.G., F.J.v.S., W.R., T.v.Z., J.C.M.

Abbreviations

The following abbreviations are used in this manuscript:

5caC	5-carboxylcytosine
5fC	5-formylcytosine
5hmC	5-hydroxymethycytosine
5mC	5-methylcytosine
8-oxodG	8-oxo-7,8-dihydro-2′-deoxyguanosine
BER	base excision repair
Mecp2	methyl-CpG binding protein 2
Mutyh	mutY DNA glycosylase
nDNA	nuclear DNA
NER	nucleotide excision repair
Neil1	Nei-like DNA glycosylase 1
mtDNA	mitochondrial DNA
Ogg1	DNA glycosylase oxoguanosine 1
ROS	reactive oxygen species;
TDG	thymine DNA glycosylase
TEMPO	2,2,6,6-tetramethylpiperidine-*N*-oxyl
TET	ten-eleven translocation enzymes
TF	transcription factors
Xrcc1	X-ray repair cross-complementing protein 1

References

1. Cooke, M.S.; Evans, M.D.; Dizdaroglu, M.; Lunec, J. Oxidative DNA damage: Mechanisms, mutation, and disease. *FASEB J.* **2003**, *17*, 1195–1214. [CrossRef]

2. Lopez-Otin, C.; Blasco, M.A.; Partridge, L.; Serrano, M.; Kroemer, G. The hallmarks of aging. *Cell* **2013**, *153*, 1194–1217. [CrossRef] [PubMed]

3. Langie, S.A.S.; Lara, J.; Mathers, J.C. Early determinants of the ageing trajectory. *Best Pract. Res. Clin. Endocrinol. Metab.* **2012**, in press. [CrossRef] [PubMed]

4. Parsons, J.L.; Zharkov, D.O.; Dianov, G.L. NEIL1 excises 3′ end proximal oxidative DNA lesions resistant to cleavage by NTH1 and OGG1. *Nucleic Acids Res.* **2005**, *33*, 4849–4856. [CrossRef] [PubMed]

5. Robertson, A.B.; Klungland, A.; Rognes, T.; Leiros, I. DNA repair in mammalian cells: Base excision repair: The long and short of it. *Cell. Mol. Life Sci.* **2009**, *66*, 981–993. [CrossRef]

6. Marsin, S.; Vidal, A.E.; Sossou, M.; Menissier-de Murcia, J.; Le Page, F.; Boiteux, S.; de Murcia, G.; Radicella, J.P. Role of XRCC1 in the coordination and stimulation of oxidative DNA damage repair initiated by the DNA glycosylase hOGG1. *J. Biol. Chem.* **2003**, *278*, 44068–44074. [CrossRef] [PubMed]

7. Osterod, M.; Hollenbach, S.; Hengstler, J.G.; Barnes, D.E.; Lindahl, T.; Epe, B. Age-related and tissue-specific accumulation of oxidative DNA base damage in 7,8-dihydro-8-oxoguanine-DNA glycosylase (*Ogg1*) deficient mice. *Carcinogenesis* **2001**, *22*, 1459–1463. [CrossRef]

8. Chen, S.K.; Hsieh, W.A.; Tsai, M.H.; Chen, C.C.; Hong, A.I.; Wei, Y.H.; Chang, W.P. Age-associated decrease of oxidative repair enzymes, human 8-oxoguanine DNA glycosylases (hOgg1), in human aging. *J. Radiat. Res.* **2003**, *44*, 31–35. [CrossRef] [PubMed]

9. Jacob, K.D.; Noren Hooten, N.; Tadokoro, T.; Lohani, A.; Barnes, J.; Evans, M.K. Alzheimer's disease-associated polymorphisms in human OGG1 alter catalytic activity and sensitize cells to DNA damage. *Free Radic. Biol. Med.* **2013**, *63*, 115–125. [CrossRef] [PubMed]

10. Mao, G.; Pan, X.; Zhu, B.B.; Zhang, Y.; Yuan, F.; Huang, J.; Lovell, M.A.; Lee, M.P.; Markesbery, W.R.; Li, G.M.; et al. Identification and characterization of OGG1 mutations in patients with alzheimer's disease. *Nucleic Acids Res.* **2007**, *35*, 2759–2766. [CrossRef] [PubMed]

11. Cabelof, D.C.; Raffoul, J.J.; Yanamadala, S.; Ganir, C.; Guo, Z.; Heydari, A.R. Attenuation of DNA polymerase beta-dependent base excision repair and increased DMS-induced mutagenicity in aged mice. *Mutat. Res.* **2002**, *500*, 135–145. [CrossRef]

12. Noren Hooten, N.; Fitzpatrick, M.; Kompaniez, K.; Jacob, K.D.; Moore, B.R.; Nagle, J.; Barnes, J.; Lohani, A.; Evans, M.K. Coordination of DNA repair by NEIL1 and PARP-1: A possible link to aging. *Aging* **2012**, *4*, 674–685. [CrossRef] [PubMed]

13. Esteller, M. Cancer epigenomics: DNA methylomes and histone-modification maps. *Nat. Rev. Genet.* **2007**, *8*, 286–298. [CrossRef] [PubMed]

14. Fang, M.Z.; Chen, D.; Sun, Y.; Jin, Z.; Christman, J.K.; Yang, C.S. Reversal of hypermethylation and reactivation of *p16^{INK4a}*, *RARbeta*, and *MGMT* genes by genistein and other isoflavones from soy. *Clin. Cancer Res.* **2005**, *11*, 7033–7041. [CrossRef] [PubMed]

15. Fang, M.Z.; Wang, Y.; Ai, N.; Hou, Z.; Sun, Y.; Lu, H.; Welsh, W.; Yang, C.S. Tea polyphenol (−)-epigallocatechin-3-gallate inhibits DNA methyltransferase and reactivates methylation-silenced genes in cancer cell lines. *Cancer Res.* **2003**, *63*, 7563–7570. [PubMed]

16. Langie, S.A.; Koppen, G.; Desaulniers, D.; Al-Mulla, F.; Al-Temaimi, R.; Amedei, A.; Azqueta, A.; Bisson, W.H.; Brown, D.G.; Brunborg, G.; et al. Causes of genome instability: The effect of low dose chemical exposures in modern society. *Carcinogenesis* **2015**, *36* (Suppl. S1), S61–S88. [CrossRef] [PubMed]

17. Langie, S.A.; Kowalczyk, P.; Tomaszewski, B.; Vasilaki, A.; Maas, L.M.; Moonen, E.J.; Palagani, A.; Godschalk, R.W.; Tudek, B.; van Schooten, F.J.; et al. Redox and epigenetic regulation of the *APE1* gene in the hippocampus of piglets: The effect of early life exposures. *DNA Repair* **2014**, *18*, 52–62. [CrossRef] [PubMed]

18. Mehler, M.F. Epigenetic principles and mechanisms underlying nervous system functions in health and disease. *Prog. Neurobiol.* **2008**, *86*, 305–341. [CrossRef] [PubMed]

19. Jones, P.A.; Takai, D. The role of DNA methylation in mammalian epigenetics. *Science* **2001**, *293*, 1068–1070. [CrossRef] [PubMed]

20. Calvanese, V.; Lara, E.; Kahn, A.; Fraga, M.F. The role of epigenetics in aging and age-related diseases. *Ageing Res. Rev.* **2009**, *8*, 268–276. [CrossRef] [PubMed]

21. Mathers, J.C.; Strathdee, G.; Relton, C.L. Induction of epigenetic alterations by dietary and other environmental factors. In *Advances in Genetics*; Herceg, Z., Ushijima, T., Eds.; Academic Press: Burlington, MA, USA, 2010; Volume 71, pp. 1–39.

22. Nabel, C.S.; Kohli, R.M. Molecular biology. Demystifying DNA demethylation. *Science* **2011**, *333*, 1229–1230. [CrossRef] [PubMed]

23. Van den Hove, D.L.; Chouliaras, L.; Rutten, B.P. The role of 5-hydroxymethylcytosine in aging and Alzheimer's disease: Current status and prospects for future studies. *Curr. Alzheimer Res.* **2012**, *9*, 545–549. [CrossRef] [PubMed]

24. Rasmussen, K.D.; Helin, K. Role of TET enzymes in DNA methylation, development, and cancer. *Genes Dev.* **2016**, *30*, 733–750. [CrossRef] [PubMed]

25. Fraga, M.F.; Ballestar, E.; Paz, M.F.; Ropero, S.; Setien, F.; Ballestar, M.L.; Heine-Suner, D.; Cigudosa, J.C.; Urioste, M.; Benitez, J.; et al. Epigenetic differences arise during the lifetime of monozygotic twins. *Proc. Natl. Acad. Sci. USA* **2005**, *102*, 10604–10609. [CrossRef] [PubMed]

26. Arai, T.; Kasahara, I.; Sawabe, M.; Honma, N.; Aida, J.; Tabubo, K. Role of methylation of the *hMLH1* gene promoter in the development of gastric and colorectal carcinoma in the elderly. *Geriatr. Gerontol. Int.* **2010**, *10* (Suppl. S1), S207–S212. [CrossRef] [PubMed]

27. Wheeler, J.M. Epigenetics, mismatch repair genes and colorectal cancer. *Ann. R. Coll. Surg. Engl.* **2005**, *87*, 15–20. [CrossRef] [PubMed]

28. Agrelo, R.; Cheng, W.H.; Setien, F.; Ropero, S.; Espada, J.; Fraga, M.F.; Herranz, M.; Paz, M.F.; Sanchez-Cespedes, M.; Artiga, M.J.; et al. Epigenetic inactivation of the premature aging werner syndrome gene in human cancer. *Proc. Natl. Acad. Sci. USA* **2006**, *103*, 8822–8827. [CrossRef] [PubMed]

29. Butterfield, D.A.; Reed, T.; Newman, S.F.; Sultana, R. Roles of amyloid beta-peptide-associated oxidative stress and brain protein modifications in the pathogenesis of alzheimer's disease and mild cognitive impairment. *Free Radic. Biol. Med.* **2007**, *43*, 658–677. [CrossRef] [PubMed]

30. McKinnon, P.J. DNA repair deficiency and neurological disease. *Nat. Rev. Neurosci.* **2009**, *10*, 100–112. [CrossRef] [PubMed]

31. Nouspikel, T.; Hanawalt, P.C. When parsimony backfires: Neglecting DNA repair may doom neurons in alzheimer's disease. *BioEssays* **2003**, *25*, 168–173. [CrossRef] [PubMed]

32. Rowlatt, C.; Chesterman, F.C.; Sheriff, M.U. Lifespan, age changes and tumour incidence in an ageing C57BL mouse colony. *Lab. Anim.* **1976**, *10*, 419–442. [CrossRef] [PubMed]

33. Godschalk, R.W.; Maas, L.M.; van Zandwijk, N.; van't Veer, L.J.; Breedijk, A.; Borm, P.J.; Verhaert, J.; Kleinjans, J.C.; van Schooten, F.J. Differences in aromatic-DNA adduct levels between alveolar macrophages and subpopulations of white blood cells from smokers. *Carcinogenesis* **1998**, *19*, 819–825. [CrossRef] [PubMed]

34. European Standards Committee on Oxidative DNA Damage. Comparison of different methods of measuring 8-oxoguanine as a marker of oxidative DNA damage. *Free Radic. Res.* **2000**, *32*, 333–341.

35. Langie, S.A.; Kowalczyk, P.; Tudek, B.; Zabielski, R.; Dziaman, T.; Olinski, R.; van Schooten, F.J.; Godschalk, R.W. The effect of oxidative stress on nucleotide-excision repair in colon tissue of newborn piglets. *Mutat. Res.* **2010**, *695*, 75–80. [CrossRef] [PubMed]

36. Wang, Y.; Leung, F.C. An evaluation of new criteria for cpg islands in the human genome as gene markers. *Bioinformatics* **2004**, *20*, 1170–1177. [CrossRef]

37. Law, J.A.; Jacobsen, S.E. Establishing, maintaining and modifying DNA methylation patterns in plants and animals. *Nat. Rev. Genet.* **2010**, *11*, 204–220. [CrossRef] [PubMed]

38. Langie, S.A.; Cameron, K.M.; Waldron, K.J.; Fletcher, K.P.; von Zglinicki, T.; Mathers, J.C. Measuring DNA repair incision activity of mouse tissue extracts towards singlet oxygen-induced DNA damage: A comet-based in vitro repair assay. *Mutagenesis* **2011**, *26*, 461–471. [CrossRef]

39. Borgesius, N.Z.; de Waard, M.C.; van der Pluijm, I.; Omrani, A.; Zondag, G.C.; van der Horst, G.T.; Melton, D.W.; Hoeijmakers, J.H.; Jaarsma, D.; Elgersma, Y. Accelerated age-related cognitive decline and neurodegeneration, caused by deficient DNA repair. *J. Neurosci.* **2011**, *31*, 12543–12553. [CrossRef] [PubMed]

40. De Waard, M.C.; van der Pluijm, I.; Zuiderveen Borgesius, N.; Comley, L.H.; Haasdijk, E.D.; Rijksen, Y.; Ridwan, Y.; Zondag, G.; Hoeijmakers, J.H.; Elgersma, Y.; et al. Age-related motor neuron degeneration in DNA repair-deficient Ercc1 mice. *Acta Neuropathol.* **2010**, *120*, 461–475. [CrossRef] [PubMed]

41. Mathers, J.C.; Coxhead, J.M.; Tyson, J. Nutrition and DNA repair—Potential molecular mechanisms of action. *Curr. Cancer Drug Targets* **2007**, *7*, 425–431. [CrossRef] [PubMed]

42. Zawia, N.H.; Lahiri, D.K.; Cardozo-Pelaez, F. Epigenetics, oxidative stress, and alzheimer disease. *Free Radic. Biol. Med.* **2009**, *46*, 1241–1249. [CrossRef] [PubMed]

43. Hochberg, Z.; Feil, R.; Constancia, M.; Fraga, M.; Junien, C.; Carel, J.C.; Boileau, P.; Le Bouc, Y.; Deal, C.L.; Lillycrop, K.; et al. Child health, developmental plasticity, and epigenetic programming. *Endocr. Rev.* **2011**, *32*, 159–224. [CrossRef] [PubMed]

44. Jung, M.; Pfeifer, G.P. Aging and DNA methylation. *BMC Biol.* **2015**, *13*, 7. [CrossRef] [PubMed]

45. Wagner, M.; Steinbacher, J.; Kraus, T.F.; Michalakis, S.; Hackner, B.; Pfaffeneder, T.; Perera, A.; Muller, M.; Giese, A.; Kretzschmar, H.A.; et al. Age-dependent levels of 5-methyl-, 5-hydroxymethyl-, and 5-formylcytosine in human and mouse brain tissues. *Angew. Chem. Int. Ed. Engl.* **2015**, *54*, 12511–12514. [CrossRef] [PubMed]

46. Madrid, A.; Papale, L.A.; Alisch, R.S. New hope: The emerging role of 5-hydroxymethylcytosine in mental health and disease. *Epigenomics* **2016**, *8*, 981–991. [CrossRef] [PubMed]

47. Munzel, M.; Globisch, D.; Carell, T. 5-hydroxymethylcytosine, the sixth base of the genome. *Angew. Chem. Int. Ed. Engl.* **2011**, *50*, 6460–6468. [CrossRef] [PubMed]

48. Kohli, R.M.; Zhang, Y. TET enzymes, TDG and the dynamics of DNA demethylation. *Nature* **2013**, *502*, 472–479. [CrossRef] [PubMed]

49. Tan, L.; Shi, Y.G. TET family proteins and 5-hydroxymethylcytosine in development and disease. *Development* **2012**, *139*, 1895–1902. [CrossRef] [PubMed]

50. Santiago, M.; Antunes, C.; Guedes, M.; Sousa, N.; Marques, C.J. TET enzymes and DNA hydroxymethylation in neural development and function—How critical are they? *Genomics* **2014**, *104*, 334–340. [CrossRef] [PubMed]

51. Crawford, D.J.; Liu, M.Y.; Nabel, C.S.; Cao, X.J.; Garcia, B.A.; Kohli, R.M. *Tet2* catalyzes stepwise 5-methylcytosine oxidation by an iterative and de novo mechanism. *J. Am. Chem. Soc.* **2016**, *138*, 730–733. [CrossRef] [PubMed]

52. Mahfoudhi, E.; Talhaoui, I.; Cabagnols, X.; Della Valle, V.; Secardin, L.; Rameau, P.; Bernard, O.A.; Ishchenko, A.A.; Abbes, S.; Vainchenker, W.; et al. *Tet2*-mediated 5-hydroxymethylcytosine induces genetic instability and mutagenesis. *DNA Repair* **2016**, *43*, 78–88. [CrossRef] [PubMed]

53. Meng, H.; Cao, Y.; Qin, J.; Song, X.; Zhang, Q.; Shi, Y.; Cao, L. DNA methylation, its mediators and genome integrity. *Int. J. Biol. Sci.* **2015**, *11*, 604–617. [CrossRef] [PubMed]

54. Pal, S.; Tyler, J.K. Epigenetics and aging. *Sci. Adv.* **2016**, *2*, e1600584. [CrossRef] [PubMed]

55. Booth, M.J.; Branco, M.R.; Ficz, G.; Oxley, D.; Krueger, F.; Reik, W.; Balasubramanian, S. Quantitative sequencing of 5-methylcytosine and 5-hydroxymethylcytosine at single-base resolution. *Science* **2012**, *336*, 934–937. [CrossRef] [PubMed]

56. Branco, M.R.; Ficz, G.; Reik, W. Uncovering the role of 5-hydroxymethylcytosine in the epigenome. *Nature Rev. Genet.* **2012**, *13*, 7–13. [CrossRef] [PubMed]

57. Wakasugi, T.; Izumi, H.; Uchiumi, T.; Suzuki, H.; Arao, T.; Nishio, K.; Kohno, K. ZNF143 interacts with p73 and is involved in cisplatin resistance through the transcriptional regulation of DNA repair genes. *Oncogene* **2007**, *26*, 5194–5203. [CrossRef] [PubMed]

58. Ishiguro, A.; Aruga, J. Functional role of Zic2 phosphorylation in transcriptional regulation. *FEBS Lett.* **2008**, *582*, 154–158. [CrossRef] [PubMed]

59. Riccio, A. Dynamic epigenetic regulation in neurons: Enzymes, stimuli and signaling pathways. *Nat. Neurosci.* **2010**, *13*, 1330–1337. [CrossRef] [PubMed]

60. Gorniak, J.; Langie, S.A.S.; Cameron, K.; von Zglinicki, T.; Mathers, J.C. The effect of ageing and short-term dietary restriction on the epigenetic, transcriptomic and phenotypic profile of base excision repair in mouse brain and liver. *Proc. Nutr. Soc.* **2012**, *71*, E56. [CrossRef]

61. Cabelof, D.C.; Yanamadala, S.; Raffoul, J.J.; Guo, Z.; Soofi, A.; Heydari, A.R. Caloric restriction promotes genomic stability by induction of base excision repair and reversal of its age-related decline. *DNA Repair* **2003**, *2*, 295–307. [CrossRef]

62. Imam, S.Z.; Karahalil, B.; Hogue, B.A.; Souza-Pinto, N.C.; Bohr, V.A. Mitochondrial and nuclear DNA-repair capacity of various brain regions in mouse is altered in an age-dependent manner. *Neurobiol. Aging* **2006**, *27*, 1129–1136. [CrossRef]

63. Xu, G.; Herzig, M.; Rotrekl, V.; Walter, C.A. Base excision repair, aging and health span. *Mech. Ageing Dev.* **2008**, *129*, 366–382. [CrossRef] [PubMed]

64. Rao, K.S. Dietary calorie restriction, DNA-repair and brain aging. *Mol. Cell. Biochem.* **2003**, *253*, 313–318. [CrossRef] [PubMed]

65. Gedik, C.M.; Grant, G.; Morrice, P.C.; Wood, S.G.; Collins, A.R. Effects of age and dietary restriction on oxidative DNA damage, antioxidant protection and DNA repair in rats. *Eur. J. Nutr.* **2005**, *44*, 263–272. [CrossRef] [PubMed]

66. Moller, P.; Lohr, M.; Folkmann, J.K.; Mikkelsen, L.; Loft, S. Aging and oxidatively damaged nuclear DNA in animal organs. *Free Radic. Biol. Med.* **2010**, *48*, 1275–1285. [CrossRef] [PubMed]

67. Jurk, D.; Wang, C.; Miwa, S.; Maddick, M.; Korolchuk, V.; Tsolou, A.; Gonos, E.S.; Thrasivoulou, C.; Saffrey, M.J.; Cameron, K.; et al. Postmitotic neurons develop a p21-dependent senescence-like phenotype driven by a DNA damage response. *Aging Cell* **2012**, *11*, 996–1004. [CrossRef] [PubMed]

68. Langie, S.A.; Achterfeldt, S.; Gorniak, J.P.; Halley-Hogg, K.J.; Oxley, D.; van Schooten, F.J.; Godschalk, R.W.; McKay, J.A.; Mathers, J.C. Maternal folate depletion and high-fat feeding from weaning affects DNA methylation and DNA repair in brain of adult offspring. *FASEB J.* **2013**, *27*, 3323–3334. [CrossRef] [PubMed]

69. Kulis, M.; Queiros, A.C.; Beekman, R.; Martin-Subero, J.I. Intragenic DNA methylation in transcriptional regulation, normal differentiation and cancer. *Biochim. Biophys. Acta* **2013**, *1829*, 1161–1174. [CrossRef] [PubMed]

70. Doi, A.; Park, I.H.; Wen, B.; Murakami, P.; Aryee, M.J.; Irizarry, R.; Herb, B.; Ladd-Acosta, C.; Rho, J.; Loewer, S.; et al. Differential methylation of tissue- and cancer-specific cpg island shores distinguishes human induced pluripotent stem cells, embryonic stem cells and fibroblasts. *Nat. Genet.* **2009**, *41*, 1350–1353. [CrossRef]

71. Hon, G.C.; Hawkins, R.D.; Caballero, O.L.; Lo, C.; Lister, R.; Pelizzola, M.; Valsesia, A.; Ye, Z.; Kuan, S.; Edsall, L.E.; et al. Global DNA hypomethylation coupled to repressive chromatin domain formation and gene silencing in breast cancer. *Genome Res.* **2012**, *22*, 246–258. [CrossRef] [PubMed]

72. Berman, B.P.; Weisenberger, D.J.; Aman, J.F.; Hinoue, T.; Ramjan, Z.; Liu, Y.; Noushmehr, H.; Lange, C.P.; van Dijk, C.M.; Tollenaar, R.A.; et al. Regions of focal DNA hypermethylation and long-range hypomethylation in colorectal cancer coincide with nuclear lamina-associated domains. *Nat. Genet.* **2011**, *44*, 40–46. [CrossRef] [PubMed]

73. Hansen, K.D.; Timp, W.; Bravo, H.C.; Sabunciyan, S.; Langmead, B.; McDonald, O.G.; Wen, B.; Wu, H.; Liu, Y.; Diep, D.; et al. Increased methylation variation in epigenetic domains across cancer types. *Nat. Genet.* **2011**, *43*, 768–775. [CrossRef] [PubMed]

74. Gredilla, R.; Bohr, V.A.; Stevnsner, T. Mitochondrial DNA repair and association with aging—An update. *Exp. Gerontol.* **2010**, *45*, 478–488. [CrossRef] [PubMed]

75. Blanch, M.; Mosquera, J.L.; Ansoleaga, B.; Ferrer, I.; Barrachina, M. Altered mitochondrial DNA methylation pattern in alzheimer disease-related pathology and in parkinson disease. *Am. J. Pathol.* **2016**, *186*, 385–397. [CrossRef] [PubMed]

76. Byun, H.M.; Barrow, T.M. Analysis of pollutant-induced changes in mitochondrial DNA methylation. *Methods Mol. Biol.* **2015**, *1265*, 271–283. [PubMed]

77. Castegna, A.; Iacobazzi, V.; Infantino, V. The mitochondrial side of epigenetics. *Physiol. Genom.* **2015**, *47*, 299–307. [CrossRef] [PubMed]

12

Transgenic Tobacco Expressing the TAT-Helicokinin I-CpTI Fusion Protein Show Increased Resistance and Toxicity to *Helicoverpa armigera* (Lepidoptera: Noctuidae)

Zhou Zhou [1,2,†], Yongli Li [2,†], Chunyan Yuan [2], Yongan Zhang [1] and Liangjian Qu [1,*]

1   Key Laboratory of Forest Protection, State Forestry Administration, Research Institute of Forest Ecology, Environment and Protection, Chinese Academy of Forestry, Beijing 100091, China; zhouzhouhaust@163.com (Z.Z.); Zhangyab59@gmail.com (Y.Z.)

2   College of Forestry, Henan University of Science and Technology, Luoyang 471003, China; yonglili1978@163.com (Y.L.); haustforestry@yeah.net (C.Y.)

*   Correspondence: qulj2001@caf.ac.cn

†   These authors contributed equally to this work.

Academic Editor: Paolo Cinelli

Abstract: Insect kinins were shown to have diuretic activity, inhibit weight gain, and have antifeedant activity in insects. In order to study the potential of the TAT-fusion approach to deliver diuretic peptides per os to pest insects, the HezK I peptide from *Helicoverpa zea*, as a representative of the kinin family, was selected. The fusion gene TAT-HezK I was designed and was used to transform tobacco plants. As a means to further improve the stability of TAT-HezK I, a fusion protein incorporating HezK I, transactivator of transcription (TAT), and the cowpea trypsin inhibitor (CpTI) was also designed. Finally, the toxicity of the different tobacco transgenic strains toward *Helicoverpa armigera* was compared. The results demonstrated that TAT-HezK I had high toxicity against insects via transgenic expression of the peptide in planta and intake through larval feeding. The toxicity of the fusion TAT-HezK I and CpTI was higher than the CpTI single gene in transgenic tobacco, and the fusion TAT-HezK I and CpTI further enhanced the stability and bioavailability of agents in oral administration. Our research helps in targeting new genes for improving herbivore tolerance in transgenic plant breeding.

Keywords: helicokinin I; cowpea trypsin inhibitor (CpTI); transactivator of transcription (TAT); protein transduction domain (PTD); *Helicoverpa armigera*

1. Introduction

The insect kinins are multifunctional neuropeptides found in several arthropod and invertebrate groups [1]. Insect kinins were shown to have diuretic activity on isolated Malpighian tubules of the cricket *Acheta* [2] and the yellow fever mosquito *Aedes aegypti* [3]. Insect neuropeptides of the insect kinin class share a common C-terminal pentapeptide sequence $Phe^1-Xaa_1^2-Xaa_2^3-Trp^4-Gly^5-NH_2$ (Xaa_1^2 = His, Asn, Phe, Ser, or Tyr; Xaa_2^3 = Pro, Ser, or Ala). Insect kinins or their synthetic analogs have been reported to inhibit weight gain when fed to, or injected in larvae of tobacco budworm *Heliothis virescens* and corn earworm *Helicoverpa zea*, both serious agricultural pests [4,5]. Interestingly, antifeedant activity and high mortality were found in the pea aphid *Acyrthosiphon pisum* [6].

Seinsche et al. [5] demonstrated that the weight gain inhibition observed in *H. virescens* is accompanied by an increase in the excretion of water in the feces, consistent with the diuretic activity previously observed in crickets [7], flies [8,9], as well as the lepidopteran *H. virescens* [5]. The authors further speculated that the insect kinins could have induced a starvation signal in the *Heliothis* larvae, resulting in mobilization of energy stores and a decreased efficiency in exploiting digested nutrients [5]. Together with the increased excretion of fluid and the induction of a starvation response, an inhibition of digestive enzyme release may have led to the weight losses observed in both *H. virescens* [5] and *H. zea* [8] treated with insect kinins and/or analogs. The diuretic activity of the helicokinins I, II, and III (HezK I, II, and III) from *H. zea* was tested on *H. virescens* larvae. All three kinins increased fluid secretion in isolated Malpighian tubules in a dose-dependent manner. Injections into the haemolymph caused a significant reduction in weight gain after 24 h and, in the case of HezK I, led to an increased mortality of 43% within six days, which was the most efficient of the three helicokinins [5]. No oral activity data has been reported for the insect kinin class of neuropeptides. Generally, oral activity for unmodified insect neuropeptides is poor to nonexistent, in large part due to the inability of these molecules to cross the gut epithelium.

Cell penetrating peptides (CPP) have been studied to facilitate the non-invasive (e.g., oral, transdermal) delivery of macromolecular therapeutic agents in animal models such as the mouse [10]. One of the best characterized CPP is the protein transduction domain (PTD) of the human immunodeficiency virus-1 (HIV-1) transactivator of transcription (TAT) protein. TAT-PTD is a lysine- and arginine-rich peptide of the sequence YGRKKRRQRRR that is required for TAT cell membrane transduction, and has been successfully used to transduce a variety of protein cargos into mammalian cells [11–13]. TAT-PTD has also been demonstrated to be useful for the systemic delivery of a galactosidase-TAT-PTD fusion protein cargo in mice via intraperitoneal injection [14], and this phenomenon could possibly be replicated in other organisms, including invertebrates. Recently, the use of TAT-PTD for the oral delivery of the *Clostera anastomosis* diapause hormone (caDH) to larvae of the moth *Helicoverpa armigera* was demonstrated [15]. The fusion peptide TAT-caDH was able to penetrate *H. armigera* midgut tissues after ingestion by third-instar larvae, and larvae exhibited pronounced growth retardation under conditions that either promoted development (27 °C, 14L:10D) or induced diapause (20 °C, 10L:14D). Under development-promoting conditions, larvae exhibited an 8% reduction in pupation rate and the duration of larval development was on average 3 days longer compared to untreated controls. Under diapause-inducing conditions, larvae fed a diet containing TAT-caDH exhibited a 14% reduction in pupation rate and the duration of larval development exhibited an increase of 12 days longer compared to controls. Fusion of the TAT peptide to a broader range of insect peptides may open up their use in pest management.

In this report, we selected the HezK I peptide from *H. zea* as a representative of the kinin family to study the potential of the TAT-fusion approach to deliver diuretic peptides per os to pest insects. We first designed the fusion gene *TAT-HezK I* and used it to transform tobacco plants. As a means to further improve the stability of the TAT-HezK I, A fusion protein incorporating HezK I, TAT, and the Cowpea Trypsin Inhibitor (CpTI) was designed. CpTI can inhibit protein degradation in the digestive tract [16]. Finally, we compared the toxicity of the different tobacco transgenic strains (*TAT-HezK I*, *TAT-HezK I-CpTI*, and controls) toward *H. armigera*. Our results may help in improving the management of insect pests of economically important crop plants, by targeting kinin-regulated physiological processes.

2. Materials and Methods

2.1. Expression Vector Construction

The *pBin438-TAT-HezK I*, *pBin438-CpTI*, and *pBin438-TAT-HezK I-CpTI* constructs were generated for plant transformation. First, the sequences for *TAT-HezK I*, *CpTI*, and *TAT-HezK I-CpTI* were obtained. The *TAT-HezK I* (GenBank: KX492908), *CpTI* (GenBank: KX492909), and *TAT-HezK I-CpTI* (GenBank: KX492910) coding sequences were synthesized by Sangon Biotech (Shanghai, China) with restriction enzyme sites added at each end (*Bam*HI at the 5′ end and *Sal*I at the 3′ end). The synthetic constructs were cloned into pUC57 plasmid by a commercial service provider (Sangon). The *HezK I* was based on the helicokinin I coding sequence from *H. zea* [5]. The TAT sequence contained the codons encoding residues 47–57 (PTD: YGRKKRRQRRR) [17]. A nucleotide sequence encoding a flexible linker (FL, amino acids GGGGS) was inserted between the TAT and HezK I coding sequences [18]. The *CpTI* sequence was based on the modified cowpea trypsin inhibitor coding sequence [16]. The TAT-HezK I-CpTI fusion construct comprised the *TAT, HezK I,* and *CpTI* sequences, each separated by an FL linker coding sequence.

The *pUC57-TAT-HezK I, pUC57-CpTI, pUC57-TAT-HezK I-CpTI,* and pBin438 plasmids were digested using *Bam*HI and *Sal*I (Sangon) at 37 °C for 3 h. The digested DNA fragments were purified using a PCR product purification kit (Sangon, Shanghai, China) and ligated using T4 DNA ligase (Sangon). The ligation mixture was used to transform competent Top10 *Escherichia coli*. Recombinant *pBin438-TAT-HezK I, pBin438-CpTI,* and *pBin438-TAT-HezK I-CpTI* plasmids sequences were confirmed by restriction endonuclease digestion and DNA sequencing (Sangon).

For the purpose of overexpression, the *TAT-HezK I, CpTI,* and *TAT-HezK I-CpTI* genes were cloned into the plant binary expression vector pBin438, under the control of the double CaMV 35S promoters with a nopaline synthase (NOS) terminator sequence (Figure 1A). The neomycin phosphotransferase II (nptII) gene was used as a selectable marker gene. Recombinant *pBin438-TAT-HezK I, pBin438-CpTI,* and *pBin438-TAT-HezK I-CpTI* plasmids were then transformed into *Agrobacterium tumefaciens* stain LBA4404 by electroporation.

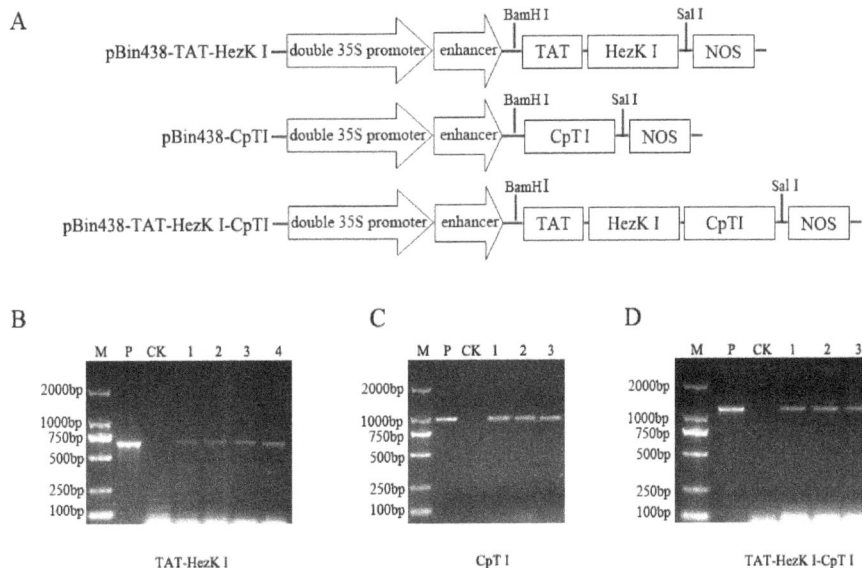

Figure 1. Molecular analysis of transgenic tobaccos. (**A**) Schematic representations of the *TAT-HezK I* (transactivator of transcription, TAT), *CpTI,* and *TAT-HezK I-CpTI* expression constructs, based on the pBin438 plasmid. Each fusion protein coding sequence was cloned downstream of a 2× CaMV 35S promoter and an Ω enhancer of TMV (Tobacco mosaic virus, TMV) to drive gene expression. The fusion protein sequence is also located upstream of a nopaline synthase (NOS) terminator sequence; (**B**) PCR analysis of genomic DNA from transgenic tobacco expressing TAT-HezK I. M, marker. P, Plasmid pBin438-TAT-HezK I as template. CK, DNA of wild-type plant as template. 1, 2, 3, and 4, DNA of TAT-HezK I transgenic line 1, 2, 3, and 4 as template; (**C**) PCR analysis of genomic DNA from transgenic tobacco expressing CpTI. M, marker. P, Plasmid pBin438-CpTI as template. CK, DNA of wild-type plant as template. 1, 2, and 3, DNA of CpTI transgenic line 1, 2, and 3 as template; (**D**) PCR analysis of genomic DNA from transgenic tobacco expressing TAT-HezK I-CpTI. M, Marker. P, Plasmid pBin438-TAT-HezK I-CpTI as template. CK, DNA of wild-type plant as template. 1, 2, and 3, DNA of TAT-HezK I-CpTI transgenic line 1, 2 and 3 as template.

2.2. Tobacco Transformation

Gene transfer experiments were performed on the *Nicotiana tabacum* cv. K326 strain. Tobacco leave discs were transformed with the *A. tumefaciens* strain LBA4404 by suspension in MS (Murashige and Skoog) liquid medium solution [19]. Leaves of 15- to 25-day-old plantlets K326 clone at tissue culture stage were used as explants for transformation. *A. tumefaciens* strain LBA4404 harboring the *pBin438-TAT-HezK I, pBin438-CpTI*, or *pBin438-TAT-HezK I-CpTI* were used to transform K326 via the leaf-disc method [20]. Briefly, infected leaf discs were grown in MS basal medium supplemented with 1.0 mg/L of 6-benzyl aminopurine (6-BA) and 0.1 mg/L of naphthaleneacetic acid (NAA) in the dark for two days at 25 °C, and then transferred on the same medium with 50 mg/L kanamycin under a 16 h/8 h light/dark regime. Individual regenerated shoots were removed and induced for rooting on 1/2 MS medium supplemented with 0.1 mg/L NAA and 50 mg/L kanamycin. The plants were grown in a greenhouse at 27 °C under constant illumination.

2.3. PCR Analysis of Genomic DNA from Transgenic Tobacco Plants

Primary rooted transformants at the tissue culture stage were screened by genomic PCR using a 35S promoter and Nos terminator specific primer pairs. Genomic DNA of T1 generations of transgenic tobacco were isolated from transgenic plants using a Rapid Plant Genomic DNA Isolation Kit B518231-0100 (Sangon). To confirm the insertion of transgenic constructs expressing *TAT-HezK I, CpTI*, and *TAT-HezK I-CpTI*, PCR analysis of genomic DNA was performed with two primer pairs as follows: forward 35S promoter primer: 35S-F, 5′-GGAAACCTCCTCGGATTCCAT-3′ and backward Nos terminator primer: NOSTR, 5′-CTCATAAATAACGTCATGCATTAC-3′.

2.4. Target Gene Expression Analysis by Quantitative Real-Time PCR (qRT-PCR)

Total RNA of T1 generation of transgenic tobacco was extracted using the Spin Column Plant total RNA Purification Kit (Sangon) according to the manufacturer's instructions. First-strand cDNA was generated from 4 µg RNA primed with oligo (dT)$_{18}$ using the AMV First Strand cDNA Synthesis Kit (Sangon). qPCR analysis was performed in a 20 µL volume using ABI SybrGreen PCR Master Mix (ABI, Carlsbad, CA, USA), and the results were analyzed using the Bio-Rad iQTM5 Real-Time PCR Detection system (Bio-Rad, Hercules, CA, USA), according to the manufacturer's instructions. The qRT-PCR reactions were initiated with a predenaturation step at 95 °C for 30 s, followed by 40 cycles of denaturation (95 °C for 20 s), annealing (59 °C for 20 s), extension (72 °C for 20 s), and a final stage of 55–95 °C to determine dissociation curves of the amplified products. *Actin* was used as the endogenous control for normalization. The value for the lowest expressing *TAT-HezK I-CpTI* transgenic line was normalised to one. More than three tobaccos were used for each transgenic line (*TAT-HezK I, CpTI*, or *TAT-HezK I-CpTI*) and three PCR reaction replicates were set up for each line. Primer sequences were as follows: *TAT-HezK I* or *TAT-HezK I-CpTI* genes checking primers: *TAT-HezK I*-F, 5′-ATGTACGGCCGCAAGAAGA-3′ and *TAT-HezK I*-R, 5′-ACGCCCCAGGGGCTGAAG-3′. *CpTI* or *TAT-HezK I-CpTI* genes checking primers: *CpTI*-F, 5′-TCATACCTACCTTCAGCCATCC-3′ and *CpTI*-R, 5′-AAGACTCAGAAGGTTCATCGCT-3′. Tobacco *Actin* primers: *Actin*-F, 5′-CCCCTTGTC TGTGATAACGG-3′ and *Actin*-R, 5′- AGAATACCCCTTTTGGACTGAG-3′.

2.5. TAT-HezK I and CpTI Polyclonal Antibody Generation and Detection of Protein in the Transgenic Plants

Polypeptides of TAT-HezK I (YGRKKRRQRRRGGGGSYFSPWGa) and middle amino acids of CpTI and (GSNHI IDDSSDEPSESSEPCCDSCa) were synthesized (Genscript, Nanjing, China), and polyclonal antibodies (anti-TAT-HezK I and anti-CpTI) were generated in rabbits, as described previously [21]. Western blot analysis of the transgenic plant materials was performed. Total soluble protein (TSP) was extracted from leaves harvested from six week old plants, 3–5 nodes below the apex. Tissues were ground to a fine powder in liquid nitrogen with a chilled mortar and pestle. TSP was extracted using the One Step Plant Active Protein Extraction Kit (Sangon) according to the manufacturer's instructions. Concentration of TSP was determined by the BCA Protein Assay Kit (Sangon) according to the manufacturer's protocol. For the Western blot, 20 micrograms of TSP was separated by 12% sodium dodecyl sulfate-polyacrylamide gel electrophoresis (SDS-PAGE) in $1\times$ SDS running buffer (Sangon) at 80 V for 2 h. Proteins were electrophoretically transferred to polyvinylidene fluoride (PVDF) membranes (Millipore, Bedford, MA, USA). The primary antibodies (anti-TAT-HezK I and anti-CpTI) were used at 1:5000 dilution, and the IgG-AP was used at a 1:20,000 dilution. Detection was performed using the NPT/BCIP kit (CWBIO company, Beijing, China) as described by the manufacturer.

2.6. Tobacco Plants Culture

To maintain transgenic and non-transgenic tobacco lines, plants were propagated by vegetative multiplication. Leaf discs were grown in MS basal medium supplemented with 1.0 mg/L 6-benzyl aminopurine (6-BA) and 0.1 mg/L naphthaleneacetic acid (NAA) at 25 °C, under 16L:8D (photophase:scotophase). Individual regenerated shoots were removed and induced for rooting on 1/2 MS medium supplemented with 0.1 mg/L NAA. After the shoots rooted, tissue culture seedlings were transplanted and grown in pre-sterilized soil at 25 ± 1 °C and 60%–80% relative humidity under a 16 h light/8 h dark photoperiod in a greenhouse.

2.7. Insect Rearing

H. armigera and diet were purchased from Baiyun Industry Co., Ltd. (Henan, China). Larvae were reared at 27 °C with a 14 h light and 10 h dark (14L:10D) cycle under $65\% \pm 5\%$ relative humidity. *H. armigera* larvae were reared in groups until they reached the third-instar and were fed individually thereafter [22,23]. Similar third-instar larvae were individually transferred to a 9 cm Petri dish containing a tobacco leaf. Fresh leaves were supplied every 24 h.

2.8. Feeding Bioassays for H. armigera on Tobacco Leaves

To investigate whether transgenic plants displayed any differences against insect attack, feeding bioassays were conducted with detached tobacco leaves as previously described [23] with minor modifications. Leaves from two-month-old tobacco clones of transgenic lines expressing TAT-HezK I (TAT-caDH 1–4), CpTI (CpTI 1–3), or TAT-HezK I-CpTI (TAT-HezK I-CpTI 1–3), as well as age-matched wild-type controls (CK) were used. Leaves were rinsed with sterile distilled water, air dried, cut to square (5–6 cm × 5–6 cm), and were placed on a moist filter paper disc in a 90-mm-diameter Petri dish. Five hundred microlitres of sterile distilled water was provided each day to maintain relative humidity. Thirty leaves of each of the transgenic lines were taken and a third-instar larva was released on each of these leaves. The larvae were starved for 2 h prior to release to increase propensity for feeding. Three biological replicates of 30 larvae per treatment were performed and larvae were kept on fresh foliage which were replaced with leaves of the same transgenic line each day for six days in a controlled greenhouse at 27 °C with a 14L:10D photoperiod and 75% relative humidity. Larval fresh weights were recorded at d 0, d 2, d 4, and d 6 after exposure to the leaves, and mortality was recorded at d 6.

2.9. Statistical Analysis

All of the data were subjected to a one-way analysis of variance using the DPS software, version 7.05 (Zhejiang University, Hangzhou, China). The Student-Newman-Keuls test was used to evaluate the intergroup differences. The results of larvae weight in which "*" denotes $p < 0.05$ and "**" denotes $p < 0.01$ were considered to represent statistically significant differences to control. The data are presented as the mean $\pm$ standard error (SE).

3. Results

3.1. Generation of Transgenic Plants Expressing TAT-HezK I, CpTI, or TAT-HezK I-CpTI

The TAT-HezK I, CpTI, and TAT-HezK I-CpTI recombinant proteins were expressed via transgenic tobacco plants to assess the impact of oral delivery in *H. armigera* larvae. Transgenic tobacco plants were used in feeding assays, along with untransformed wild-type plants as controls. The *TAT-HezK I, CpTI*, and *pBin438-TAT-HezK I-CpTI* constructs were cloned in the pBin438 vector plasmid which drives expression with a 2× 35S promoter and an Ω enhancer of TMV (Tobacco Mosaic Virus) (Figure 1A). At least three independent transgenic T1 lines were selected for each construct. To confirm the successful insertion of each construct, genomic DNA from the transgenic T1 plants was subjected to PCR screening with specific primers of 35S-F and NOSTR (Figure 1B: TAT-HezK I 1, 2, 3, and 4; Figure 1C: CpTI 1, 2, and 3; Figure 1D: TAT-HezK I-CpTI 1, 2, and 3). PCR analysis of plant genomic DNA showed that four *pBin438-TAT-HezK* I-transformed, three *pBin438-CpTI*-transformed, and three *pBin438-TAT-HezK I-CpTI*-transformed tobacco plants were obtained.

We used qRT-PCR to quantify the expression of each transgene. Our results indicate that the levels of *TAT-HezK I, CpTI*, or *TAT-HezK I-CpTI* mRNAs expressed in each of the lines were not similar to one another. Using the TAT-HezK I-CpTI 1 line as a baseline for transgene expression comparisons, we found that the TAT-HezK I-CpTI 2 and TAT-HezK I-CpTI 3 lines expressed 1.2 and 2.5 times more of the transgene, respectively. The CpTI 1, CpTI 2, and CpTI 3 lines had levels of transgene expression of 1.4, 1.5, and 3.3 times that of TAT-HezK I-CpTI 1. The TAT-HezK I 1, TAT-HezK I 2, TAT-HezK I 3, and TAT-HezK I 4 lines had levels of transgene expression of 2.5, 4.1, 24.2, and 30.3 times that of TAT-HezK I-CpTI 1, respectively (Figure 2).

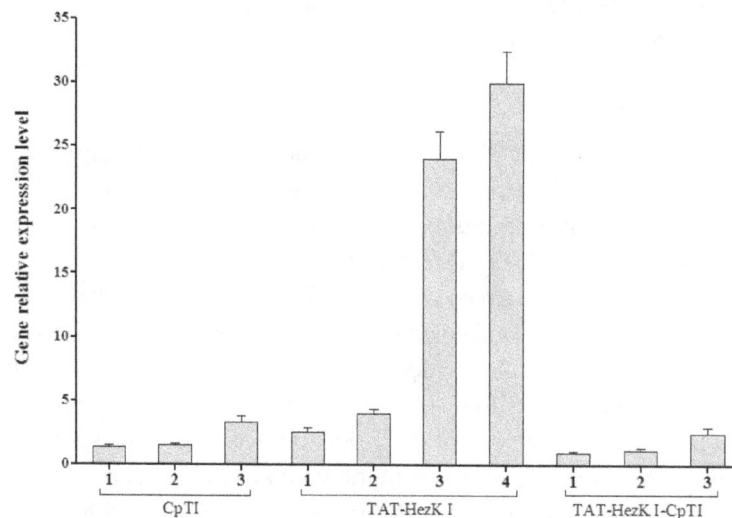

Figure 2. mRNA expression of CpTI, TAT-HezK I, and TAT-HezK I-CpTI in leaves of transgenic tobacco.

3.2. Analysis of the Protein Expression in Transgenic Lines

When probed with antiserum raised against the TAT-HezK I protein, three TAT-HezK I transgenic lines showed a major immunoreactive band of approximately 3 KDa, while three TAT-HezK I-CpTI lines showed a major immunoreactive band of approximately 18 KDa. When probed with antiserum raised against the CpTI protein, three CpTI transgenic lines expressed a single fusion protein of ca. 15 KDa, and TAT-HezK I-CpTI line 3 showed a major immunoreactive band of approximately 18 KDa. As a control, no band could be detected in the wild type (WT) plant. The evidence for the presence of immunoreactive bands corresponding in size to the predicted molecular masses for *TAT-HezK I, CpTI*, and *TAT-HezK I-CpTI* suggests that the introduced gene cassettes were successfully translated in planta (Figure 3). TAT-HezK I-CpTI 1–3 lines expressed similar component polypeptides. Using the

TAT-HezK I-CpTI 1 line as a baseline for transgene expression comparisons, the TAT-HezK I-CpTI 2, TAT-HezK I-CpTI 3, CpTI 1, CpTI 2, CpTI 3, TAT-HezK I 1, TAT-HezK I 2, TAT-HezK I 3, and TAT-HezK I 4 lines had levels of polypeptides expression of 0.9, 1.1, 0.7, 0.9, 1.0, 0.8, 2.4, 3.6, and 3.5 times that of TAT-HezK I-CpTI 1 in the plant tissue, according to the band intensity as measured by ImageJ software 1.8.0 (National Institutes of Health, Bethesda, MD, USA).

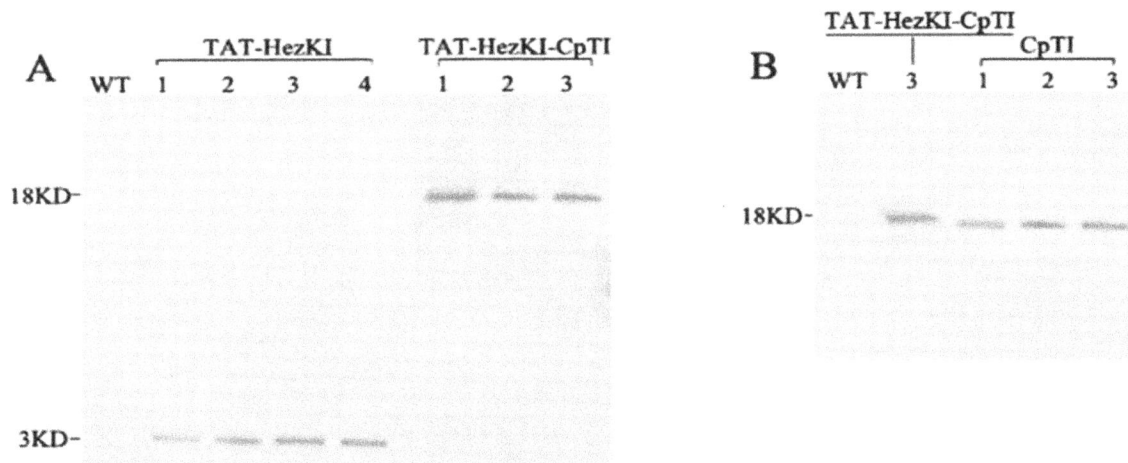

Figure 3. Western blotting analysis of the protein expression of TAT-HezK I, CpTI, and TAT-HezK I-CpTI in the transgenic plants. (**A**) Western blotting analysis of TAT-HezK I/TAT-HezK I-CpTI expression in transgenic lines using anti-TAT-HezK I. WT, protein extracted from WT tobacco plants. In every lane, 20 µg TSP was loaded; (**B**) Western blotting analysis of TAT-HezK I-CpTI/CpTI expression in transgenic lines using anti-CpTI. WT, Protein extracted from WT tobacco plants. In every lane, 20 µg TSP was loaded.

3.3. Increased Resistance to, and Toxicity toward H. armigera in Transgenic Tobacco Plants Expressing CpTI, TAT-HezK I, and TAT-HezK I-CpTI

Two-month-old tobacco plants were used for the feeding bioassays. Larvae feeding upon transgenic plants expressing CpTI, TAT-HezK I, and TAT-HezK I-CpTI exhibited a reduced survival rate. Larval mortality was assayed at d 6. Transgenic lines TAT-HezK I 3, TAT-HezK I 4, and TAT-HezK I-CpTI3 were observed with the highest levels of mortality (57%, 57%, and 50%) by feeding detached mature leaves to third-instar larvae. The larval mortality rates associated with feeding upon transgenic lines CpTI 3, TAT-HezK I 3, TAT-HezK I 4, TAT-HezK I-CpTI 1, TAT-HezK I-CpTI 2, and TAT-HezK I-CpTI 3 increased to approximately 15%, 47%, 47%, 17%, 23%, and 40% compared to the control, respectively (Figure 4A).

Larvae feeding on transgenic tobacco expressing CpTI, TAT-HezK I, and TAT-HezK I-CpTI achieved lower weight gains from d 2 to d 6 than the controls. Average larval weight was 221.7 mg when larval feeding occurred on untransformed tobacco controls. However, larval feeding on transgenic lines TAT-HezK I 3, TAT-HezK I 4, and TAT-HezK I-CpTI3 were observed with the lowest weights, and the average larval weights were 95.1 mg, 80.1 mg, and 90.3 mg at d 6, respectively (Figure 4B). When detached mature leaves were fed to *H. armigera* larvae for six days, larval average weight feeding on transgenic tobacco lines CpTI 3, TAT-HezK I 3, TAT-HezK I 4, TAT-HezK I-CpTI1, TAT-HezK I-CpTI2, and TAT-HezK I-CpTI3 were 42%, 57%, 64%, 37%, 43%, and 59% statistically less than ($p < 0.05$) feeding on untransformed tobacco controls, respectively.

A

B

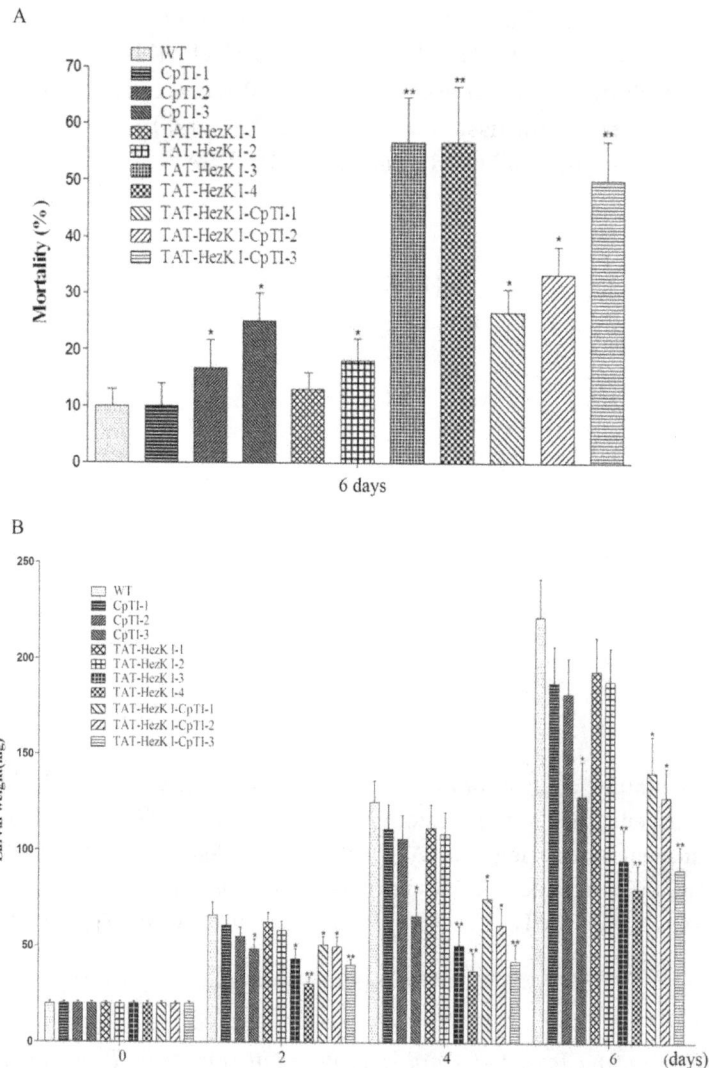

Figure 4. Effect of transgenic tobacco expressing CpTI, TAT-HezK I, or TAT-HezK I-CpTI proteins on larval mortality (**A**) and weight gain (**B**) as measured in feeding bioassays. Detached mature leaves of similar sizes and weights from two-month-old tobacco plants were individually placed in 9-cm Petri dishes and covered with filter paper to retain proper moisture. One third-instar were placed in each Petri dish containing a single leaf, with thirty larvae in one group. After three and six days of feeding, the larval mortality (**A**) and weight gain (**B**) were recorded. For both assays, all treatments were compared to the untransformed control (CK) (Student-Newman-Keuls test: "*" denotes $p < 0.05$ and "**" denotes $p < 0.01$).

4. Discussion

Injection of HezK I into the haemolymph of *H. virescens* larvae caused a significant reduction in weight gain after 24 h and led to an increased mortality of 43% within six days, which was the most efficient in the three described helicokinins; helicokinins I (YFSPWG-amide), II (VRFSPWG-amide), and III (KVKFSAWG-amide) [5]. For this reason, we decided to study the potential insecticidal activity of HezK I delivered in planta to *H. armigera* larvae. *H. armigera* (Lepidoptera: Noctuidae) is closely related to *H. zea*, and is a widely distributed insect pest of high agricultural importance. Here, we first designed the fusion gene *TAT-HezK I* that was transformed and expressed in tobacco used to manipulate development in agriculturally important insects.

Members of the insect kinin family are hydrolyzed and therefore inactivated by tissue-bound peptidases found in insects. Two susceptible hydrolysis sites in insect kinins have been reported.

The primary site is located between the Pro[3] and the Trp[4] residues, with a secondary site N-terminal to the Phe[1] residue in natural extended insect kinin sequences. Experiments demonstrate that the angiotensin converting enzyme (ACE) from the housefly can cleave at the primary hydrolysis site, whereas neprilysin (NEP) can cleave insect kinins at both the primary and secondary hydrolysis sites [4,8,24,25]. Prior to this report, no oral activity data has been reported for the insect kinin class of neuropeptides. Generally, activity for unmodified insect neuropeptides is poor to nonexistent when delivered via an oral route. The exploitation of insect feeding behavior using peptides is a valuable approach in pest control; diapause hormone (DH) were previously tested for its potential function. To reduce potential problems related to DH degradation in the insect digestive tract, the TAT-PTD peptide were used to increase the absorption of the caDH fusion protein in the insect gut. A fusion peptide that included TAT, DH, and GFP (TAT-caDH-eGFP) passed across the midgut wall and entered the hemolymph, after which it was transported to other tissues. In the previous experiments, *H. armigera* larvae feeding on a diet containing synthetic TAT-caDH were affected in their development [15]. TAT fusion partner enhanced the stability and bioavailability of the fusion protein to increase the absorption of the caDH fusion protein in the insect gut, and TAT-PTD fusion expression provides a method for the delivery of agents by oral administration. Here, our report demonstrates that TAT-HezK I can be delivered to insects via a second method, by transgenic expression of the peptide in planta and intake through larval feeding. *H. armigera* larval survival rates decreased to approximately 47% when feeding on transgenic tobacco lines of TAT-HezK I 3 and TAT-HezK I 4 at d 6. In addition, the weights of the larvae were 57% and 64% less than untransformed tobacco controls.

Previous reports indicated that the expression of the foreign CpTI protein in transgenic tobacco (*N. tabacum* L.) confers high resistance to larvae of the cotton boll-worm *H. armigera* due to the increased accumulation level of foreign CpTI protein [26]. The strategy of CpTI can offset potential problems related to protein degradation in the digestive tract and can be widely applied to other related research fields in plant genetic engineering [16]. In our research, *CpTI* transgenic tobacco were used as the positive control, and the *TAT-HezK I-CpTI* fusion gene was designed to try to further enhance the stability and bioavailability of agents in oral administration. Using suitable antibodies to detect the protein levels of CpTI and TAT-HezK I-CpTI, the expression levels of the fusion protein in transgenic lines were normalized. Though three CpTI lines and three TAT-HezK I-CpTI lines had similar target protein levels, larvae of *H. armigera* feeding on transgenic lines expressing TAT-HezK I-CpTI achieved lower weight gains; TAT-HezK I-CpTI 3 transgenic line had statistically higher mortality rates and lower weight gains on larvae than transgenic lines expressing CpTI. The *TAT-HezK I-CpTI* fusion gene was a better insecticidal fusion gene than the *CpTI* single gene.

TAT-HezK I 3, TAT-HezK I 4, and TAT-HezK I-CpTI 3 were the most effective lines toxic to *H. armigera* larvae. *TAT-HezK I* can be used as a kind of new gene toxic to herbivores in transgenic plant breeding. Considering mRNA expression levels, transgenic line TAT-HezK I 3 and TAT-HezK I 4 had 9.7 and 12.1 times that of TAT-HezK I-CpTI 3 in the plant tissue. However, the protein expression levels of transgenic lines TAT-HezK I 3 and TAT-HezK I 4 were 3.3 and 3.2 times that of TAT-HezK I-CpTI 3 in the plant tissue. In our opinion, the presence of the CpTI gene might help the translation of the TAT-HezK I genes when *TAT-HezK I* and *CpTI* are used. For TAT-HezK I 3, TAT-HezK I 4, and TAT-HezK I-CpTI 3, the lines had similar toxic effects to the *H. armigera* larvae; the higher TAT-HezK I protein expression level in transgenic lines TAT-HezK I 3 and TAT-HezK I 4 was the key factor cause higher toxicity. As the lower component of the TAT-HezK I-CpTI protein in the TAT-HezK I-CpTI 3 line had similar toxicity to TAT-HezK I 3 and TAT-HezK I 4 lines, the fusion TAT-HezK I and CpTI enhanced the stability and bioavailability of agents in oral administration.

5. Conclusions

The results demonstrated that TAT-HezK I had high toxicity against insects via transgenic expression of the peptide in planta and from intake through larval feeding. The *TAT-HezK I-CpTI* fusion gene was a better insecticidal fusion gene than a *CpTI* single gene. In conclusion, the studies presented here have led to the identification of interesting tools for the development of selective, environmentally friendly arthropod pest control genes capable of disrupting insect kinin regulated processes. In order to comprehensively demonstrate that the technique has value for enhancing plant resistance in field crops, bioassays will need to be repeated on intact plants in the field, and biological safety evaluations will also need to be performed in the future.

Acknowledgments: We appreciate Daniel Doucet for revising an earlier draft of this manuscript. This work was supported by the Fundamental Research Funds for the Central Non-profit Research Institute of CAF (No. CAFRIFEEP201501) of China, Key Scientific Research Projects of Colleges and Universities in Henan Province (15A210005) and Backbone Teachers Research Projects of Colleges and Universities in Henan Province (No. 2015GGJS-055).

Author Contributions: Liangjian Qu, Zhou Zhou, and Yongan Zhang conceived and designed the experiments; Yongli Li, Zhou Zhou, and Liangjian Qu performed the experiments; Zhou Zhou, Yongli Li, and Chunyan Yuan analyzed the data; Liangjian Qu, Chunyan Yuan, and Yongan Zhang contributed reagents/materials/analysis tools; Zhou Zhou and Yongli Li wrote the paper.

References

1. Gäde, G. Regulation of intermediary metabolism and water balance of insects by neuropeptides. *Annu. Rev. Entomol.* **2004**, *49*, 93–113. [CrossRef] [PubMed]

2. Hayes, T.K.; Pannabecker, T.L.; Hinckley, D.J.; Holman, G.M.; Nachman, R.J.; Petzel, D.H.; Beyenbach, K.W. Leucokinins, a new family of ion transport stimulators and inhibitors in insect malpighian tubules. *Life Sci.* **1989**, *44*, 1259–1266. [CrossRef]

3. Coast, G.M.; Holman, G.M.; Nachman, R.J. The diuretic activity of a series of cephalomyotropic neuropeptides, the achetakinins, on isolated Malpighian tubules of the house cricket, *Acheta domesticus. J. Insect Physiol.* **1990**, *36*, 481–488. [CrossRef]

4. Nachman, R.J.; Strey, A.; Isaac, E.; Pryor, N.; Lopez, J.D.; Deng, J.G.; Coast, G.M. Enhanced in vivo activity of peptidase-resistant analogs of the insect kinin neuropeptide family. *Peptides* **2002**, *23*, 735–745. [CrossRef]

5. Seinsche, A.; Dyker, H.; Losel, P.; Backhaus, D.; Scherkenbeck, J. Effect of helicokinins and ACE inhibitors on water balance and development of *Heliothis. virescens* larvae. *J. Insect Physiol.* **2000**, *46*, 1423–1431. [CrossRef]

6. Smagghe, G.; Mahdian, K.; Zubrzak, P.; Nachman, R.J. Antifeedant activity and high mortality in the pea aphid *Acyrthosiphon. pisum* (Hemiptera: Aphidae) induced by biostable insect kinin analogs. *Peptides* **2010**, *31*, 498–505. [CrossRef] [PubMed]

7. Coast, G.M. Fluid secretion by single isolated Malpighian tubules of the house cricket, *Acheta domesticus* and their response to diuretic hormone. *Physiol. Entomol.* **1988**, *13*, 381–391. [CrossRef]

8. Nachman, R.J.; Isaac, R.E.; Coast, G.M.; Holman, G.M. Aib-containing analogues of the insect kinin neuropeptide family demonstrate resistance to an insect angiotensin-converting enzyme and potent diuretic activity. *Peptides* **1997**, *18*, 53–57. [CrossRef]

9. Coast, G.M. Diuresis in the housefly (*Musca domestica*) and its control by neuropeptides. *Peptides* **2001**, *22*, 153–160. [CrossRef]

10. Walker, L.R.; Ryu, J.S.; Perkins, E.; McNally, L.R.; Raucher, D. Fusion of cell-penetrating peptides to thermally responsive biopolymer improves tumor accumulation of p21 peptide in a mouse model of pancreatic cancer. *Drug Des. Dev. Ther.* **2014**, *8*, 1649–1658. [CrossRef]

11. Dietz, G.P.H.; Bohr, M. Delivery of bioactive molecules into the cell: The Trojan horse approach. *Mol. Cell. Neurosci.* **2004**, *27*, 85–131. [CrossRef] [PubMed]

12. Simon, M.J.; Kang, W.H.; Gao, S.; Banta, S.; Morrison, B. TAT is not capable of transcellular delivery across an intact endothelial monolayer in vitro. *Ann. Biomed. Eng.* **2011**, *39*, 394–401. [CrossRef] [PubMed]

13. Yu, J.Y.; Meng, X.L.; Xu, J.P.; Chen, D.D.; Meng, M.X.; Ni, Y.W. Fusion of TAT-PTD to the C-terminus of catfish growth hormone enhances its cell uptakes and growth-promoting effects. *Aquaculture* **2013**, *392*, 84–93. [CrossRef]

14. Schwarze, S.R.; Ho, A.; Vocero-Akbani, A.; Dowdy, S.F. In vivo protein transduction: Delivery of a biologically active protein into the mouse. *Science* **1999**, *285*, 1569–1572. [CrossRef] [PubMed]

15. Zhou, Z.; Li, Y.L.; Yuan, C.Y.; Zhang, Y.A.; Qu, L.J. Oral administration of TAT-PTD–diapause hormone fusion protein interferes with *Helicoverpa. armigera* (Lepidoptera: Noctuidae) development. *J. Insect Sci.* **2015**. [CrossRef] [PubMed]

16. Deng, C.Y.; Song, G.S.; Xu, J.W.; Zhu, Z. Increasing accumulation level of foreign protein in transgenic plants through protein targeting. *Acta Bot. Sin.* **2003**, *45*, 1084–1088.

17. Vivès, E.; Brodin, P.; Lebleu, B.A. Truncated HIV-1 Tat protein basic domain rapidly translocates through the plasma membrane and accumulates in the nucleus. *J. Biol. Chem.* **1997**, *272*, 16010–16017. [CrossRef] [PubMed]

18. Zhao, H.L.; Yao, X.Q.; Xue, C.; Wang, Y.; Xiong, X.H.; Liu, Z.M. Increasing the homogeneity, stability and activity of human serum albumin and interferon-alpha2b fusion protein by linker engineering. *Protein Expr. Purif.* **2008**, *61*, 73–77. [CrossRef] [PubMed]

19. Murashige, T.; Skoog, F. A revised medium for rapid growth and bio assays with tobacco tissue cultures. *Physiol. Plant.* **1962**, *15*, 473–497. [CrossRef]

20. Zhou, Z.; Wang, M.J.; Hu, J.J.; Lu, M.Z.; Wang, J.H. Improve freezing tolerance in transgenic poplar by over expressing a ω-3 fatty acid desaturase gene. *Mol. Breed.* **2010**, *25*, 571–579. [CrossRef]

21. Zhou, Z.; Li, Y.L.; Yuan, C.Y.; Qu, L.J. Prokaryotic expression of TAT-Cloan-DH-EGFP fusion protein and its transduction in larvae of the gypsy moth, *Lymantria. Dispar* (Lepidoptera: Lymantriidae) by oral administration. *Acta Entomol. Sin.* **2014**, *57*, 1361–1367.

22. Mao, Y.B.; Tao, X.Y.; Xue, X.Y.; Wang, L.J.; Chen, X.Y. Cotton plants expressing *CYP6AE14* double-stranded RNA show enhanced resistance to bollworms. *Transgenic Res.* **2011**, *20*, 665–673. [CrossRef] [PubMed]

23. Zhu, J.Q.; Liu, S.; Ma, Y.; Zhang, J.Q.; Qi, H.S.; Wei, Z.J.; Yao, Q.; Zhang, W.Q.; Li, S. Improvement of pest resistance in transgenic tobaccos expressing dsRNA of an insect-specific gene *EcR*. *PLoS ONE* **2012**, *7*, e38572.

24. Cornell, M.J.; Williams, T.A.; Lamango, N.S.; Coates, D.; Corvol, P.; Soubrier, F.; Hoheisel, J.; Lehrach, H.; Isaac, R.E. Cloning and expression of an evolutionary conserved single-domain angiotensin converting enzyme from *Drosophila melanogaster*. *J. Biol. Chem.* **1995**, *270*, 13613–13619. [CrossRef] [PubMed]

25. Lamango, N.S.; Sajid, M.; Isaac, R.E. The endopeptidase activity and the activation by Cl⁻ of angiotensin-converting enzyme is evolutionarily conserved: Purification and properties of an antiotensin-converting enzyme from the housefly, *Musca Domestica*. *Biochem. J.* **1996**, *314*, 639–646. [CrossRef]

26. Feng, P.Z.; Wang, Y.; Cao, Y.; Liu, B.L.; Zhu, Z.; Li, X.H. On the resistance of CpTI transgenic tobacco plants to cotton boll worm. *Acta Phytophylacica Sin.* **1997**, *24*, 331–335.

13

AAC as a Potential Target Gene to Control *Verticillium dahliae*

Xiaofeng Su, Latifur Rehman, Huiming Guo, Xiaokang Li, Rui Zhang * and Hongmei Cheng *

Biotechnology Research Institute, Chinese Academy of Agricultural Sciences, Beijing 100081, China; suxiaofeng@caas.cn (X.S.); latif_ibge@yahoo.com (L.R.); guohuiming@caas.cn (H.G.); lixiaokang2016@163.com (X.L.)
* Correspondence: zhangrui@caas.cn (R.Z.); chenghongmei@caas.cn (H.C.)

Academic Editors: Wenyi Gu and Paolo Cinelli

Abstract: *Verticillium dahliae* invades the roots of host plants and causes vascular wilt, which seriously diminishes the yield of cotton and other important crops. The protein AAC (ADP, ATP carrier) is responsible for transferring ATP from the mitochondria into the cytoplasm. When *V. dahliae* protoplasts were transformed with short interfering RNAs (siRNAs) targeting the *VdAAC* gene, fungal growth and sporulation were significantly inhibited. To further confirm a role for *VdAAC* in fungal development, we generated knockout mutants (Δ*VdACC*). Compared with wild-type *V. dahliae* (Vd wt), Δ*VdAAC* was impaired in germination and virulence; these impairments were rescued in the complementary strains (Δ*VdAAC*-C). Moreover, when an RNAi construct of *VdAAC* under the control of the 35S promoter was used to transform *Nicotiana benthamiana*, the expression of *VdAAC* was downregulated in the transgenic seedlings, and they had elevated resistance against *V. dahliae*. The results of this study suggest that *VdAAC* contributes to fungal development, virulence and is a promising candidate gene to control *V. dahliae*. In addition, RNAi is a highly efficient way to silence fungal genes and provides a novel strategy to improve disease resistance in plants.

Keywords: *Verticillium dahliae*; *VdAAC*; RNAi; growth; virulence

1. Introduction

Verticillium dahliae is one of the most destructive soil-borne fungi, infecting many important economic crops, fruit trees and ornamental flowers [1,2]. This fungus can cause typical disease symptoms, including stunted growth, necrosis, wilt and defoliation, which severely decrease the yield and quality of crops [3]. Each year, *Verticillium* wilt is reported to cause extensive economic losses to the crop industry [4]. The fungus can survive in soil for many years and infect the roots of its hosts. Its mycelium then abundantly colonizes the vascular bundle to block the transportation of nutrients [5]. Once the fungus is established in the host, *Verticillium* wilt is an intractable disease because of the intricate pathogenic mechanism of *V. dahliae* [6]. Currently, no fungicide is available to cure the disease caused by this fungus [7]. Previous studies on *V. dahliae* have thus focused on identifying genes that are crucial for fungal development and virulence [8–10], inestimable knowledge for crop breeding programs.

RNA interference (RNAi) is an effective tool to investigate gene function and elevate plant resistance against a fungus [11–13]. In *V. longisporum*, the expression of *Vlaro2* was reduced via RNAi, resulting in a bradytrophic mutant [14]. In vitro cultures of *Fusarium graminearum*, the introduction of double-stranded (ds)RNA that targeted cytochrome P450 lanosterol C14α-demethylase-encoding genes inhibited fungal growth. Similarly, expressing the same region of dsRNA into susceptible *Arabidopsis thaliana* and *Hordeum vulgare* conferred high resistance to fungal infection [15]. Transgenic banana plants with siRNAs targeted against velvet and *Fusarium* transcription factor 1 were protected against *Fusarium oxysporum* f. sp. *cubense* (Foc) [16], as were transgenic cotton plants against *V. dahliae* when the fungal *VdH1* gene was silenced [17].

Genes for essential cellular components may prove to be likely effective targets. For example, mitochondrial carriers are a series of proteins that transport nucleotides, amino acids, fatty acids, and so on across the inner mitochondrial membrane of eukaryotes [18]. Of these carriers, the highly conserved AAC is the most abundant protein [19,20]. AAC is essential for maintaining fluxes in energy and mediating the exchange of ADP and ATP between the mitochondria and cytoplasm [21]. AAC consists of six transmembrane helices embedded in the inner mitochondrial membrane with its N- and C-terminals exposed to the cytosolic side [22]. The C-terminal structure of yeast AAC is predicted to be involved in regulating the accessibility of the transmembrane core to water [23]. Silencing of the *AAC* gene of *Blumeria graminis* by biolistically bombarding barley cells with RNAi constructs led to the formation of fewer haustoria in barley cells [24]. In *Saccharomyces cerevisiae*, AAC might transmit a signal and facilitate permeabilization of the outer mitochondrial membrane to accelerate mitochondrial degradation followed by cytochrome C release during acetic acid-induced apoptosis [25]. Thylakoid ADP/ATP carrier (TAAC), apart from regulating ADP and ATP balance, has an additional role in transporting 3′-Phosphoadenosine 5′-phosphosulfate as the high-energy sulfate donor through the plastid envelope in Arabidopsis [26]. AAC also increases mitochondrial proton conductance for adapting to cold water stress in king penguins [27]. A decrease in the expression of *Trypanosoma brucei* AAC resulted in a reduced level of cytosolic ATP and mitochondrial oxygen consumption, severe growth defects and elevation in the amount of reactive oxygen species [28].

Although the functions of the *AAC* gene in development, resistance and signal transduction pathways have been explored in other species, its role in the development and virulence of *V. dahliae* has not yet been reported. In the present study, we used siRNA-induced silencing of the *VdAAC* gene in *V. dahliae* to establish the relationship between *VdAAC* and fungal development. Deletion of *VdAAC* resulted in reduced colony growth and sporulation. Virulence was significantly decreased in the Δ*VdAAC* mutants compared with the wild type (Vd wt) and complemented strains (Δ*VdAAC*-C). Confocal microscopic observations revealed that conidial germination of Δ*VdAAC* was significantly impaired. Moreover, transgenic *N. benthamiana* expressing dsRNA against *VdAAC* showed strong resistance against *V. dahliae*. Our results indicate that *VdAAC* contributes to fungal germination, development and sporulation, which are requisite for the fungus to invade the plants and induce full virulence in the host. For potential exploitation of this gene to protect crops against *V. dahliae*, its biological function needs to be further elaborated.

2. Materials and Methods

2.1. Fungal Strains, Plants and Inoculation with V. dahliae

Strain V991, a highly toxic and defoliating wild-type pathogenic strain of *V. dahliae*, was kindly gifted by Prof. Guiliang Jian of the Institute of Plant Protection, Chinese Academy of Agricultural Sciences (CAAS). *V. dahliae* strain Vd-GFP that expresses *GFP*, is from our laboratory culture collection [29]. Single-conidium cultures of all *V. dahliae* strains were grown in complete medium broth (CM) at 25 °C. After 1 week, the conidia were harvested for inoculation.

Two-week-old seedlings of *N. benthamiana* were transplanted from Murashige-Skoog (MS) agar into disinfested soil and incubated in the greenhouse at 23 ± 2 °C, $75\% \pm 5\%$ relative humidity, and a photoperiod of 16 h day/8 h night. After 1 month, seedlings with 6–8 leaves were inoculated by dipping the roots in a suspension of 10^6 conidia·mL^{-1} for 2 min.

2.2. Disease Index

Disease severity was evaluated using a five-grade scale based on a previous study with modifications [30]: grade 0, no wilt; grade 1, less than two leaves wilting; grade 2, three to five leaves wilting; grade 3, more than five leaves wilting or chlorotic; and grade 4, plant death or near death.

Each respective experiment comprised 5 seedlings and was independently repeated three times for each assessment. Symptoms were recorded and the disease index (DI) calculated at 10 days post inoculation (dpi), 11 dpi and 12 dpi using the formula: DI = [Σ (number × grade)/(5 × 4)] × 100 [30,31].

2.3. Bioinformatics Analysis

The whole sequence of *VdAAC* was obtained from the *Verticillium* genomic database (www.broadinstitute.org). Amino acid sequences homologous to AAC in other species, identified with a Blastp search of the Protein Data Base (PDB), were used to construct a phylogenetic tree in MEGA software (version 6.06) (http://www.megasoftware.net/).

2.4. siRNA Design and Transformation of V. dahliae Protoplasts

The siRNAs targeting different regions of the *VdAAC* gene (siRNA-1, siRNA-2, siRNA-3 and siRNA-4) were designed using BLOCK-iT™ (Invitrogen, Carlsbad, CA, USA) RNAi Designer and synthesized by Oligobio, Beijing, China. The siRNAs sequences are given in Table 1, and the locations of these siRNAs in *VdAAC* are displayed in Figure S1A.

Table 1. siRNA sequences designed against VdAAC.

Name	Sense Sequence	Antisense Sequence
Control	UUCUCCGAACGUGUCACGUTT	ACGUGACACGUUCGGAGAATT
siRNA-1	UCAAGCUCCUCAUCCAGAATT	UUCUGGAUGAGGAGCUUGATT
siRNA-2	GCAACACUGCCAACGUCAUTT	AUGACGUUGGCAGUGUUGCTT
siRNA-3	GCUUUCCGUGACAAGUUCATT	UGAACUUGUCACGGAAAGCTT
siRNA-4	GCAUGUACGACUCCAUCAATT	UUGAUGGAGUCGUACAUGCTT

V. dahliae protoplasts isolated from fresh mycelia were transformed with the siRNAs as described in our previous study [29]. After 72 h in TB3 broth, the mycelia were collected to extract RNA with an RNA Extraction Kit (YPHBio, Tianjin, China). First strand cDNA was synthesized using a Reverse Transcription Kit (TransGen, Beijing, China) based on the manufacturer's instructions. qRT-PCR was carried out with a 7500 Real Time PCR System (ABI, Foster City, CA, USA) [29]. *Vdactin* was used as a housekeeping gene [32]. The relative expression level of *VdAAC* was analyzed using the $2^{-\Delta\Delta Ct}$ method. The standard curve met the experimental requirements ($R^2 > 0.99$, E > 95%) [33]. Transformed protoplasts were also cultured for 2 weeks on the center of PDA (Potato Dextrosa Agar: potato infusion 200 g, dextrose 20 g and agar 20 g in 1 L H_2O) plates to measure colony diameter and count the conidia produced to assess the effect of silencing on fungal growth and sporulation.

2.5. Plasmid Construction and Fungal Transformation

For creating a knockout-infused gene fragment, flanking regions (1 kbp upstream and downstream) of the *VdAAC* gene and a hygromycin resistance (HPT) expression cassette were amplified and fused via the overlapping sequences.

For *GFP* (Green Fluorescence Protein) disruption mutants (Δ*VdAAC-GFP*), the neomycin resistance (Neo^R) cassette containing XbaI and BstEII restriction sites was cloned into the pCAMBIA1302 vector. After that, the *GFP* expression cassette was introduced into the plasmid via XbaI and KpnI restriction sites to generate pCAMBIA1302::Neo::GFP. Meanwhile, the *VdAAC* ORF was substituted for the *GFP* open reading frame (ORF) of the recombinant plasmid as pCAMBIA1302::Neo::VdAAC for complementary strains.

The respective constructs (knockout-infused fragment, pCAMBIA1302::Neo::GFP, pCAMBIA1302::Neo::VdAAC) were used to transform *V. dahliae* protoplasts [29]. Transformants were selected based on antibiotic resistance and confirmed by RT-PCR. The primers are listed in Table 2.

Table 2. Primers and their sequences used in this study.

Primers	Sequence
qRT-AAC	TTGCCGAGTGCTTCAAGCGTAC GGCGTAGTCGAGGGAGTAGACG
qRT-Vdactin	GGCTTCCTCAAGGTCGGCTATG GCTGCATGTCATCCCACTTCTTC
qRT-VdITS	CCGCCGGTCCATCAGTCTCTCTGTTTATAC CGCCTGCGGGACTCCGATGCGAGCTGTAAC
qRT-Nbactin	GGACCTTTATGGAAACATTGTGCTCAGT CCAAGATAGAACCTCCAATCCAGACAC
qRT-VA	GGGTATTCAGACCCTATTGGACG CGAACTTCTTGTACTCAGCCTCC
qRT-ATP6	CTAGACCAATTTGAAATAAGA AAAGATTCTTGGCTAATAGAT
qRT-VdAC	TCTCCATCGTCTTCACCGACATCA TCTGCACGGCGAAACACCACA
qRT-VdATP-PRT	CGACGCCAACGTGCGGTCCTACAA GCCCGAGAAGCTCGTGCCAAT
HPT expression cassette	TTGAAGGAGCATTTTTGGGC TTATCTTTGCGAACCCAGGG
ΔAAC	CTTGGTGAAGGAGAGCGTTGAAAGT GCCCAAAAATGCTCCTTCAATGACAAGTTCAAGGCCATGTTCGGC CCCTGGGTTCGCAAAGATAACTCCGTTGCTGGTATCGTTGTCTAC GGTTCCTCGTCGCTGTCAATGACC
Neo expression cassette	aat*TCTAGA*GTTTGCGGGCTGTCTTGACG ata*GGTCACC*TACCTGTGCATTCTGGGTAA
GFP expression cassette	ggc*TCTAGA*CTTTCGACACTGAAATACGTCG ata*GGTACC*GCATCAGAGCAGATTGTACTGAGAG
ΔAAC-C	aaa*AGTACT*ATGTCCGTCGAGAAGCAG aaa*CTGCAG*TTATTTGAAGGCCTTGCC
Trans-AAC	GGGGACAAGTTTGTACAAAAAAGCAGGCTGTGCTTCAAGCGTAC GGGGACCACTTTGTACAAGAAAGCTGGGTCCCTTGAAGAGAGAC
Det-trans	CGTCATCCGTTACTTCCCTACCCA AGACCGGCAATACCGTCAGAGGC
Det-GFP	CGACGTAAACGGCCACAAGTT TCTTTGCTCAGGGCGGACTGG
Det-AAC	GCGCCAGTTCAACGGTCTTGTCG TCACCAGAGGTCATCATCATGCGAC
Det-Neo	GTTGTCACTGAAGCGGGAAGGG GCGATACCGTAAAGCACGAGGAA
Det-HPT	TTCGACAGCGTCTCCGACCTGA AGATGTTGGCGACCTCGTATTGGG

Restriction enzyme sites are indicated using bold and italic fonts.

2.6. Stress Treatments of V. dahliae Strains

For characterizing the development and morphology of the wild type and mutant strains of *V. dahliae*, 10 μL samples of 1×10^6 conidia·mL^{-1} of the respective strains were cultured on Czapek-Dox (3 g NaNO$_3$, 1 g K$_2$HPO$_4$, 0.5 g MgSO$_4$.7H$_2$O, 0.5 g KCl, 0.01g FeSO$_4$ and 30 g Sucrose in 1 L H$_2$O) agar with either 0.5 M NaCl or sorbitol. Similarly, plates containing conidia of the respective strains were exposed to UV light for 10 s in a Gel doc system (Syngene, Cambridge, UK) to assess the impact of UV-stress on the survival of these conidia [34]. The plates were then incubated at 25 °C and the colony diameter on each plate was measured after 2 weeks. For estimating conidia production, 3 mL of sterilized water was added to each plate, which was then gently shaken to release the conidia [35]. The conidia were then counted using a light microscope (BX52, OLYMPUS, Tokyo, Japan).

2.7. Confocal Microscopy

Confocal microscopy was used to facilitate the observation of the infection process of both wild type and mutant strains. *N. benthamiana* seedlings were inoculated with Vd-GFP and Δ*VdAAC*-GFP strains respectively. At 7 dpi, roots of the infected plants were collected, washed with water for 3 times and then observed under confocal microscope (LSM 700, Carl Zeiss, Jena, Germany) [36].

2.8. Plasmid Construction and Plant Transformation

A pair of primers was designed based on the *VdAAC* ORF [37] (Table 2). The targeted fragment (648 bp) in *VdAAC*, shown in Figure S2A, was amplified with partial BP adaptors. The whole sequence was cloned using BP site primers and inserted into pDONR207 by a BP recombinant reaction (Invitrogen, Carlsbad, CA, USA). Then, the targeted fragment was cloned into the pK7GWIWG2(I) vector using an LR recombinant reaction (Invitrogen, Carlsbad, CA, USA). The recombinant plasmid was named pK7GWIWG2(I)-VdAAC (Figure S2B), confirmed by sequencing, and then used to transform *Agrobacterium tumefaciens* strain LBA4404 using electroporation [38].

Sterile leaves of *N. benthamiana* were immersed in *A. tumefaciens* strain LBA4404 containing the recombinant plasmid and transferred to MS agar. After 3 days, the leaves were cultured on MS agar containing 100 mg·L^{-1} kanamycin [39]. Seedlings were confirmed by PCR to be transgenic (Figure S2C). Two transgenic lines (Trans-1 and Trans-2), expressing dsRNA against *VdAAC* were selected for further analysis. The primer sequences for detection are listed in Table 2.

2.9. Analysis of Fungal Biomass

Colonization of *V. dahliae* in seedlings was quantified at 12 dpi by isolating DNA from the roots, stems (0–3 cm above the soil line) and leaves, respectively, using the Plant Genomic DNA Kit (TIANGEN, Beijing, China). Fungal biomass was quantified via qRT-PCR by amplifying ITS1 and ITS2 of rDNA (Z29511) of *V. dahliae* [35]. The *N. benthamiana* housekeeping gene (*Nbactin*) was used as an internal control [40]. The primers are listed in Table 2.

2.10. qRT-PCR Analysis of the Expression Level of Targeted Genes

The silencing effect of *VdAAC* in the infected seedlings was assessed by extracting RNA from roots at 12 dpi for *qRT-PCR* as described in Section 2.4. The housekeeping gene *Vdactin* was used as a control [32]. The relative expression of targeted gene was analyzed using the $2^{-\Delta\Delta Ct}$ method. The standard curve met experimental requirements ($R^2 > 0.99$, E > 95%) [33]. The primers are listed in Table 2.

2.11. Statistical Analysis

All experiments were independently repeated thrice, and data was analyzed for significant differences among the groups using Duncan's multiple range test ($p < 0.05$) and SPSS Statistics 17.0 software (SPSS, Chicago, IL, USA).

3. Results

3.1. Bioinformatics Analysis of VdAAC

The ORF of *VdAAC* (VDAG_07535.1) contains 930 bp, which encodes a protein with 310 amino acids (GenBank NO.: XP_009654735.1). The neighbor-joining phylogenetic tree for the *VdAAC* sequences from *V. dahliae* and other fungi constructed using MEGA (bootstraps: 1000) demonstrated that the AAC sequences are relatively conserved among these fungal species (Figure 1).

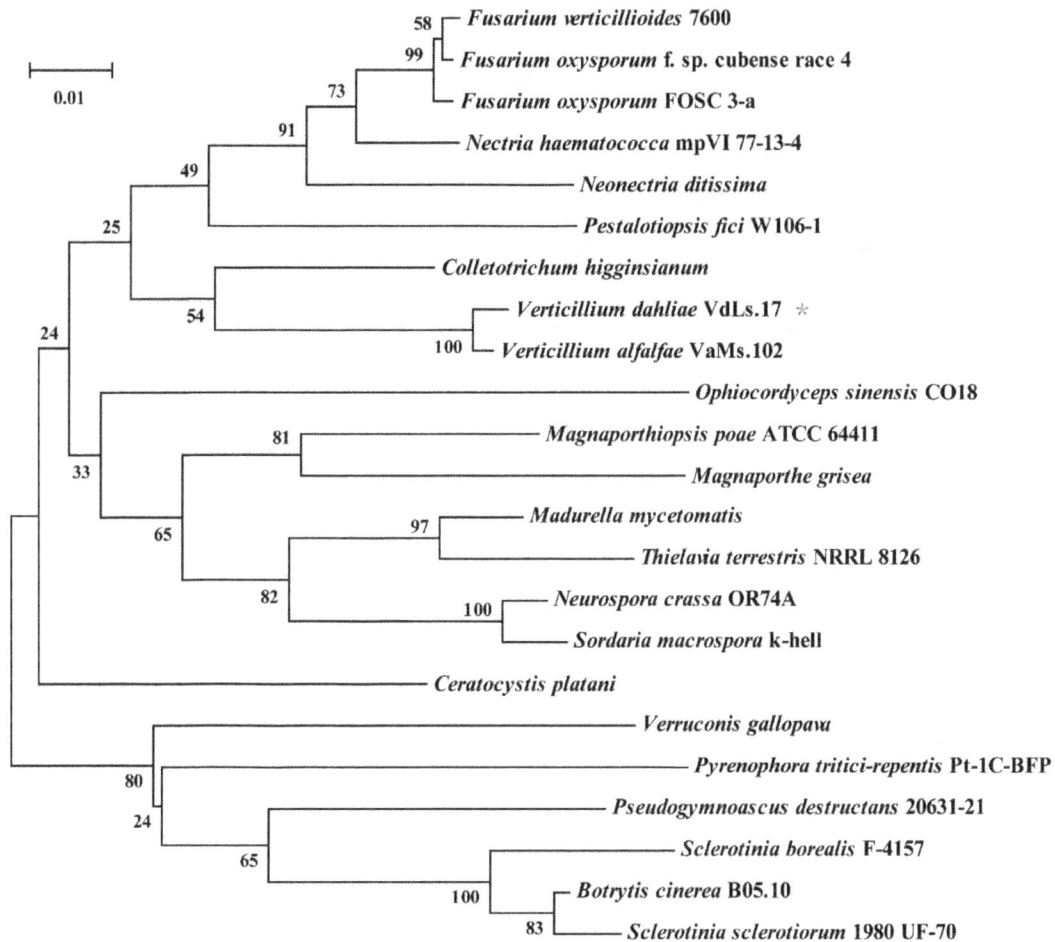

Figure 1. Phylogenetic analysis of AAC amino acid sequences from different fungal species. The phylogenetic tree was constructed by MEGA software (version 6.06; bootstraps: 1000). Fungal species and protein accession numbers: *Verticillium dahliae* VdLs.17 (XP_009654735.1); *Nectria haematococca* mpVI 77-13-4 (XP_003051617.1); *Fusarium oxysporum* FOSC 3-a (EWZ02370.1); *Verticillium alfalfae* VaMs.102 (XP_003004480.1); *Fusarium verticillioides* 7600 (EWG42987.1); *Pestalotiopsis fici* W106-1 (XP_007835574.1); *Fusarium oxysporum* f. sp. *cubense* race 4 (EMT68221.1); *Magnaporthiopsis poae* ATCC 64411 (KLU91586.1); *Neonectria ditissima* (KPM44251.1); *Madurella mycetomatis* (KOP45712.1); *Colletotrichum higginsianum* (CCF32866.1); *Ceratocystis platani* (KKF94862.1); *Pseudogymnoascus destructans* 20631-21 (XP_012739498.1); *Botrytis cinerea* B05.10 (XP_001559435.1); *Sclerotinia sclerotiorum* 1980 UF-70 (XP_001598713.1); *Verruconis gallopava* (KIW06207.1); *Sclerotinia borealis* F-4157 (ESZ96107.1); *Neurospora crassa* OR74A (XP_011393638.1); *Sordaria macrospora* k-hell (XP_003351160.1); *Pyrenophora tritici-repentis* Pt-1C-BFP (XP_001934086.1); *Ophiocordyceps sinensis* CO18 (EQK99145.1); *Magnaporthe grisea* (AAX07662.1); *Thielavia terrestris* NRRL 8126 (XP_003658188.1). * represents the query sequence.

3.2. Silencing of VdAAC Effectively Inhibited Fungal Growth and Sporulation

In our previous study [29], we showed that the siRNAs can enter *V. dahliae* protoplasts to silence the targeted genes. Thus, siRNAs designed against *VdAAC* were used to transform the protoplasts. After 2 weeks, the mean colony diameter of the siRNA-1 group (14.2 mm) and siRNA-3 (16 mm) were distinctly smaller than that of the siRNA-control group (24.5 mm) (Figure 2A and Figure S1B). To further confirm whether the silencing of *VdAAC* gene led to the reduced colony growth, qRT-PCR was carried out to determine the relative expression level of *VdAAC* in all the groups. The data was consistent with the colony diameter assessment (Figure 2B). Similarly, the siRNA-1 and siRNA-3 groups produced fewer conidia than the other groups did (Figure 2C). Taken together, these results demonstrate that inhibition of *VdAAC* expression impairs the fungal growth and sporulation.

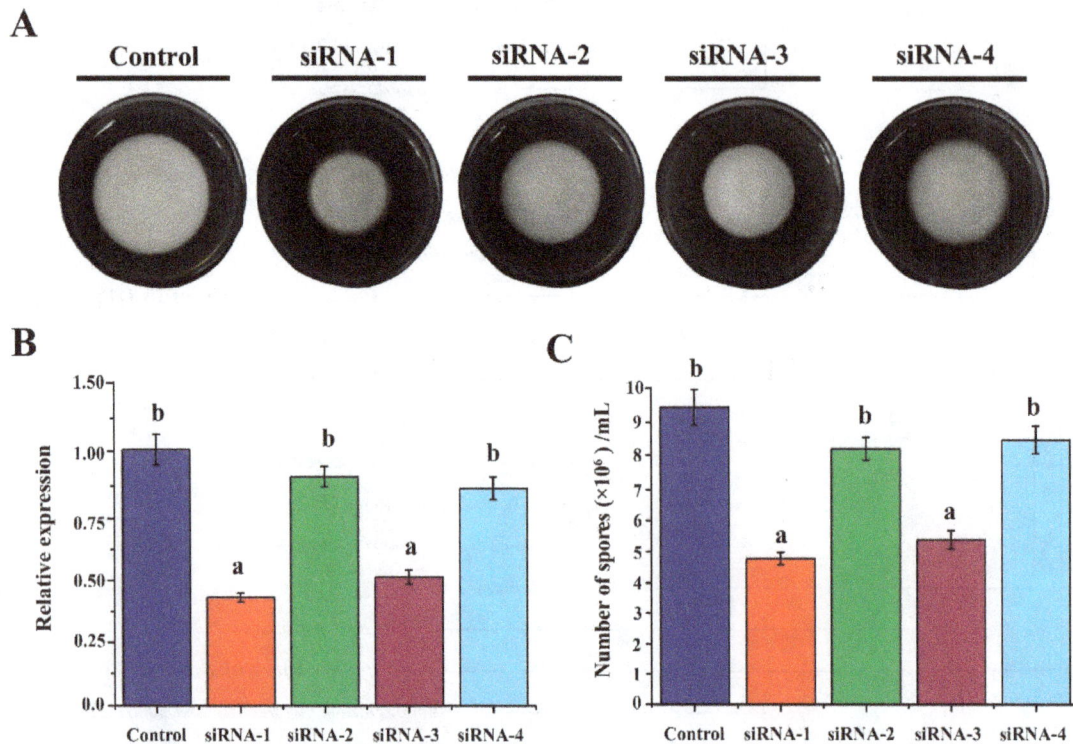

Figure 2. Assay for siRNA inhibition of the *VdAAC* gene. *V. dahliae* protoplasts were transformed with 10 μM siRNA-1, siRNA-2, siRNA-3, siRNA-4 or control, respectively. After regenerating for 18 h in TB3 broth, the protoplasts were cultured in the center of a PDA plate. (**A**) Colony morphology on PDA after 2 weeks; (**B**) Relative expression levels of *VdAAC* in different RNAi-treated groups. RNA was extracted from mycelia 72 h after transformation. First strand cDNA was synthesized, and qRT-PCR was carried out; (**C**) Number of conidia produced by the control and siRNA groups after 2 weeks. Error bars represent standard deviation (SD) calculated from means for three independent replicates. Significant differences ($p < 0.05$) among means for the different incubation times in Duncan's multiple range test are indicated with different letters.

3.3. Generation of the VdAAC Mutant

To further explore the function of *VdAAC*, we used the knockout-infused fragment to transform Vd wt protoplasts to generate the *VdAAC* deletion mutants (Figure 3A). pCAMBIA1302::Neo::GFP was used to transform the gene deletion strains to facilitate confocal microscopy while pCAMPIA1302::Neo::VdAAC for complementation assays (Figure 3B,C).

Subsequently, transformants of Δ*VdAAC* (*VdAAC* deletion mutant), Δ*VdAAC-C* (*VdAAC* complementation mutant, obtained from transforming pCAMPIA1302::Neo::VdAAC into Δ*VdAAC*) and Δ*VdAAC-GFP* (VdAAC mutant transformed with GFP plasmid) strains were selected randomly and anlayzed by PCR (Figure 3D). Two gene deletion strains Δ*VdAAC-1* and Δ*VdAAC-2* and two complementary strains Δ*VdAAC-C-1* (derived from the transformation of Δ*VdAAC-1*) and Δ*VdAAC-C-2* (derived from the transformation of Δ*VdAAC-2*) were selected for further work. As expected, *VdAAC* expression was only detected in Vd wt, Δ*VdAAC-C-1* and Δ*VdAAC-C-2*, and not in Δ*VdAAC-1* and Δ*VdAAC-2*. Moreover, *GFP* expression was detected in Δ*VdAAC-GFP*. The transformants were then further analyzed for the role of *VdAAC* in the development and virulence of *V. dahliae*.

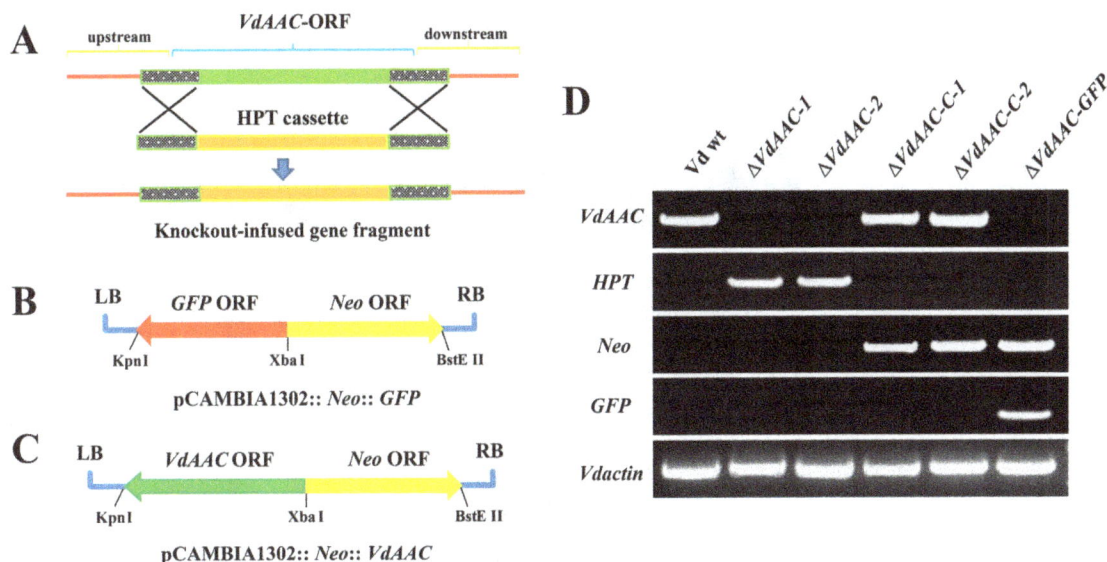

Figure 3. Disruption of the *VdAAC* gene and confirmation of *V. dahliae* mutants. (**A**) Construction of the knockout-infused fragment for gene disruption. The fragment was obtained by fusing about 1 kb upstream and downstream of the *VdAAC* gene and hygromycin resistance (HPT) cassette; (**B**) GFP expression cassette (GFP) and neomycin resistance (Neo) cassette were introduced into pCAMBIA1302 to generate pCAMBIA1302::Neo::GFP; (**C**) GFP expression cassette (GFP) was repalced by VdAAC expression cassette (VdAAC) to produce pCAMBIA1302::Neo::VdAAC; (**D**) Confirmation of transformants. RNA was isolated from mycelia of mutants cultured in CM broth. The first strand cDNA was synthesized, and RT-PCR was carried out to confirm the transformants. *Vdactin* gene was used as a housekeeping gene.

3.4. Stress Response Assay

The function of *VdAAC* in stress responses was analyzed by exposing conidia of Vd wt, $\Delta VdAAC$, and $\Delta VdAAC$-C strains to UV light, high NaCl or sorbitol. The phenotype of all the strains in the absence of stress was much better (Figure 4A). Exposure to each stress resulted in no significant reduction in the colony diameters and conidial number of $\Delta VdAAC$ when compared with the effect of the stress on Vd wt and $\Delta VdAAC$-C (Figure 4A,B). Although the effect of NaCl and sorbitol stresses was significant on the growth of all the strains, however the ration of growth reduction for gene deletion mutants with wild type and complementary strains was similar to no stress conditions (Figure 4C). In brief, *VdAAC* might not have a significant contribution in stress tolerance.

3.5. Expression of Genes Involved in Energy Metabolism in $\Delta VdAAC$

To understand the regulation of target genes relevant to energy metabolism, we analyzed the expression of vacuolar ATPase (VDAG_05626.1, *VdVA*), a gene that is involved in generating electrochemical potential across the vacuolar membrane [41–43], ATP synthase F0 subunit 6 (VDAG_17005.1, *VdATP6*), responsible for phosphorylating ADP [44,45], adenylate cyclase (VDAG_04508.1, *VdAC*), required for the conversion of ATP to cAMP [46,47] and ATP phosphoribosyltransferase (VDAG_08760.1, *VdATP-PRT*), necessary for the formation of phosphoribosyl-ATP and inorganic pyrophosphate [48–51] in Vd wt, $\Delta VdAAC$ and $\Delta VdAAC$-C. The expression of these genes in $\Delta VdAAC$ strains increased >2-fold as compared with Vd wt and $\Delta VdAAC$-C (Figure 5). Collectively, these results suggest that the upregulation of these genes might be due to the disturbance in ADP/ATP levels.

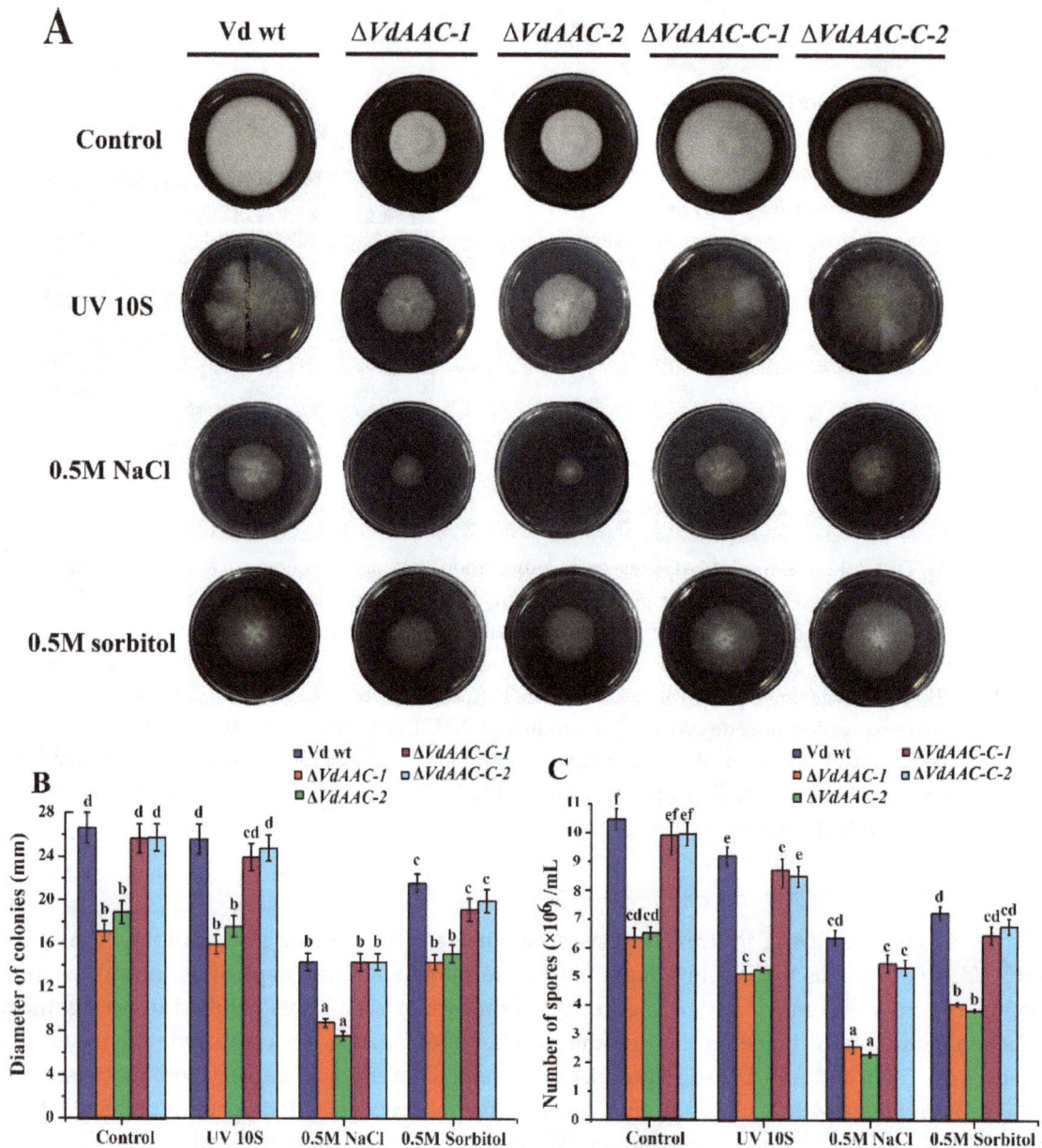

Figure 4. Colony morphology, diameter and conidia number of Δ*VdAAC*, Δ*VdAAC-C* and wild-type *V. dahliae* (Vd wt) strains exposed to stresses and without stress. Conidia from the respective strains were exposed to UV light and cultured in the center of Czapek-Dox agar plates. Conidia without UV light exposure were cultured on media supplemented with either NaCl or sorbitol. After 2 weeks, fungal traits were assessed: (**A**) colony morphology; (**B**) colony diameter; and (**C**) conidia number of mutants and Vd wt strain. Different letters (a–f) above the bars represent significant differences among the treatment groups ($p < 0.05$) as determined by the Duncan's multiple range test.

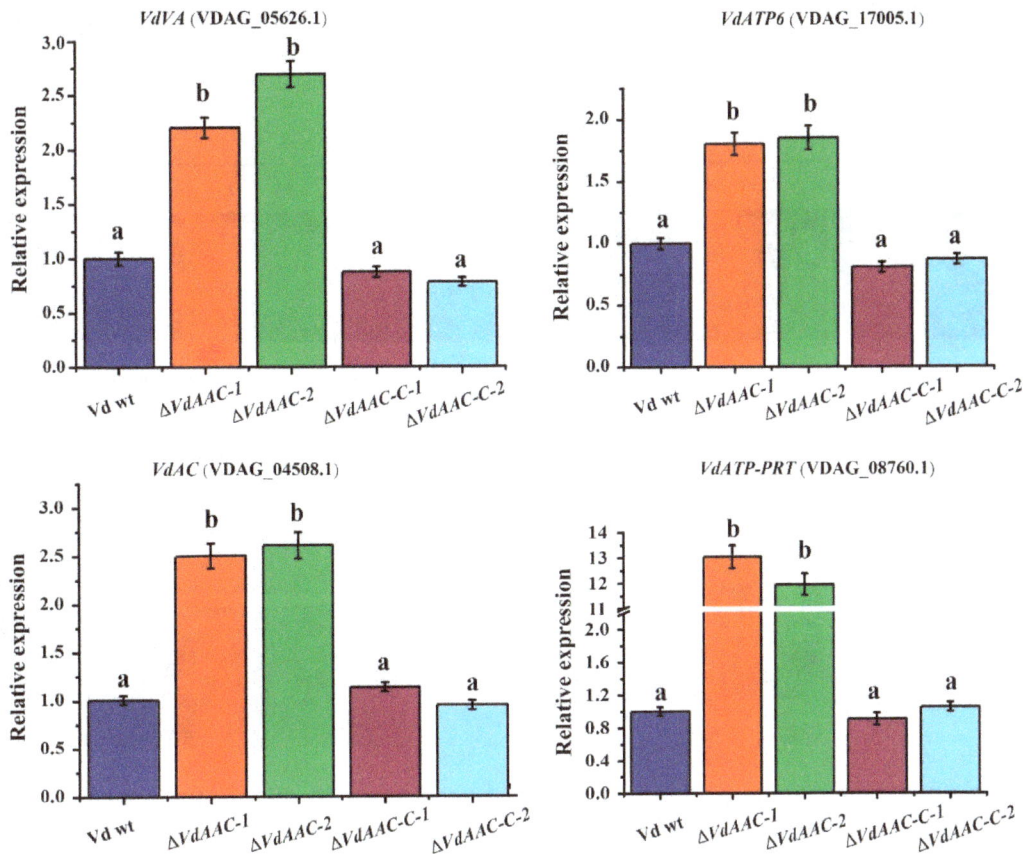

Figure 5. Relative expression of related genes involved in energy metabolism. Conidia from the respective strains were cultured in CM broth. After 4 days, the mycelia were collected for RNA extraction and the cDNA was synthesized. Expression patterns of four genes, vacuolar ATPase (VDAG_05626.1, *VdVA*), ATP synthase F0 subunit 6 (VDAG_17005.1, *VdATP6*), adenylate cyclase (VDAG_04508.1, *VdAC*), ATP phosphoribosyltransferase (VDAG_08760.1, *VdATP-PRT*), were determined by qRT-PCR. *Vdactin* gene was used as the reference gene for the expression analysis. Significant differences among the treatment groups ($p < 0.05$) are indicated by different letters (a, b) above the bars determined by the Duncan's multiple range test.

3.6. VdAAC Is Involved in Fungal Virulence

For evaluating the consequences of *VdAAC* deletion on fungal virulence, wild-type *N. benthamiana* (Nb wt) seedlings were inoculated with Vd wt, Δ*VdAAC*, and Δ*VdAAC-C*. At 12 dpi, seedlings inoculated with the Vd wt exhibited typical wilting symptoms and seemed nearly dead. In contrast, the disease index of seedlings inoculated with Δ*VdAAC* remained at a low level and was 70%–80% lower than that of the Vd wt group. The lower leaves of plants displayed a mild necrosis. The symptoms and disease index of seedlings inoculated with Δ*VdAAC-C* were similar to that of the Vd wt group (Figure 6A and Figure S3A). Fungal biomass in the various tissues of the plants inoculated with the different strains was then analyzed using qRT-PCR (Figure 6B). Fungal biomass of Δ*VdAAC* was significantly lower than that of the Vd wt and Δ*VdAAC*. These results were consistent with the phenotype and disease index data.

To investigate the *VdAAC* effect on fungal germination, 10^3 conidia of Vd-GFP and Δ*VdAAC-GFP* strains were added to PDA plates. After 48 h, the germination of Δ*VdAAC-GFP* conidia was nearly half that of Vd-GFP (Figure S3B). Furthermore, when the infection process of Vd-GFP and Δ*VdAAC-GFP* strains was examined microscopically (Figure 6C), at 7 dpi, many conidia of Vd-GFP had germinated, and hyphae were growing on the root surface. In contrast, fewer Δ*VdAAC-GFP* conidia had germinated compared with Vd-GFP. Meanwhile, the fungal biomass was lower on the root surface. All these results demonstrate that *VdAAC* contributes to fungal germination and growth, requisite for invasion and full virulence.

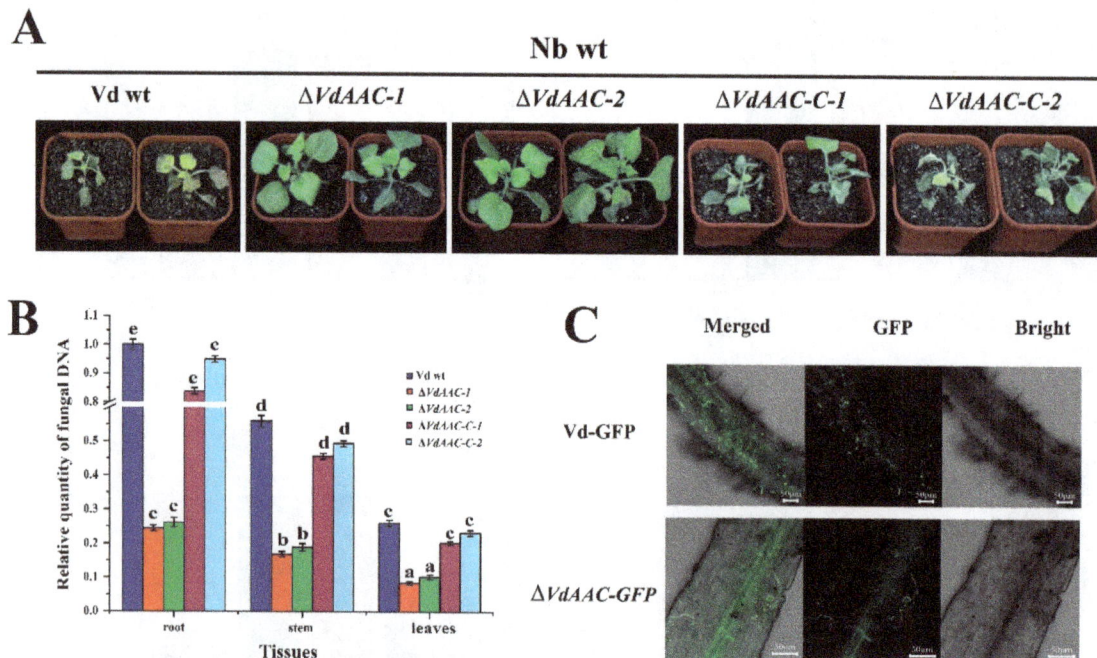

Figure 6. Virulence analysis of ΔVdAAC, ΔVdAAC-C and wild-type *V. dahliae* (Vd wt) strains on the wild-type *Nicotiana benthamiana* (Nb wt). The Nb wt seedlings were inoculated with 10^6 conidia·mL^{-1}. (**A**) Virulence phenotypes of Nb wt seedlings and (**B**) fungal biomass of different tissues of plants at 12 days after inoculation with the different strains. For the relative quantitative analysis, ITS1 and ITS2 of rDNA (Z29511) of *V. dahliae* were quantified relative to *N. benthamiana* housekeeping gene (*Nbactin*) for equilibration. (**C**) GFP fluorescence detection in roots of plants 7 days after inoculation with *Vd-GFP* or *ΔVdAAC-GFP*. Duncan's multiple range test was applied to determine significant differences among the treatment groups ($p < 0.05$) indicated by different letters (a–e).

3.7. DsRNA of VdAAC Confers Resistance against Vd in Transgenic Lines

Transgenic seedlings (Trans-1 and Trans-2) were inoculated with fungal Vd wt conidia to validate whether dsVdAAC can confer resistance against *V. dahliae* (Figure 7A and Figure S2D). From 10 dpi, the Nb wt group displayed typical wilt symptoms, and the disease index was more than 80. At 12 dpi, the seedlings of the Nb wt group were nearly dead, and the disease index was approximately 100. In contrast, at 10 dpi, seedlings of the transgenic groups had weak symptoms. At 12 dpi, the disease index was 70% lower in the Trans-1 group and 36% lower in the Trans-2 group than in the Nb wt group.

On the basis of the qRT-PCR to estimate fungal biomass in the root, stem and leaves of different groups at 12 dpi (Figure 7B), fungal biomass was significantly lower in transgenic seedlings than in Nb wt. To further examine whether the decreasing disease index in transgenic seedlings resulted from the silencing of *VdAAC*, we used qRT-PCR to analyze the relative expression level of *VdAAC* in transgenic and wild-type seedlings (Figure 7C). With the expression level of *VdAAC* in the Nb wt group estimated as 1, the Trans-1 group had better silencing efficiency (up to 47%) compared with 29% in Trans-2 group. The relative quantitative results, including the level of *V. dahliae* biomass and *VdAAC* expression, were strongly in accordance with the phenotypes of different groups.

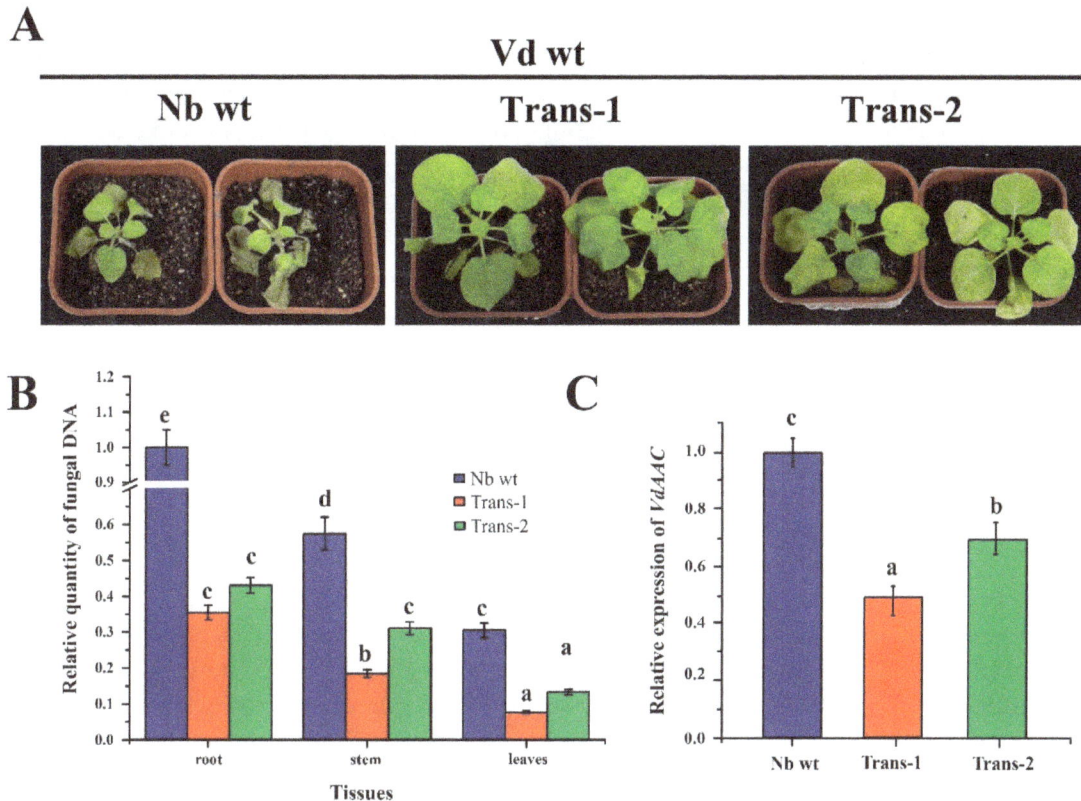

Figure 7. Assessment of the transgenic *Nicotiana benthamiana* for resistance against wild-type *V. dahliae*. One-month-old wild-type *Nicotiana benthamiana* (Nb wt) and transgenic lines (Trans-1 and -2) seedlings were inoculated with Vd wt and analyzed at 12 dpi. (**A**) Phenotypes of seedlings; (**B**) fungal biomass and (**C**) relative expression level of *VdAAC* determined by qRT-PCR. The bars with different letters (a–e) are significantly different ($p < 0.05$), based on the Duncan's multiple range test.

4. Discussion

Because the membrane protein AAC is needed to maintain a balance between ADP and ATP to generate energy for cells, we postulated that siRNAs designed against the *VdAAC* gene could be introduced into *V. dahliae* to decrease the expression of *VdAAC*; silencing of *VdAAC* could ultimately inhibit the mycelial growth and sporulation. By transforming fungal protoplasts in this way, we confirmed this hypothesis. Further support for this data comes from a previous study in which the silencing of a single functional *AAC* gene (*TbAAC*), in *Trypanosoma brucei*, by RNAi resulted in a severe growth defect, mainly due to reduced mitochondrial ATP synthesis [28]. Consistent with our RNAi results, the Δ*VdAAC* mutant had significant reduced colony diameter, conidia number and virulence as compared with Vd wt and Δ*VdAAC*-C.

Previous studies indicated that genes related to energy metabolism are upregulated in response to adverse environments [52–54]. In litchi fruit, when exposed to cold temperature, short-term anaerobic and pure oxygen conditions, genes related to energy metabolism including *LcAAC* were found to be upregulated [55]. Under reduced oxygen tension, the *AAC* gene deletion mutants of *Schizosaccharomyces pombe* exhibited impaired growth and were also unable to grow on a nonfermentable carbon source [56]. In our study, however, we found that *VdAAC* gene does not have similar role of adapting the fungus to adverse conditions. When exposed to UV light and high osmotic stress, the ratio of reduction in mycelial growth and sporulation for Δ*VdAAC* with Vd wt and Δ*VdAAC*-C strains were not significant from no stress conditions. Overall, the gene deletion mutants (Δ*VdAAC*) grows at roughly 60% of the wild type (Vd wt) and complementary strains (Δ*VdAAC*-C) rate under both stress and no stress conditions.

The *AAC* gene has a vital role in maintaining ADP/ATP balance in vivo, transporting ATP to the cytoplasm and ADP to mitochondria [57,58]. The movement of adenine-nucleotide (ADP/ATP) across the inner membrane of mitochondria is dependent on the concentration of internal ATP and on the energy state of mitochondria. The knockout of this gene can have a significant effect on genes that are putatively involved in energy metabolism e.g., vacuolar ATPase (VDAG_05626.1, *VdVA*) was upregulated in gene deletion mutants. The deletion of *AAC* gene also had a significant impact on the expression ATP synthase F0 subunit 6 (VDAG_17005.1, *VdATP6*). Similarly adenylate cyclase (VDAG_04508.1, *VdAC*) and ATP phosphoribosyltransferase (VDAG_08760.1, *VdATP-PRT*), were also upregulated in mutants as compared to the wild type and complementary strains.

Under favorable conditions, conidia germinate and produce a germ tube as the first step to invade a host and initiate wilt disease [59]. As we discussed earlier, sporulation is also a requisite factor for this fungus to infect the host [60]. The genes involved in germination and sporulation have become a target to control this fungus [61]. Sporulation is significantly impaired in *VdPR3* deletion mutants, which had decreased virulence on cotton [62]. The disruption of *VdRNS/ER* downregulated glycan synthesis, leading to the inhibition of conidia germination and infection by *V. dahliae* [63]. In this study, germination and sporulation were significantly reduced in ΔVdAAC, suggesting these reductions were the main reason for reduced fungal biomass and virulence in Nb wt.

Transgenic plants harboring dsRNA against appropriate target genes can have improved resistance [64,65]. Expression of 16D10 dsRNA in *Arabidopsis* improved resistance against the four major species of root-knot nematodes [66]. Transgenic tomato plants expressing a hairpin construct exhibited resistance against potato spindle tuber viroid infection [67]. Wheat plants transformed with RNAi constructs against three targeted fungal genes exhibited strong resistance against *Puccinia triticina* and thus a suppressed disease phenotype [36]. In our study, in the transgenic lines of *N. benthamiana* that expressed dsVdAAC, expression of the targeted gene and fungal biomass were reduced in the plant, which also had a lower disease index than Nb wt did.

5. Conclusions

In this study, the siRNAs transformed into *V. dahliae* protoplasts silenced *VdAAC*, and mycelial growth and sporulation were inhibited. Gene knockout mutants, as compared with wild-type and complementary strains, were impaired in mycelial growth, conidia production, stress tolerance and virulence. Moreover, the transgenic plants expressing ds*VdAAC* showed enhanced resistance against *V. dahliae*. Taken together, the present data demonstrates that *VdAAC* has potential as a target gene for an RNAi-based strategy to protect crops against *V. dahliae*.

Supplementary Materials: The following are available online at www.mdpi.com/2073-4425/8/1/25/s1, Figure S1: Position of siRNAs designed from different regions of the *VdAAC* gene of *V. dahliae* and colony diameter in different RNAi-treated groups. (A) Position of siRNAs along the *VdAAC* gene. siRNAs were designed and synthesized by Oligobio, Beijing, China; (B) Colony diameters of control and siRNA groups observed 2 weeks after transformation on PDA agar plates. Figure S2: Evaluation of resistance for wild-type (Nb wt) and transgenic *N. benthamiana* against *V. dahliae*. (A) Region (189–836 bp) of *VdAAC* gene was amplified and cloned into pK7GW1WG2(I) by LR recombination reaction. Numbers indicate nucleotide positions; (B) Schematic representation of the pK7GWIWG2(I)-VdAAC construction containing the sense and antisense partial ORF of VdAAC; (C) Confirmation of transgenic plants transformed with pK7GWIWG2(I)-VdAAC by PCR. The Nb wt seedling served as the negative control. (D) Disease index for seedlings from 10 to 12 days post inoculation with V. dahliae. Figure S3: Virulence and germination analysis of mutants and wild-type *V. dahliae* (Vd wt). (A) Disease index for *N. benthamiana* seedlings at 10 to 12 dpi with ΔVdAAC, ΔVdAAC-C and Vd wt; (B) Percentage germination of conidia produced by Vd-GFP or ΔVdAAC-GFP after 18 h on PDA.

Acknowledgments: This work was funded by a grant from National Nonprofit Industry Research (201503109).

Author Contributions: Hongmei Cheng, Rui Zhang and Huiming Guo conceived and designed the experiments. Xiaofeng Su, Latifur Rehman and Xiaokang Li performed experiments and analysed data.

Abbreviations

The following abbreviations are used in this manuscript:

AAC	ADP, ATP carrier
siRNAs	short interfering RNAs
RNAi	RNA interference
CM	complete medium broth
MS	Murashige-Skoog
dpi	days post inoculation
DI	disease index
ORF	open reading frame

References

1. Klosterman, S.J.; Atallah, Z.K.; Vallad, G.E.; Subbarao, K.V. Diversity, pathogenicity, and management of Verticillium species. *Annu. Rev. Phytopathol.* **2009**, *47*, 39–62. [CrossRef] [PubMed]
2. Pang, J.; Zhu, Y.; Li, Q.; Liu, J.; Tian, Y.; Liu, Y.; Wu, J. Development of Agrobacterium-mediated virus-induced gene silencing and performance evaluation of four marker genes in *Gossypium barbadense*. *PLoS ONE* **2013**, *8*, e73211. [CrossRef] [PubMed]
3. Pegg, G.F.; Brady, B.L. *Verticillium Wilts*; CABI Publishing: Oxford, UK, 2002.
4. Wang, Y.; Liang, C.; Wu, S.; Zhang, X.; Tang, J.; Jian, G.; Jiao, G.; Li, F.; Chu, C. Significant improvement of cotton Verticillium wilt resistance by manipulating the expression of gastrodia antifungal proteins. *Mol. Plant* **2016**, *9*, 1436–1439. [CrossRef] [PubMed]
5. Tsror, L.; Levin, A.G. Vegetative compatibility and pathogenicity of *Verticillium dahliae* Kleb. Isolates from Olive in Israel. *J. Phytopathol.* **2003**, *151*, 451–455. [CrossRef]
6. Duressa, D.; Anchieta, A.; Chen, D.; Klimes, A.; Garcia-Pedrajas, M.D.; Dobinson, K.F.; Klosterman, S.J. RNA-seq analyses of gene expression in the microsclerotia of *Verticillium dahliae*. *BMC Genom.* **2013**, *14*, 607. [CrossRef] [PubMed]
7. Fradin, E.F.; Thomma, B.P. Physiology and molecular aspects of *Verticillium* wilt diseases caused by *V. dahliae* and *V. albo-atrum*. *Mol. Plant Pathol.* **2006**, *7*, 71–86. [CrossRef] [PubMed]
8. Zhang, D.D.; Wang, X.Y.; Chen, J.Y.; Kong, Z.Q.; Gui, Y.J.; Li, N.Y.; Bao, Y.M.; Dai, X.F. Identification and characterization of a pathogenicity-related gene *VdCYP1* from *Verticillium dahliae*. *Sci. Rep.* **2016**. [CrossRef] [PubMed]
9. Xiong, D.; Wang, Y.; Tang, C.; Fang, Y.; Zou, J.; Tian, C. VdCrz1 is involved in microsclerotia formation and required for full virulence in *Verticillium dahliae*. *Fungal Genet. Biol.* **2015**, *82*, 201–212. [CrossRef] [PubMed]
10. Tian, L.; Xu, J.; Zhou, L.; Guo, W. VdMsb regulates virulence and microsclerotia production in the fungal plant pathogen *Verticillium dahliae*. *Gene* **2014**, *550*, 238–244. [CrossRef] [PubMed]
11. Nakayashiki, H.; Hanada, S.; Nguyen, B.Q.; Kadotani, N.; Tosa, Y.; Mayama, S. RNA silencing as a tool for exploring gene function in ascomycete fungi. *Fungal Genet. Biol.* **2005**, *42*, 275–283. [CrossRef] [PubMed]
12. Deshmukh, R.; Purohit, H.J. siRNA mediated gene silencing in *Fusarium* sp. HKF15 for overproduction of bikaverin. *Bioresour. Technol.* **2014**, *157*, 368–371. [CrossRef] [PubMed]
13. Mumbanza, F.M.; Kiggundu, A.; Tusiime, G.; Tushemereirwe, W.K.; Niblett, C.; Bailey, A. In vitro antifungal activity of synthetic dsRNA molecules against two pathogens of banana, *Fusarium oxysporum* f. sp. *cubense* and *Mycosphaerella fijiensis*. *Pest Manag. Sci.* **2013**, *69*, 1155–1162. [CrossRef] [PubMed]
14. Singh, S.; Braus-Stromeyer, S.A.; Timpner, C.; Tran, V.T.; Lohaus, G.; Reusche, M.; Knufer, J.; Teichmann, T.; von Tiedemann, A.; Braus, G.H. Silencing of *Vlaro2* for chorismate synthase revealed that the phytopathogen *Verticillium longisporum* induces the cross-pathway control in the xylem. *Appl. Microbiol. Biotechnol.* **2010**, *85*, 1961–1976. [CrossRef] [PubMed]
15. Koch, A.; Kumar, N.; Weber, L.; Keller, H.; Imani, J.; Kogel, K.H. Host-induced gene silencing of cytochrome P450 lanosterol C14alpha-demethylase-encoding genes confers strong resistance to *Fusarium* species. *Proc. Natl. Acad. Sci. USA* **2013**, *110*, 19324–19329. [CrossRef] [PubMed]
16. Ghag, S.B.; Shekhawat, U.K.; Ganapathi, T.R. Host-induced post-transcriptional hairpin RNA-mediated gene silencing of vital fungal genes confers efficient resistance against *Fusarium* wilt in banana. *Plant Biotechnol. J.* **2014**, *12*, 541–553. [CrossRef] [PubMed]

17. Zhang, T.; Jin, Y.; Zhao, J.H.; Gao, F.; Zhou, B.J.; Fang, Y.Y.; Guo, H.S. Host-induced gene silencing of target gene in fungal cells confers effective resistance to cotton wilt disease pathogen *Verticillium dahliae*. *Mol. Plant* **2016**, *9*, 939–942. [CrossRef] [PubMed]

18. Nury, H.; Dahout-Gonzalez, C.; Trezeguet, V.; Lauquin, G.; Brandolin, G.; Pebay-Peyroula, E. Structural basis for lipid-mediated interactions between mitochondrial ADP/ATP carrier monomers. *FEBS Lett.* **2005**, *579*, 6031–6036. [CrossRef] [PubMed]

19. Klingenberg, M. Molecular aspects of the adenine nucleotide carrier from mitochondria. *Arch. Biochem. Biophys.* **1989**, *270*, 1–14. [CrossRef]

20. Fiore, C.; Le, S.A.; Roux, P.; Schwimmer, C.; Dianoux, A.N.F.; Gjm, L.; Brandolin, G.; Pv, V.; Trezeguet, V. The mitochondrial ADP/ATP carrier: Structural, physiological and pathological aspects. *Biochimie* **1998**, *80*, 137–150. [CrossRef]

21. Pebay-Peyroula, E.; Dahout-Gonzalez, C.; Kahn, R.; Trezeguet, V.; Lauquin, G.J.; Brandolin, G. Structure of mitochondrial ADP/ATP carrier in complex with carboxyatractyloside. *Nature* **2003**, *426*, 39–44. [CrossRef] [PubMed]

22. Hatanaka, T.; Kihira, Y.; Shinohara, Y.; Majima, E.; Terada, H. Characterization of loops of the yeast mitochondrial ADP/ATP carrier facing the cytosol by site-directed mutagenesis. *Biochem. Biophys. Res. Commun.* **2001**, *286*, 936–942. [CrossRef] [PubMed]

23. Ohkura, K.; Hori, H.; Shinohara, Y. Role of C-terminal region of yeast ADP/ATP carrier 2 protein: Dynamics of flexible C-terminal arm. *Anticancer Res.* **2009**, *29*, 4897–4900. [PubMed]

24. Nowara, D.; Gay, A.; Lacomme, C.; Shaw, J.; Ridout, C.; Douchkov, D.; Hensel, G.; Kumlehn, J.; Schweizer, P. HIGS: Host-induced gene silencing in the obligate biotrophic fungal pathogen *Blumeria graminis*. *Plant Cell* **2010**, *22*, 3130–3141. [CrossRef] [PubMed]

25. Pereira, C.; Chaves, S.; Alves, S.; Salin, B.; Camougrand, N.; Manon, S.; Sousa, M.J.; Corte-Real, M. Mitochondrial degradation in acetic acid-induced yeast apoptosis: The role of Pep4 and the ADP/ATP carrier. *Mol. Microbiol.* **2010**, *76*, 1398–1410. [CrossRef] [PubMed]

26. Gigolashvili, T.; Geier, M.; Ashykhmina, N.; Frerigmann, H.; Wulfert, S.; Krueger, S.; Mugford, S.G.; Kopriva, S.; Haferkamp, I.; Flugge, U.I. The *Arabidopsis* thylakoid ADP/ATP carrier TAAC has an additional role in supplying plastidic phosphoadenosine 5′-phosphosulfate to the cytosol. *Plant Cell* **2012**, *24*, 4187–4204. [CrossRef] [PubMed]

27. Talbot, D.A.; Duchamp, C.; Rey, B.; Hanuise, N.; Rouanet, J.L.; Sibille, B.; Brand, M.D. Uncoupling protein and ATP/ADP carrier increase mitochondrial proton conductance after cold adaptation of king penguins. *J. Physiol.* **2004**, *558*, 123–135. [CrossRef] [PubMed]

28. Gnipova, A.; Subrtova, K.; Panicucci, B.; Horvath, A.; Lukes, J.; Zikova, A. The ADP/ATP carrier and its relationship to oxidative phosphorylation in ancestral protist *Trypanosoma brucei*. *Eukaryot. Cell* **2015**, *14*, 297–310. [CrossRef] [PubMed]

29. Rehman, L.; Su, X.; Guo, H.; Qi, X.; Cheng, H. Protoplast transformation as a potential platform for exploring gene function in *Verticillium dahliae*. *BMC Biotechnol.* **2016**, *16*, 57–65. [CrossRef] [PubMed]

30. Wang, H.M.; Lin, Z.X.; Zhang, X.L.; Chen, W.; Guo, X.P.; Nie, Y.C.; Li, Y.H. Mapping and quantitative trait loci analysis of *Verticillium* wilt resistance genes in cotton. *J. Integr. Plant Biol.* **2008**, *50*, 174–182. [CrossRef] [PubMed]

31. Tian, J.; Zhang, X.; Liang, B.; Li, S.; Wu, Z.; Wang, Q.; Leng, C.; Dong, J.; Wang, T. Expression of baculovirus anti-apoptotic genes *p35* and op-iap in cotton (*Gossypium hirsutum* L.) enhances tolerance to *Verticillium* wilt. *PLoS ONE* **2010**, *5*, e14218. [CrossRef] [PubMed]

32. Yang, X.; Ben, S.; Sun, Y.; Fan, X.; Tian, C.; Wang, Y. Genome-wide identification, phylogeny and expression profile of vesicle fusion components in *Verticillium dahliae*. *PLoS ONE* **2013**, *8*, e68681. [CrossRef] [PubMed]

33. Bustin, S.A.; Benes, V.; Garson, J.A.; Hellemans, J.; Huggett, J.; Kubista, M.; Mueller, R.; Nolan, T.; Pfaffl, M.W.; Shipley, G.L.; et al. The MIQE guidelines: Minimum information for publication of quantitative real-time PCR experiments. *Clin. Biochem.* **2009**, *55*, 611–622. [CrossRef] [PubMed]

34. Hoppenau, C.E.; Tran, V.-T.; Kusch, H.; Aßhauer, K.P.; Landesfeind, M.; Meinicke, P.; Popova, B.; Braus-Stromeyer, S.A.; Braus, G.H. *Verticillium dahliae* VdTHI4, involved in thiazole biosynthesis, stress response and DNA repair functions, is required for vascular disease induction in tomato. *Environ. Exp. Bot.* **2014**, *108*, 14–22. [CrossRef]

35. Tzima, A.K.; Paplomatas, E.J.; Tsitsigiannis, D.I.; Kang, S. The G protein beta subunit controls virulence and multiple growth- and development-related traits in *Verticillium dahliae*. *Fungal Genet. Biol.* **2012**, *49*, 271–283. [CrossRef] [PubMed]

36. Panwar, V.; McCallum, B.; Bakkeren, G. Endogenous silencing of *Puccinia triticina* pathogenicity genes through in planta-expressed sequences leads to the suppression of rust diseases on wheat. *Plant J.* **2013**, *73*, 521–532. [CrossRef] [PubMed]

37. Ellendorff, U.; Fradin, E.F.; de Jonge, R.; Thomma, B.P. RNA silencing is required for *Arabidopsis* defence against *Verticillium* wilt disease. *J. Exp. Bot.* **2009**, *60*, 591–602. [CrossRef] [PubMed]

38. Zhang, Z.; Song, Y.; Liu, C.M.; Thomma, B.P. Mutational analysis of the Ve1 immune receptor that mediates *Verticillium* resistance in tomato. *PLoS ONE* **2014**, *9*, e99511. [CrossRef] [PubMed]

39. Mao, Y.; Cai, W.; Wang, J.; Hong, G.; Tao, X.; Wang, L.; Huang, Y.; Chen, X. Silencing a cotton bollworm P450 monooxygenase gene by plant-mediated RNAi impairs larval tolerance of gossypol. *Nat. Biotechnol.* **2007**, *25*, 1307–1313. [CrossRef] [PubMed]

40. Obrepalska-Steplowska, A.; Wieczorek, P.; Budziszewska, M.; Jeszke, A.; Renaut, J. How can plant virus satellite RNAs alter the effects of plant virus infection? A study of the changes in the *Nicotiana benthamiana* proteome after infection by peanut stunt virus in the presence or absence of its satellite RNA. *Proteomics* **2013**, *13*, 2162–2175. [CrossRef] [PubMed]

41. Klionsky, D.J.; Nelson, H.; Nelson, N.; Yaver, K. Mutations in the yeast vacuolar ATPase result in the mislocalization of vacuolar proteins. *J. Exp. Biol.* **1992**, *172*, 83–92. [PubMed]

42. Clague, M.J.; Urbe, S.; Aniento, F.; Gruenberg, J. Vacuolar ATPase activity is required for endosomal carrier vesicle formation. *J. Biol. Chem.* **1994**, *269*, 21–24. [PubMed]

43. Klionsky, D.J.; Herman, P.K.; Emr, S.D. The fungal vacuole: Composition, function, and biogenesis. *Microbiol. Rev.* **1990**, *54*, 266–292. [PubMed]

44. Deckers-Hebestreit, G.; Schmid, R.; Kiltz, H.H.; Altendorf, K. F0 portion of *Escherichia coli* ATP synthase: Orientation of subunit c in the membrane. *Biochemistry* **1987**, *26*, 5486–5492. [CrossRef] [PubMed]

45. Van Walraven, H.S.; Scholts, M.J.; Lill, H.; Matthijs, H.C.; Dilley, R.A.; Kraayenhof, R. Introduction of a carboxyl group in the loop of the F0 c-subunit affects the H^+/ATP coupling ratio of the ATP synthase from *Synechocystis 6803*. *J. Bioenerg. Biomembr.* **2002**, *34*, 445–454. [CrossRef] [PubMed]

46. Salomon, Y.; Londos, C.; Rodbell, M. A highly sensitive adenylate cyclase assay. *Anal. Biochem.* **1974**, *58*, 541–548. [CrossRef]

47. Klimpel, A.; Gronover, C.S.; Williamson, B.; Stewart, J.A.; Tudzynski, B. The adenylate cyclase (BAC) in *Botrytis cinerea* is required for full pathogenicity. *Mol. Plant Pathol.* **2002**, *3*, 439–450. [CrossRef] [PubMed]

48. Piszkiewicz, D.; Tilley, B.E.; Rand-Meir, T.; Parsons, S.M. Amino acid sequence of ATP phosphoribosyltransferase of *Salmonella typhimurium*. *Proc. Natl. Acad. Sci. USA* **1979**, *76*, 1589–1592. [CrossRef] [PubMed]

49. Cho, Y.; Sharma, V.; Sacchettini, J.C. Crystal structure of ATP phosphoribosyltransferase from *Mycobacterium tuberculosis*. *J. Biol. Chem.* **2003**, *278*, 8333–8339. [CrossRef] [PubMed]

50. Brenner, M.; Ames, B.N. The histidine operon and its regulation. In *Metabolic Regulation*; Vogel, H.J., Ed.; Academic Press: New York, NY, USA, 1971; Volume 5, pp. 349–387.

51. Goldberger, R.F.; Kovach, J.S. Regulation of histidine biosynthesis in *Salmonella typhimurium*. *Curr. Top. Cell. Regul.* **1972**, *5*, 285–308. [PubMed]

52. Voncken, F.; Gao, F.; Wadforth, C.; Harley, M.; Colasante, C. The phosphoarginine energy-buffering system of *Trypanosoma brucei* involves multiple arginine kinase isoforms with different subcellular locations. *PLoS ONE* **2013**, *8*, e65908. [CrossRef] [PubMed]

53. Pereira, C.A. Arginine kinase: A potential pharmacological target in trypanosomiasis. *Infect. Disord. Drug Targets* **2014**, *14*, 30–36. [CrossRef] [PubMed]

54. Miranda, M.R.; Canepa, G.E.; Bouvier, L.A.; Pereira, C.A. Trypanosoma cruzi: Oxidative stress induces arginine kinase expression. *Exp. Parasitol.* **2006**, *114*, 341–344. [CrossRef] [PubMed]

55. Liu, T.; Wang, H.; Kuang, J.; Sun, C.; Shi, J.; Duan, X.; Qu, H.; Jiang, Y. Short-term anaerobic, pure oxygen and refrigerated storage conditions affect the energy status and selective gene expression in litchi fruit. *LWT-Food Sci. Technol.* **2015**, *60*, 1254–1261. [CrossRef]

56. Trezeguet, V.; Zeman, I.; David, C.; Lauquin, G.J.; Kolarov, J. Expression of the ADP/ATP carrier encoding genes in aerobic yeasts; phenotype of an ADP/ATP carrier deletion mutant of *Schizosaccharomyces pombe*. *Biochim. Biophys. Acta* **1999**, *1410*, 229–236. [CrossRef]

57. Miura, K.; Inouye, S.; Sakai, K.; Takaoka, H.; Kishi, F.; Tabuchi, M.; Tanaka, T.; Matsumoto, H.; Shirai, M.; Nakazawa, T.; Nakazawa, A. Cloning and characterization of adenylate kinase from *Chlamydia pneumoniae*. *J. Biol. Chem.* **2001**, *276*, 13490–13498. [CrossRef] [PubMed]

58. Claypool, S.M.; Oktay, Y.; Boontheung, P.; Loo, J.A.; Koehler, C.M. Cardiolipin defines the interactome of the major ADP/ATP carrier protein of the mitochondrial inner membrane. *J. Cell Biol.* **2008**, *182*, 937–950. [CrossRef] [PubMed]

59. Hawke, M.A.; Lazarovits, G. Production and manipulation of individual microsclerotia of *Verticillium dahliae* for use in studies of survival. *Phytopathology* **1994**, *23*, 582–584.

60. Isaac, I. Verticillium wilt of sainfoin. *Ann. Appl. Biol.* **1946**, *33*, 28–34. [CrossRef] [PubMed]

61. Debode, J.; De Maeyer, K.; Perneel, M.; Pannecoucque, J.; De Backer, G.; Hofte, M. Biosurfactants are involved in the biological control of *Verticillium microsclerotia* by *Pseudomonas* spp. *J. Appl. Microbiol.* **2007**, *103*, 1184–1196. [CrossRef] [PubMed]

62. Zhang, Y.L.; Li, Z.F.; Feng, Z.L.; Feng, H.J.; Zhao, L.H.; Shi, Y.Q.; Hu, X.P.; Zhu, H.Q. Isolation and functional analysis of the pathogenicity-related gene *VdPR3* from *Verticillium dahliae* on cotton. *Curr. Genet.* **2015**, *61*, 555–566. [CrossRef] [PubMed]

63. Santhanam, P.; Boshoven, J.C.; Salas, O.; Bowler, K.; Islam, T.; Keykha Saber, M.; van den Berg, G.C.; Bar-Peled, M.; Thomma, B.P. Rhamnose synthase activity is required for pathogenicity of the vascular wilt fungus *Verticillium dahliae*. *Mol. Plant Pathol.* **2016**. [CrossRef] [PubMed]

64. Novina, C.D.; Sharp, P.A. The RNAi revolution. *Nature* **2004**, *430*, 161–164. [CrossRef] [PubMed]

65. Kalantidis, K.; Psaradakis, S.; Tabler, M.; Tsagris, M. The occurrence of CMV-specific short RNAS in transgenic tobacco expressing virus-derived double-stranded RNA is indicative of resistance to the virus. *MPMI* **2002**, *15*, 826–833. [CrossRef] [PubMed]

66. Huang, G.; Allen, R.; Davis, E.L.; Baum, T.J.; Hussey, R.S. Engineering broad root-knot resistance in transgenic plants by RNAi silencing of a conserved and essential root-knot nematode parasitism gene. *Proc. Natl. Acad. Sci. USA* **2006**, *103*, 14302–14306. [CrossRef] [PubMed]

67. Schwind, N.; Zwiebel, M.; Itaya, A.; Ding, B.; Wang, M.B.; Krczal, G.; Wassenegger, M. RNAi-mediated resistance to Potato spindle tuber viroid in transgenic tomato expressing a viroid hairpin RNA construct. *Mol. Plant Pathol.* **2009**, *10*, 459–469. [CrossRef] [PubMed]

PERMISSIONS

LIST OF CONTRIBUTORS

Ana M. Vélez and Natalie Matz
Department of Entomology, University of Nebraska, 103 Entomology Hall, Lincoln, NE 68583, USA

Elane Fishilevich, Nicholas P. Storer and Kenneth E. Narva
Dow AgroSciences, 9330 Zionsville Road, Indianapolis, IN 46268, USA

Blair D. Siegfried
Entomology and Nematology Department, University of Florida, Charles Steinmetz Hall, Gainesville, FL 32611, USA

Anna Moszczynska and Dongyue Yu
Department of Pharmaceutical Sciences, Eugene Applebaum College of Pharmacy and Health Sciences, Wayne State University, Detroit, MI 48201, USA

Kyle J. Burghardt
Department of Pharmacy Practice, Eugene Applebaum College of Pharmacy and Health Sciences, Wayne State University, Detroit, MI 48201, USA

Mi Ae Kim, Tae Ha Kim and Young Chang Sohn
Department of Marine Molecular Bioscience, Gangneung-Wonju National University, Gangneung 25457, Korea

Jae-Sung Rhee
Department of Marine Science, College of Natural Sciences, Incheon National University, Incheon 22012, Korea

Jung Sick Lee
Department of Aqualife Medicine, Chonnam National University, Yeosu 59626, Korea

Ah-Young Choi and Beom-Soon Choi
Phyzen Genomics Institute, Seongnam 13558, Korea

Ik-Young Choi
Department of Agriculture and Life Industry, Kangwon National University, Chuncheon 24341, Korea

Quin NeeWong
Biotechnology Research Centre, School of Biosciences, Faculty of Science, University of Nottingham Malaysia Campus, Jalan Broga, 43500 Semenyih, Selangor Darul Ehsan, Malaysia

Sean Mayes
Biotechnology Research Centre, School of Biosciences, Faculty of Science, University of Nottingham Malaysia Campus, Jalan Broga, 43500 Semenyih, Selangor Darul Ehsan, Malaysia
Crops For the Future, Jalan Broga, 43500 Semenyih, Selangor Darul Ehsan, Malaysia
School of Biosciences, Faculty of Science, University of Nottingham Sutton Bonington Campus, Sutton Bonington, Leicestershire LE12 5RD, UK

Alberto Stefano Tanzi, Wai Kuan Ho and Festo Massawe
Biotechnology Research Centre, School of Biosciences, Faculty of Science, University of Nottingham Malaysia Campus, Jalan Broga, 43500 Semenyih, Selangor Darul Ehsan, Malaysia
Crops For the Future, Jalan Broga, 43500 Semenyih, Selangor Darul Ehsan, Malaysia

Asha Karunaratne
Crops For the Future, Jalan Broga, 43500 Semenyih, Selangor Darul Ehsan, Malaysia
Department of Export Agriculture, Faculty of Agricultural Sciences, Sabaragamuwa University of Sri Lanka, Belihuloya 70140, Sri Lanka

Sunir Malla and Martin Blythe
Deep Seq, Faculty of Medicine and Health Sciences, Queen's Medical Centre, University of Nottingham, Nottingham NG7 2UH, UK
School of Biosciences, Faculty of Science, University of Nottingham Sutton Bonington Campus, Sutton Bonington, Leicestershire LE12 5RD, UK

Heather Holl and Samantha Brooks
Department of Animal Sciences, UF Genomics Institute, University of Florida, Gainesville, FL 32610, USA

Ramiro Isaza
Department of Small Animal Clinical Sciences, College of Veterinary Medicine, University of Florida, Gainesville, FL 32608, USA

Yasmin Mohamoud and Ayeda Ahmed
Department of Genetic Medicine, Weill Cornell Medical College in Qatar, Doha, Qatar

Faisal Almathen
Veterinary Public Health and Animal Husbandry, College of Veterinary Medicine and Animal Resources, King Faisal University, Al-Ahsa 31982, Saudi Arabia

Cherifi Youcef and Semir Gaouar
Department of Biology, University of Abou Bekr Belkaïd, Tlemcen 13000, Algeria

Douglas F. Antczak
Baker Institute for Animal Health, College of Veterinary Medicine, Cornell University, Ithaca, NY 14853, USA

Shuyun Zeng, Tao Zhou, Kai Han, Yanci Yang, Jianhua Zhao and Zhan-Lin Liu
Key Laboratory of Resource Biology and Biotechnology in Western China (Ministry of Education), College of Life Science, Northwest University, Xi'an, 710069, China

Chang-Duck Koo
Department of Forest Science, Chungbuk National University, Cheongju 28644, Korea

Hwa-Yong Lee
Department of Forest Science, Chungbuk National University, Cheongju 28644, Korea
Department of Biology, Chungbuk National University, Cheongju 28644, Korea

Suyun Moon and Hojin Ryu
Department of Biology, Chungbuk National University, Cheongju 28644, Korea

Donghwan Shim
Department of Forest Genetic Resources, National Institute of Forest Science, Suwon 16631, Korea

Chang Pyo Hong
Theragen Etex Bio Institute, Suwon 16229, Korea

Yi Lee and Jong-Wook Chung
Department of Industrial Plant Science and Technology, Chungbuk National University, Cheongju 28644, Korea

Xudong Zhu, Mengqi Wang, Xiaopeng Li, Songtao Jiu, Chen Wang and Jinggui Fang
Nanjing Agricultural University, Weigang 1 hao, 210095 Nanjing, China

Yue Li, Liqiang Wan, Shuyi Bi, Xiufu Wan, Zhenyi Li, Jing Cao, Zongyong Tong, Hongyu Xu, Feng He and Xianglin Li
Institute of Animal Sciences, Chinese Academy of Agricultural Sciences, Beijing 100193, China

Faraz Khan
School of Biosciences, University of Nottingham, Sutton Bonington Campus, Nottingham LE12 5RD, UK

Sean Mayes
School of Biosciences, University of Nottingham, Sutton Bonington Campus, Nottingham LE12 5RD, UK

Crops for the Future, Jalan Broga, 43500 Semenyih, Selangor Darul Ehsan, Malaysia

Hui Hui Chai
Crops for the Future, Jalan Broga, 43500 Semenyih, Selangor Darul Ehsan, Malaysia

Ishan Ajmera and Charlie Hodgman
Centre for Plant Integrative Biology, University of Nottingham, Sutton Bonington Campus, Nottingham LE12 5RD, UK

Chungui Lu
School of Animal Rural and Environmental Sciences, Nottingham Trent University, Clifton Campus, Nottingham NG11 8NS, UK

Bartłomiej Tomaszewski, Joanna P. Gorniak and John C. Mathers
Centre for Ageing and Vitality, Human Nutrition Research Centre, Institute of Cellular Medicine, Newcastle University, Campus for Ageing and Vitality, Newcastle upon Tyne NE4 5PL, UK

Sabine A. S. Langie
Centre for Ageing and Vitality, Human Nutrition Research Centre, Institute of Cellular Medicine, Newcastle University, Campus for Ageing and Vitality, Newcastle upon Tyne NE4 5PL, UK
Environmental Risk and Health unit, Flemish Institute of Technological Research (VITO), Boeretang 200, 2400 Mol, Belgium

Kerry M. Cameron and Thomas von Zglinicki
The Ageing Biology Centre and Institute for Cell and Molecular Biology, Newcastle University, Campus for Ageing and Vitality, Newcastle upon Tyne NE4 5PL, UK

Gabriella Ficz
Barts Cancer Institute, Queen Mary University, London EC1M 6BQ, UK

David Oxley
Mass Spectrometry Laboratory, Babraham Institute, Cambridge CB22 3AT, UK

Lou M. Maas, Roger W. L. Godschalk and Frederik J. van Schooten
Department of Pharmacology and Toxicology, School for Nutrition and Translational Research in Metabolism (NUTRIM), Maastricht University, 6200 MD Maastricht, The Netherlands

Wolf Reik
Epigenetics Programme, Babraham Institute, Cambridge CB22 3AT, UK
Wellcome Trust Sanger Institute, Hinxton CB10 1SA, UK

Yongan Zhang and Liangjian Qu
Key Laboratory of Forest Protection, State Forestry
Administration, Research Institute of Forest Ecology,
Environment and Protection, Chinese Academy of
Forestry, Beijing 100091, China

Zhou Zhou
Key Laboratory of Forest Protection, State Forestry
Administration, Research Institute of Forest Ecology,
Environment and Protection, Chinese Academy of
Forestry, Beijing 100091, China

College of Forestry, Henan University of Science and
Technology, Luoyang 471003, China

Yongli Li and Chunyan Yuan
College of Forestry, Henan University of Science and
Technology, Luoyang 471003, China

**Xiaofeng Su, Latifur Rehman, Huiming Guo,
Xiaokang Li, Rui Zhang and Hongmei Cheng**
Biotechnology Research Institute, Chinese Academy of
Agricultural Sciences, Beijing 100081, China

Index